PLANAR ANTENNAS

PLANAR ANTENNAS

Design and Applications

Edited by

Praveen Kumar Malik

CRC Press
Taylor & Francis Group
Boca Raton London New York

CRC Press is an imprint of the
Taylor & Francis Group, an **informa** business

First edition published 2022
by CRC Press
6000 Broken Sound Parkway NW, Suite 300, Boca Raton, FL 33487-2742

and by CRC Press
2 Park Square, Milton Park, Abingdon, Oxon, OX14 4RN

CRC Press is an imprint of Taylor & Francis Group, LLC

Library of Congress Cataloging-in-Publication Data
A catalog record has been requested for this book

ISBN: 978-1-032-03446-1 (hbk)
ISBN: 978-1-032-03449-2 (pbk)
ISBN: 978-1-003-18732-5 (ebk)

DOI: 10.1201/9781003187325

Typeset in Times
by MPS Limited, Dehradun

Dedication

This book is dedicated to my late father, who taught me to be an independent and determined person, without whom I would never be able to achieve my objectives and succeed in life.

Late (Sr.) Dharamveer Singh

Contents

Preface

This edited book aims to bring together leading academic scientists, researchers, and research scholars to exchange and share their experiences and research results on all aspects of wireless communication and planer antenna design. It also provides a premier interdisciplinary platform for researchers, practitioners, and educators to present and discuss the most recent innovations, trends, and concerns as well as practical challenges encountered and solutions adopted in the fields of wireless communication and planer antenna design.

Organization of the Book

The book is organized into eight chapters. A brief description of each of the chapters follows:

1 **Overview and Introduction**
Under this chapter we have discussed an introduction to micro-strip antennas, analysis and analytical models of micro-strip antennas, advances in the design of micro-strip antennas and design challenges, reconfigurable wideband micro-strip antenna and circularly polarized wideband micro-strip antenna.

2 **Performance Analysis of Micro-Strip Antenna**
This chapter will provide the insight of micro-strip antenna design standards, micro-strip antenna design parameters, micro-strip antenna performance parameters, ultra-wideband antenna design parameters, ultra-wideband antenna design challenges and remedies, monopole based ULTRA WIDE BAND antenna and mathematical analysis of ULTRA WIDE BAND antenna.

3 **Multiple Input Multiple Output (MIMO) Antenna Design and Uses**
The third chapter includes design issues of MIMO antennas, MIMO antenna design for various applications, MIMO antenna design analysis, and performance analysis of MIMO antennas.

4 **Fractal and Defected Ground Structure Micro-Strip Antenna**
Under this chapter we have discussed defected ground structures, design of defected ground structures, fractal antenna design basics, design of fractal antenna, and recent advances in design of fractal and defected ground structure.

5 **Micro-Strip Antenna in Vehicular Communication**
Chapter 5 carried out the design of antenna for vehicular communication, design aspect and challenges in the vehicular antennas, and recent antenna design for vehicular communication.

6 **Importance and Uses of Micro-Strip Antenna in IoT**
Chapter 6 focuses on the introduction to IoT, design of micro-strip antenna for

IoT applications, design challenges of antennas for IoT applications, and current trends in the design of antennas for IoT applications.

7 **Ultra-Wide-Band Antenna Design for Wearable Applications**
Chapter 7 highlights the design of ULTRA WIDE BAND antennas, design challenges of ULTRA WIDE BAND antennas, antennas for biomedical applications and early diagnoses, wearable sensors for medical implant, and advances in the field of ULTRA WIDE BAND antenna design for biomedical applications.

8 **Micro-Strip Antenna Design for Misc Applications**
Chapter 8 presents the different other antenna used in wireless communication.

Dr. Praveen Kumar Malik

About the Editor

Dr. Praveen Kumar Malik is a Professor in the School of Electronics and Electrical Engineering, Lovely Professional University, Phagwara, Punjab, India. He received his Ph.D. with a specialization in Wireless Communication and Antenna Design. He has authored or coauthored more than 40 technical research papers published in leading journals and conferences from the IEEE, Elsevier, Springer, Wiley, etc. Some of his research findings are published in top-cited journals. He has also published three edited/authored books with international publishers. He has guided many students leading to M.E./M.Tech and guiding students leading to Ph.D. He is an Associate Editor of different journals. His current interest includes Microstrip Antenna Design, MIMO, Vehicular Communication, and IoT. He was invited as Guest Editors/Editorial Board Members of many International Journals, invited for keynote speaker in many International Conferences held in Asia, and invited as Program Chair, Publications Chair, Publicity Chair, and Session Chair in many International Conferences. He has been granted two design IPR and a few are in pipelines.

List of Contributors

T. Anilkumar
Department of ECE
K L University
Guntur, Andhra Pradesh, India

B. Bansal
Jaypee Institute of Information Technology
Noida, Uttar Pradesh, India

A. Birwal
University of Delhi
New Delhi, India

S. Prasad Jones Christydass
K. Ramakrishnan College of Technology
Trichy, Tamilnadu, India

Arpan Desai
Division of Computational Physics
Institute for Computational Science
Ton Duc Thang University
Ho Chi Minh City, Vietnam

Preksha Gandhi
Charotar University of Science and Technology
Changa, India

K. R. Gokul Anand
Assistant Professor
Department of ECE
Dr. Mahalingam College of Engineering & Technology
Coimbatore, Tamilnadu, India

Sindhu Hak Gupta
Department of Electronics and Communication Engineering
ASET
Amity University
Noida, India

Chandrasekhar Rao Jetti
Department of ECE
Bapatla Engineering College
Bapatla Andhra Pradesh, India

Nguyen Truong Khang
Faculty of Electrical and Electronics Engineering
Ton Duc Thang University
Ho Chi Minh City, Vietnam

A. J. Sharath Kumar
Assistant Professor
Department of Electronics and Communication Engineering
Vidyavardhaka College of Engineering
Mysuru, Karnataka, India

B. T. P. Madhav
Professor
Department of ECE
K L University
Guntur, Andhra Pradesh, India

Mihir Narayan Mohanty
ITER, Siksha 'O' Anusandhan
(Deemed to be University)
Bhubaneswar, Odisha, India

C. Muthu Ramya
National Institute of Technology
Puducherry, India

Venkateswara Rao Nandanavanam
Department of ECE
Bapatla Engineering College
Bapatla, Andhra Pradesh, India

Vikas Pandey
Babu Banarasi Das University
Lucknow, Uttar Pradesh, India

Shuchismita Pani
Amity School of Engineering & Technology
Amity University
Noida, India

Neha Parmar
Antenna and RF Engineer
CoreIoT Technologies
Andhra Pradesh, India

Yesha Patel
Charotar University of Science and Technology
Changa, India

T. Poornima
Research Associate
Department of ECE
Amrita Vishwa Vidyapeetham
Coimbatore, Tamilnadu, India

P. Potey
LTCE
Navi Mumbai, India

E. L. Dhivya Priya
Assistant Professor
Department of ECE
Sri Krishna College of Technology
Coimbatore, Tamilnadu, India

N. S. Raghava
Delhi Technological University
New Delhi, India

R. Boopathi Rani
Assistant Professor
Department of ECE
National Institute of Technology
Puducherry, India

Qudsia Rubani
Department of Electronics and Communication Engineering
ASET
Amity University
Noida, India

M. Saravanan
Annamalai University
Cuddalore, Tamilnadu, India

R. Saravanakumar
Hindusthan Institute of Technology
Coimbatore, Tamilnadu, India

Parnika Saxena
Amity School of Engineering & Technology
Amity University
Noida, India

Swapnil Shah
Charotar University of Science and Technology
Changa, India

Manish Sharma
Chitkara University Institute of Engineering and Technology
Chitkara University
Punjab, India

Arun Kumar Singh
Saudi Electronic University
Saudi Arabia-KSA

T. P. Surekha
Professor
Department of Electronics and Communication Engineering
Vidyavardhaka College of Engineering
Mysuru, Karnataka, India

K. Tuckley
IIT Bombay
Mumbai, India

Trushit Upadhyaya
Charotar University of Science and Technology
Changa, India

P. Upender
Assistant Professor
VITS Deshmukhi
Hyderabad, Telangana, India

P. A. Harsha Vardhini
Professor
VITS Deshmukhi
Hyderabad, Telangana, India

Karteek Viswanadha
Delhi Technological University
New Delhi, India

Yogesh
Amity School of Engineering & Technology
Amity University
Noida, India

Part I

Overview and Introduction

1 Antenna Design for Wireless Application

Mihir Narayan Mohanty
ITER, Siksha 'O' Anusandhan (Deemed to be University), Bhubaneswar, Odisha, India

1.1 INTRODUCTION

Currently microstrip antennas design is one of the most popular research areas in wireless communication technology. Because of their minimal size they are assuming significant function in wireless communication, particularly in broadband and ultra wide band (UWB) applications. UWB communication has been turned into an appealing suggestion for research network due to its various advantages such as minimal effort, intricacy and power utilization, penetration through any kind of material, and so forth. The microstrip antenna is a basic aspect of any communication conspire and is subsequently becoming a hotly debated research topic [1]. Among various types of printed antennas, rectangular, squared, triangular, and elliptical are the most used geometry shape. This is because of the uncomplicated structure and mathematical modeling as compared to complex antenna geometry. For improving the antenna characteristics, numerous modifications in geometry were done in the last few decades. Normally, conservative printed antennas are reverberating at a single frequency equivalent to their principal mode. Low bandwidth and gain are two major limitations of these types of antennas. In advanced wireless communication framework, high data rate is one of the most enviable aspects along with the requisite of coexistence with present communication framework. This means that these antennas must be proficient of resonating on more than one resonant frequency. For resonating one or more frequency, the broadband and ultra wideband antennas can be configured by modifying the geometry. Also by adopting this design, the bandwidth can be enhanced. These aforementioned factors are the main cause behind designing and analyzing of specific planar antennas. Though bandwidth is very narrow, enhancement is very much required [2].

In UWB wireless approach, remarkable research attempts were taken to design the specific antennas that can support the communication framework. These antennas are useful for application in wideband wireless communications. With use of narrow pulses in terms of nanoseconds, it can cover wide bandwidth in the frequency domain over short distance at low power densities [3]. With the burgeoning requirement for higher data transfer speed and internet ingress, the communication systems with UWB operations are choosing more prevailing communication

DOI: 10.1201/9781003187325-1

systems and to notch the interference signal bands in the UWB frequency range. The research work deals with bandwidth enhancement of microstrip patch antenna suitable for broad band and UWB applications. The major contribution of authors in [4] was analysis of the frequency and time-domain response of the UWB antennas so that their suitability applications can be fixed for portable pulsed-UWB system. Etching narrow slots on the antenna was also proposed by them to avoid narrow band interference and their effects on a nano-second pulse.

The printed microstrip patch radiators are broadly used in different communication systems as these radiators can issue many advantages such as compact, lightweight, integral with monolithic microwave integrated circuit (MMIC), effective cost, easy fabrication, and stable radiation characteristics. Currently the stipulation of mobile communications and transportable apparatus increases. Hence, small and cost effective radiators requirement has brought the microstrip radiator to the vanguard. However, the major limitation of the microstrip patch radiator is its narrow impedance bandwidth that limits their applications. A typical microstrip patch antenna has one percent to few percent bandwidth for thin substrates. The congeniality between communication distance prolongation and multiple communication systems stipulates the bandwidth enhancement. The most recurrent approach used for improving bandwidth is to decrease the substrate dielectric constant [5].

The most auspicious technology is UWB technology for wireless communication with higher data rate of greater than or equal to 100 Mbit/s in 15 meter distance, radar applications of highest accuracy and systems of imaging. The UWB antenna is currently one of the most fascinating research fields between the researchers. As many systems stipulate more than multiband operation [6], so UWB antenna is that type of antenna which is simpler than multiband antenna. The UWB technology having very lower emission limit works with exceedingly wide band width in microwave frequency range in comparison to the traditional broadband wireless communication technology. The designs of antennas for UWB and broadband have different demanding exude because of system characteristics and distinctive applications. They are broad and wide impedance bandwidth response, radiation characteristics, gain, and also compact or small in physical size. This chapter disputes the bandwidth improvement of microstrip patch radiator for broadband and UWB technology. Then supervisory environs are concisely presented followed by a review of bandwidth enhancement of microstrip patch antenna. At last a stimulant explanation is introduced behind the present research. Various applications of UWB are in small range and wireless communications with high data speed, like ground searching radars, equipment for medical imaging, wireless local area networks (WLAN) with higher data rate, and communications for defense purposes [7].

1.2 RELATED LITERATURE

In recent years, various microstrip antennas were invented for achieving the best performance. Authors in [8,9] have designed planar UWB antenna and UWB antenna in their work. Their proposed design was of two rejected bands with a rectangular patch. In substrate they have used FR4 and they have also achieved good bandwidth from their proposed design. Similarly a novel ring microstrip antenna with rectangular polarized cut was proposed in [6]. Their proposed antenna consists of a ring radiation patch, six non-metallic column, and two different switches. From the result it was observed that the gain of the antenna was satisfactory as compared to some other designs. The result was analyzed for both simulated and measured case. Based on fractal and metamaterial, authors in [10] have designed a microstrip compact antenna for improving the data communication. The effect of fractal structure was implemented in the basic microstrip antenna. The modification in the design was applied for improving the performance in wireless application and from the result it was confirmed that the experimental result was satisfactory as compared to simulated result. A novel technique was proposed for designing a single feed circular polarized microstrip antenna [11]. The dimension of the fractal etched defected ground structure (FDGS) was adjusted for obtaining the circular polarized radiation. The proposed antenna was designed and fabricated. From the result it was observed that the gain and bandwidth of the antenna was better. A study on circular microstrip antenna for stress evaluation was conducted in [12]. In that work they have also designed an antenna with two different resonant frequencies and variations on current distributions.

The wireless communication systems are growing very fast for wide band antennas, which are of great importance. For these systems, the radiators should satisfy higher gain with larger bandwidth. The UWB varies from 3.1 to 10.6 GHz, data rate between 110 to 200 Mbps and with highest radiated power −41.3 dBm/MHz, which is approved by FCC [13]. High-speed rate of data, small interference, reliable, inexpensive, and less complexity are the main merits of the UWB communication. The applications of UWB communication are in medicine imaging, radar, and military communication. Presently the microstrip patch antenna is significant in wide scope of multifunctionality in different frameworks like UWB, WLAN, rocket, radar, and utilizations of telemetry in science. These frameworks have spurred the creation of scaling down UWB radiators with high gain, better radiation efficiency, and radiation properties. The conceivable outcomes of capacities of these applications are a direct result of well-known points of interest like light weight, planar setup, low profile, minimized in size, simple fabrication, reasonable assembling, and coordination with other microwave circuitries [14]. A restricted impedance data transmission, poor in effectiveness, is the significant disservices of microstrip antenna. There are many methods of accomplishing improved impedance bandwidth for broadband, UWB qualities and applications beating these impediments. Because of higher transmission rate, wide bandwidth, low force thickness range, it turned into an inspiration for scientists. Likewise because of exactness, less troubles, low volume, simple creation cycle and ease, the printed planar radiators were created to be acceptable contenders for UWB correspondence. It differs from 3.1 to 10.6 GHz permitted by US Federal Communications Commission (FCC) in February 2002 [13,15]. To facilitate communication in different bands, multi-band antennas can be designed. The fact that microstrip patch antennas have narrow bandwidths can be exploited to design antennas operating at multiple bands. A compact broadband antenna that can operate at 2.45 and 5.8 GHz bands is designed in [16]. The patch antenna with a U-slot was printed on the bottom side of the 1.6 mm thick FR4 substrate and was suspended on top on the ground plane with shorting pins to achieve a broadband performance. The bandwidths were obtained to be 100 MHz and 200 MHz respectively and the gains were measured to be 0.5 dBi and 1.5 dBi respectively. The antenna was evaluated on a 3-layer human phantom model. The resonant frequencies shift slightly toward the lower frequencies when in proximity to the human body. A dual-band printed antenna working in 900/1800 MHz is designed in [17] using Cordura fabric for substrate and Zelt fabric for conducting sections. The antenna exhibited gains of the order of 1.8–2.06 dBi and efficiency up to 84%. The antenna was proposed to be used for RF energy harvesting applications. For a given WBAN scenario, IRIS motes can sustain themselves for a long time if RF energy is over −10 dBm.

A miniaturized antenna for MICS (403.5 MHz) and ISM (2.45 GHz) band is designed in [18]. The antenna consists of 2 center-fed via-loaded circular patches stacked over each other with a circular ground plane. The upper patch resonated at 403.5 MHz and lower at 2.45 GHz. The antenna exhibited an omnidirectional radiation pattern. A coplanar waveguide fed folded dipole antenna design is proposed in [19], which radiated from the edges and operated at 433 MHz. The antenna had a 0.2 mm thick FR4 substrate. The antenna was tested with different surfaces and static and moving human subject. The communication link was tested with an inverted-F antenna in the upper body at 14 different positions. The antenna was found to have a SAR of around 2 W/Kg. A microstrip antenna for operating in 2.45 GHz and 5.8 GHz is designed in [20]. A quadrant of circular microstrip antenna was bisected along the fictitious magnetic wall but preserved the distribution of the electric field at 2.45 GHz and cut a quarter wavelength slot at an open end in order to excite another resonant mode at 5.8 GHz. The fractional bandwidths at the two resonant frequencies were 1.2% and 2.0% respectively. The antenna exhibited on-body radiation efficiency of around 60% against 76% for a corresponding simulation. A serpentine microstrip patch antenna for operation in 403.5 MHz (MICS band) and 2.45 GHz (ISM band) is designed in [21]. The antenna was covered by 2 mm thick household silicon as a superstreet and to prevent shorting out of the antenna. The antenna is evaluated using an FDTD solver using a 2/3rd model and a simplified 3 layer model of the upper chest. The antenna arm lengths were optimized using a genetic algorithm using S11 at −10 dB as a cost function. Acircular slot antenna, based entirely on textiles, was proposed in [22]. The antenna was fed by a substrate integrated waveguide cavity. The performance of the antenna was analyzed using a phantom, which was equivalent to a 2/3rd muscle model. The return loss (−10 dB) bandwidth, measured on the phantom, covered a 5.8 GHz ISM band. The performance was found to be significantly unaffected by the deformation of the antenna or by the presence of the human body as the antenna was fed by the SIW cavity. A multi-band UWB antenna, with a tri-monopole structure with a rectangular patch, is presented in [23]. The antenna used FR4-epoxy as a substrate and coaxial feed. The antenna resonated at 3.1 GHz, 5.1 GHz, 8.0 GHz, and 8.8 GHz. The return loss obtained were −23 dB, −24.91 dB, −29.0 dB, and −20.0 dB respectively with VSWR within 1 and 2 for these bands.

An antenna for wireless body area network is designed in [24]. The antenna operated at three bands. The proposed antenna comprised of 2 Landolt ring strips connected by a rectangular strip on FR4-epoxy substrate. Two symmetric stubs extended from the coplanar waveguide ground plane. The antenna covered 3.19–4.57 GHz, 5.37–5.6 GHz, and 7.56–10.67 GHz bands. The antenna was simulated with a three-layer human phantom model and the frequency bands for on-body scenarios were observed to be 2.49–4.48 GHz, 5.14–5.65 GHz, and 7.60–10.55 GHz. The gain and radiation efficiencies for free-space at 3.8 GHz, 5.35 GHz, and 7.8 GHz are 5.5 dBi and 89%, 3.71 dBi, and 82%, and 4.07 dBi and 92% respectively. The gain and radiation efficiencies for the on-body scenario at 3.8 GHz, 5.35 GHz, and 7.8 GHz are 3.09 dBi and 41%, 2.80 dBi, and 42%, and 5.05 dBi and 49% respectively. A dual-band wearable button type antenna is designed in [25]. The scope of application of the antenna was the wireless local area network. The antenna was designed on a Rogers 5880 dielectric disc of 16 mm diameter located on top of the textile substrate. A conductive textile constituted the ground plane. The antenna comprises two patches—one with a tuning stub and is directly coax-fed; the second was a smaller capacitively coupled and shorted to the ground plane. The performance of the antenna was measured in free-space and on arm and chest phantom models. The antenna was also tested with various tilt angles. The antenna resonated at 2.4 GHz, 5.2 GHz, and 5.8 GHz. The antenna exhibited a vertically polarized omnidirectional pattern at 2.4 GHz, suitable for on-body communication and quasi-broadside patterns with small backlobes at 5.2 GHz and 5.8 GHz, which is suitable for off-body communication. The effect of proximity to the human body was minimized due to the large conducting ground. For 2.45 GHz, 5.2 GHz, and 5.8 GHz, gains measured in free space were 1.05 dBi, 4.5 dBi, and 6.04 dBi respectively, the gains measured on the body were 0.24 dBi, 5.18 dBi, and 5.54 dBi respectively, efficiencies were 91.9%, 87.3%, and 85.3% respectively. The SAR for 2.45 GHz, 5.2 GHz, and 5.8 GHz measured on arm were 0.20 W/kg, 0.22 W/kg, and 0.10 W/kg respectively and on the chest were 0.18 W/kg, 0.12 W/kg, and 0.13 W/kg respectively. A dual-band button antenna operating in on-body and off-body modes is presented in [26]. An omnidirectional pattern in 2.45 GHz band for on-body communication is obtained from the spiral inverted-F antenna and broadside radiation in 5.8 GHz band for off-body communication is obtained as the antenna interacts with the metal reflector. The antenna is mounted on a rigid metallic flange with a large disc at the bottom and a small disc at the top at an offset from the center. The ground coplanar waveguide feed printed on a 0.305 mm thick Rogers 4003 substrate laminate is placed on the larger disc of the flange, which then transforms to coaxial feed for the antenna. The coaxial feed comprises a copper wire passing through a hole drilled in the flange and filled with a Teflon insulator. For lower and upper bands, the peak gains in free space were measured to be 1.0 and 6.4 dBi, respectively, that measured on phantom were −0.6 and 4.3 dBi, respectively, and the efficiencies as measured on the phantom were 46.3% and 69.3% respectively. The SAR values were measured with 100 mW input power averaged over 1 g and 10 g of tissue at 2.45 GHz were 0.370 W/kg and 0.199 W/kg respectively.

A dual-band printed omega-shaped monopole antenna with a modified ground on an FR4 substrate is proposed in [27]. The antenna operated in two bands—402–405 MHz MICS band on-body communications and 2.4–2.48 GHz ISM band for off-body communications. The bandwidth was enhanced and impedance matching was done by using I-shaped and T-shaped stubs in the ground plane. The bandwidths obtained on a multi-layer tissue model were 39.8% for the MICS band and 14.1% for the ISM band. The antenna was evaluated on the human arm model. The gain at 2.45 GHz was 3.11 dBi with an efficiency of 92.4% and the Front-to-Back ratio was 1.5 dB and in the presence of phantom, the gain was 0.919 dBi. The SAR at 402 MHz was 0.02 W/kg and at 2.4 GHz was 0.1 W/kg. A dual-band antenna using a polyimide substrate operating at 2.45 GHz and 5.8 GHz is presented in [28]. The polyimide substrate makes the antenna flexible enough to be wearable. A metamaterial surface was used between the antenna and the human body. The antenna has a U-shaped radiating element. The antenna was fed by a coplanar waveguide feed. The substrate was made of 50 μm thick polyimide. The metamaterial array element has 3 layers: a square peripheral metal ring, a square metal patch on the dielectric substrate, and a metallic ground plane. The metamaterial surface had a 3 × 3 array of elements. Without the metamaterial surface, the radiation efficiency and gain in free space at 2.45 GHz were 82.3% and 1.25 dB respectively, at 5.8 GHz they were 82.4% and 1.66 dB respectively. For on-body measurements, the radiation efficiency and gain were 12.9% and −4.1 dB respectively at 2.45 GHz, and 31.5% and 2.33 dB respectively at 5.8 GHz. The SAR values averaged over 1 g body tissue were 17.5 W/kg at 2.45 GHz, and 9.3 W/kg at 5.8 GHz. With the metamaterial surface, for on-body measurements, the radiation efficiency and gain were 61.3% and 5.2 dB respectively at 2.45 GHz, and 67.2%, and 7.7 dB respectively at 5.8 GHz. The SAR values averaged over 1 g body tissue, were 2.48 W/kg at 2.45 GHz, and 3.33 W/kg at 5.8 GHz. The antenna was tested on different body parts and on different people. A dual-band antenna for wireless body area network working in 3.2–3.5 GHz and 3.9–4.3 GHz is presented in [29]. The antenna has a metamaterial structure with a 2 × 2 array of H-shaped patches integrated beneath it. The antenna was a simple rectangular microstrip patch antenna and was microstrip-fed. The antenna and the metamaterial structure were fabricated on 0.254 mm thick Rogers RT Duroid 5880 substrate. In the frequency bands, 3.2–3.5 GHz and 3.9–4.3 GHz, the antenna gain was found to be 4.54 dBi and 4.71 dBi, respectively, the front-to-back ratio was 11.79 dB and 12.79 dB and 1 g averaged SAR was found to be 0.174 W/kg and 0.207 W/kg at the respective bands. A multiband wearable monopole antenna with a defected ground structure, operating in 2.12 GHz, 4.78 GHz, 5.75 GHz, and 6.11 GHz is proposed in [30]. The antenna with 2 i-shaped slots and L-slot is fabricated on a denim substrate. The ground plane also had 2+-shaped slots. The return loss was found to be of the order of −51 dB. The antenna radiated from the end-fire position. An inkjet-printed antenna for operation at 2.45 GHz and 5.8 GHz WLAN bands is presented in [31]. The antenna consisted of two parts: both of which had their origins in the analytical profile of the sinusoidal curve with exponential decay, with a small stub attached to each, are placed alongside each other to reduce mutual coupling effects, and increase bandwidth. The antenna system has polyethylene terephthalate (PET) film as a substrate, which is transparent and highly flexible. The antenna is fed by a 50 Ω coplanar waveguide structure (CPW). The impedance bandwidth, peak gain, and efficiency at 2.45 GHz were found to be 19.3%, 2.05 dBi, and 81% respectively and at 5.8 GHz, were found to be 32.3%, 3.63 dBi, and 82% respectively. The SAR levels were found using a 3-layer human arm model and were found to be 6.035 W/kg at 2.45 GHz and 7.173 W/kg at 5.8 GHz.A microstrip-fed inverted-A monopole with multiple inputs and multiple outputs for wireless body area networks is presented in [32]. The antenna exhibited 3 bands at 3.8 GHz (LTE A 43), 3.9–4.2 GHz (C-band), and 5.1–5.85 GHz (IEEE 802.11 ac). The isolation between the antenna elements was of the order of −17 dB, and gain was 6 dBi.

This chapter gives the background information and literature review associated to the bandwidth improvement of microstrip patch radiator for broadband and UWB application. The chapter is systematized as follows. A concise introduction on bandwidth enhancement for broad band applications is presented in Section 1.3. Section 1.4 presents a brief introduction on bandwidth enhancement for UWB applications. Section 1.5 presents detail information associated to the notch band characteristics in UWB radiators, followed by a conclusion in Section 1.6.

1.3 BROADBAND ANTENNA TECHNOLOGY

The bandwidth of microstrip patch radiator (MPR) varies directly in proportion to the height (h) and inversely proportional to the square root of the dielectric constant (ε_r) of the dielectric substrate. As the substrate thickness (h) is increased, then the probe inductance is also increased and the probe compensation methods have been implemented for acquiring the impedance matching. But the practical limitation of the microstrip patch radiator is that by increasing the thickness of the antenna substrate (h) above $0.1\lambda_0$, the antenna performance will be decreased due to the surface wave propagation. The thick substrate having low dielectric constant value, 10% bandwidth with VSWR ≤ 2 has been achieved experimentally. But this value is not sufficient for numerous broadband operations like radar L-band varies from 1.4 GHz to 1.7 GHz with 19% bandwidth, C-band satellite television varies from 3.7 GHz to 4.2 GHz with 12.5% bandwidth, spread spectrum applications, etc. For improving antenna bandwidth different methods have been proposed by the researchers to solve this problem. These methods are broadly classified as:

i. Stacked multi resonator microstrip patch antennas
ii. Planar multi resonator broadband microstrip antennas
iii. Regularly shaped broadband microstrip antenna
iv. Broadband planar monopole antenna

1.3.1 Stacked Multi Resonator Microstrip Patch Antennas

In this method, multiple resonators are used to increase the bandwidth. But rather than planar coupled method, two or more than two radiating elements on separate sheets of the dielectric substrates are sheaved on the top of one another. The bottom patch is fed by a coaxial probe feed or by a microstrip line feed. The top radiating element is proximity coupled to the agitated bottom radiating element. Basically, the substrate of antenna increases by this technique. But planar direction size and single patch radiator remain the same. Hence, these multi-layered structures are applicable for array components. These multilayer structures give bandwidth of 70% virtually for VSWR ≥ 2 [33,34].

1.3.2 Planar Multi Resonator Broadband Microstrip Antennas

For broadband operation, the planar multiple resonator method is used. In this method, only a single patch is excited through the microstrip line feed and the other patches are coupled parasitically. The coupling among various resonators has been acquired by a short separation among the patches, which is less than 2 h; h is height of the dielectric substrate. In planar multiple resonator method, the patch called parasitic radiating element positioned near the feed radiating element, which is being agitated by the coupling among the two radiating elements. The combination of the two resonant frequencies f_1 and f_2 of these two patches give rise to broad bandwidth.

1.3.3 Regularly Shaped Broadband Microstrip Antennas

The classification of these types of microstrip patch radiators are mainly based on different configuration shapes of their radiating elements. The resonant frequency of the radiator depends on size and shape of their radiating elements. The microstrip patch antennas achieve broadband characteristics by changing the configuration shape or size of the radiating elements or by inserting slot or slit in the radiating elements or in ground plane. Generally, the broadband performance is obtained by decreasing the quality factor (Q), since the antenna bandwidth and quality factor (Q) are inversely proportional to each other [8].

1.3.4 Broadband Planar Monopole Radiator

The monopole radiator has been reported to give the highest bandwidth among other techniques. A microstrip patch fed by microstrip line feed at its edge through a perpendicular ground plane converts to a planar monopole radiator, when the bottom ground plane is detached. Different shapes of planar monopole radiators like circular, square, rectangular, triangular, elliptical, and hexagonal disc monopoles are illustrated for every broad voltage standing wave ratio (VSWR) band widths. The easy equations are given to expect the low frequency analogous to voltage standing wave ratio (VSWR) = 2 for these radiators. The monopole radiators such as hexagonal, rectangular, square, and triangular achieve a lower bandwidth than the elliptical and circular monopole. But the bandwidths of these antennas are broadly abundant for many operations. Also because of their easy fabrication process, these radiators are glamorous.

1.4 UWB TECHNOLOGY

The wireless technology for communicating digital data by a wide frequency range of moderate power is UWB technology. This UWB technology carries an enormous amount of data by small separation. From Shannon's theorem it is cleared that by increasing the channel capacity (c), bandwidth (B) increases linearly. But the Signal-to-Noise Ratio (SNR) remains fixed.

$$c = B * \log_2(1 + \text{signal to noise ratio}) \tag{1.1}$$

The Equation 1.1 shows the communication data rate that can be improved by improving the signal to noise ratio (SNR) or bandwidth. Generally, the signal to noise ratio (SNR) can be enhanced by increasing the signal power. As several compact devices are powered by battery, so the signal power cannot be improved. Also, the potential intervention with other radio systems should be restrained. Hence to apprehend high data rate, bandwidth improvement is the best solution. Equation 1.2 is given below for the mathematical relation between distances in free space, received and transmitted power.

$$D \propto \sqrt{\frac{P_{\text{Tr}}}{P_{\text{Rr}}}}, \tag{1.2}$$

where D, P_{Tr}, P_{Rr} indicate the distance, transmitted, and received power respectively.

Equations 1.1 and 1.2 indicate that, when bandwidth increases rather than power, then it is more efficient to obtain higher capacity. However, evenlthis is complicated to obtain in long range transmission. Basically, UWB has been developed for higher data rate transmission with small range communication system. The frequency range varies from 3.1 GHz to 10.6 GHz was assigned for UWB communication of unlicensed use by United States Federal Communications Commission (FCC) in 2002 [22]. To allocate effective use of frequency spectra is the main purpose for higher data rate of small range wireless personal area network (WPAN). To yield long range lower data rate wireless network operations also visualized together with imaging systems and radar as presented in Table 1.1.

The UWB signal resides exceedingly larger bandwidth as the radio frequency (RF) energy extends above tremendous spectrum. The UWB signal spectrum is wider ranging than any obligatory narrow band wireless communications with regard to magnitude and the emitted obligatory power. The maximum restricted attainable power to UWB transmitters is 0.556 mW, since the optimum deployed allocation band for UWB is full 7.5 GHz. It is scarcely a small part of accessible transmission power in the industrial, scientific, and medical (ISM) bands, such as wireless local area network (WLAN). It adequately immures the UWB system to small distance and indoor transmission at higher data rate or medium distance transmission at lower data rate. The utilizations like Wireless Personal Area Networks (WPANs) and Universal Serial Bus (USB) have been recommended extending over hundreds of Mega bits per second (Mbps) to numerous Giga bits per second (Gbps) data rate with the covering distance from 1 meter to 4 meters [23]. The attainable data rate by UWB is inferior for behind 20 meters distances. For determination of UWB characteristics, the radiator is the crucial component. For

TABLE 1.1
Different Ultra Wide Band (UWB) Systems with Frequency Range Below −41.3 dBm Equivalent Isotropically Radiated Power (EIRP) Emanation Limit

Applications	Frequency Range
Through wall imaging systems	1.61 GHz to 10.6 GHz
Indoor communication systems	3.1 GHz to 10.6 GHz
Vehicular radar systems	22 GHz to 29 GHz
Ground penetrating radar, wall imaging	3.1 GHz to 10.6 GHz
Medical imaging systems	3.1 GHz to 10.6 GHz
Surveillance systems	1.99 GHz to 10.6 GHz

UWB applications a compact, lightweight antenna design faces many challenges Despite these features, the radiators intended in are not able to incorporate into PCBs due to oversized and manufacturing problems.

1.4.1 Antenna for UWB Systems

For economic purposes, the considerable researches have implemented the performance evaluation of UWB radiators with the advancement of UWB communications. The three distinct categories of broadband radiators are available traditionally. DC-to-daylight is the first category. These radiators are invented with maximum bandwidth. The impulse radar, ground penetrating radars, electromagnetic compatibility, or field measurement radar are the quintessential applications. To obtain as broad spectrum as possible is the main objective of these types of design of radiators.

Small radiators: Small radiators tend to be compact, quasi omnidirectional, or omnidirectional. The bow tie radiator, diamond dipole, and the Lodge's biconical radiator are the examples of small radiators.

Frequency-independent radiators: The frequency independent radiators depend on the change in configuration from small scale to large scale section. The small scale section provides upper frequency and the large scale section provides the lower frequency. The actual origin of the radiated fields varies basically in proportion with frequency, which gives a scatter manner. The log periodic radiator, spiral radiator, and conical spiral radiators are the examples of these frequency independent radiators.

Horn radiators: The horn radiator is an electromagnetic channel in which the power directs in a certain direction. The horn radiators are having higher gain and comparably narrow beams. But the horn antennas are to be enormous and largest. These radiators are applicable for point to point transmission operations that need a narrow beam width.

Speculum radiators: The speculum radiators point powers in a certain direction same as in horn radiators. Basically, the modification and adjustment of feed manipulation are easy in speculum radiators. The Hertz's parabolic cylinder reflector is the example of reflector radiator.

The radiators need to be compact enough for many applications to be incorporated within lightweight devices. Hence, the burgeoning implication of UWB transmission inspired researchers to inspect on compact radiators. Two groups of compact radiators have developed to yield these solutions. The first one is invented from biconical radiators. But the planar configurations include elliptical, diamond, bow tie, and circular disc dipoles. The second one is invented from the origination of monopole elements. Based on various planar elements such as square, elliptical, circular, etc., numbers of UWB monopoles have been exhibited to allocate UWB performances. In accordance with the traditional resonating theory, these types of radiators are no more acceptable. To account for the performance of UWB planar radiators, the higher order modes, and dipole mode are proposed by H.G. Schantz.

1.4.2 UWB Antenna Parameters

The parameters of UWB antenna for time domain or frequency domain can be elected subject to applications. The time domain is switched from the frequency domain by inverse Fourier transform. The important parameters of UWB radiator are described as follows.

Peak value of the impulse response: It is denoted by $P(\theta, \Phi)$ and is the largest value of the substantial peak of the radiator transient behavior.

Pulse width is the amount of representing the widening of the transmitted impulse. The pulse width value is restricted to a few hundred picoseconds (pc) in order that assurance of higher data rate is in transmission.

Ringing (τ_R) is an unwanted outcome in UWB radiator. The multiple reflections as energy storage in the antenna produce in oscillation of the transmitted pulse after the principal peak. There is no use of this energy and reduces the maximum value. The ringing period (τ_R) is signified as the time until the amplitude of pulse has decayed from the maximum value to beneath a definite lower limit (α). The value of ringing should be very small, which is signified in nano second (ns). That is, the ringing is less than a few pulse widths.

Group delay (τ_G) quantitatively calculates the scatter behavior of the radiator. The group delay is defined in frequency dominion as presented in Equation 1.3.

$$\tau_G(\omega) = -\frac{d\Psi(\omega)}{d(\omega)} = -\frac{d\Psi(f)}{2\pi df}. \tag{1.3}$$

Fidelity of a radiator measures how exactly the received voltage accessible at the radiator terminal emulates the performance of transient field incident on the radiator. For transmission, this measures how exactly the time integral of the radiated field emulates the performance of the voltage applied to the radiator terminals. Also, it is renowned that the fidelity of radiator is waveform precise. That means the radiator may produce a high fidelity replication for few waveforms and it is not able to produce identical fidelity for numerous other waveforms.

1.4.3 Bandwidth Enhancement of the Antenna

The important characteristic of microstrip patch radiator is its impedance bandwidth that can be considerably enhanced by using different methods. To acquire this, the importance is typically laid on bandwidth improvement methods of microstrip patch radiator, depicted for broadband and UWB applications.

1.4.4 Need for Bandwidth Enhancement

The main limitation of microstrip patch radiator is its narrow impedance bandwidth. Hence, the microstrip radiators should be designed for bandwidth enhancement in broadband and UWB applications. The impedance bandwidth of microstrip patch radiator is the order of 1% to 5%. The bandwidth enhancement is one of the areas of research in the field of microstrip patch radiator. Generally, the bandwidth is defined more concisely as a percentage $\frac{f}{f_c} \times 100\%$, where f represents the width of the range of acceptable frequencies; f_c represents the center frequency of the antenna. Center frequency, $f_c = (f_l + f_u)/2$, f_l and f_u are lower and upper cut off frequencies respectively. The parameters like return loss, voltage standing wave ratio (VSWR), and radiation efficiency are often used to define the bandwidth of microstrip patch radiator.

In literature the well-known measure of bandwidth of a radiator is determined in terms of VSWR range, which is given below:

$$1 \leq \text{VSWR} \leq 2$$

For broadband and UWB antenna design the following factors are necessary in radiator configuration.

- Lower permittivity or larger thickness of the dielectric substrate to obtain low Quality factor (Q).
- Feed impedance must be matched.
- Optimization of structure of radiating element.
- Suppression of surface waves in a thick dielectric substrate.

The impedance bandwidth of microstrip patch radiator can be enhanced by using the dielectric substrate with low dielectric constant value or by increasing the thickness of dielectric substrate. The thickness of dielectric substrate and low permitivity improves the surface waves resulting in low gain, low efficiency of the radiator, and perverts the radiated field characteristics. Hence, these effects become perceptible, when the height of dielectric substrate increases beyond 2% of the wavelength [35]. In single layer dielectric microstrip patch radiator, the impedance bandwidth may be improved by increasing the thickness of the dielectric substrate but at the expense of efficiency of the radiator because of surface wave factor. By using Wheeler's transformation [36], all the losses such as radiation, dielectric, surface wave and conduction losses, and quality factors allied with it may therefore be evaluated for calculating the impedance bandwidth. Hence, it is necessary to consider the factors that affect these losses. In the cavity model, the radiation characteristic of the microstrip patch radiator is modelled by an effective loss tangent.

The quality factor due to radiation (Q_r) is deduced as

$$Q_r \propto \frac{\varepsilon_{re}^3 c^3}{hbf_r^2}, \tag{1.4}$$

where ε_{re} is effective dielectric constant, c is velocity of light, h is height of dielectric substrate, b is width of patch, and f_r is resonant frequency. From Equation 1.4, it is deduced that:

- As h increases, quality factor (Q_r) reduces due to increase in radiated power (Pr).
- As width of the patch (b) decreases, the quality factor (Q_r) reduces due to increase in radiated power (Pr).

1.4.5 Bandwidth Enhancement and Its Use in UWB Application

The ultra-wideband antennas designed for wearable applications operate within 3–11 GHz. This band has gained immense popularity after the 2.4 GHz band. The bandwidth widening is done mostly by geometrical manipulation of the antennas. A textile antenna was designed where the conducting parts were made of metalized Nylon fabric and a 0.5 mm thick acrylic fabric was used as a substrate. The feed was used as coplanar-waveguide for the design of UWB disc monopole and a UWB annular slot antenna. As the performance measure, radiation pattern, return loss, and transfer function were evaluated. In general, textile antennas development is a critical task. It did not show much improvement with the simulated results. On the other hand, a microstrip-fed UWB annular slot antenna exhibited fairly good agreement with the simulated results barring a few discrepancies. An inverted truncated annular conical Dielectric Resonator Antenna operating within lower European UWB 3.4–5.0 GHz for Body Area Network applications is presented in [35]. The inverted configuration helped in enhancing the bandwidth. The simulated results show return losses of up to −38 dB and a bandwidth of 3.4–5.0 GHz. Three UWB antennas were presented and their performances were experimentally characterized for near/on body positions with different orientations [36]. The 3 antennas were: monopole antenna for Omni direction, directional suspended plate antenna, and printed diversity antenna. Results show that as the body moved, the omnidirectional antenna experienced increased transmission but horizontal coverage was poor. The distance from the body did not significantly affect the polarization performance of the directional antenna. The proximity to the human body had negligible effects on the performance of the diversity antenna, making it the most suitable candidate for on-body communications. A CPW-fed stair-shaped UWB antenna for Wireless Body Area Network (WBAN) is presented in [37] and impedance bandwidth of 3–11 GHz and a gain of about 6.5 dB was obtained. The antenna was evaluated in free space, on the 3-layered human phantom model, and on various on-body positions on chest, abdomen, and hands. The antenna was found to become directional when placed on the body and the resonance frequency shifted to the lower frequency side. A folded UWB antenna is presented in [38] with an aim to reduce backward radiation and human proximity effects. The antenna has a 3D structure consisting of a beveled edge feed structure for impedance matching over a broad range and a metal plate with a folded strip as the radiator. The bandwidth coverage is from 3.1 to 12 GHz with additional resonating frequency at 6 GHz. A circular microstrip patch antenna with an octagonal slot, fed by a microstrip line, is proposed in [39]. The antenna had partial ground and 1 mm thick blue jeans fabric was used as a substrate. With return loss bandwidth of around 1 GHz, the scope of application of the proposed antenna included body area network and C-band satellite communication. A UWB antenna aimed at off-body communication and works at 6.1 GHz is designed in [40]. The antenna was designed on an FR4 substrate and used a wide feed slot. Circular slots were cut in the rectangular patch antenna creating two open branches ends. The simulated and measured values of return loss were –31.47 dB and 20.16 dB with a bandwidth of more than 500 MHz. The very high radiation efficiency of more than 90% was also observed. A pentagonal monopole antenna with square slots in the defected ground and FR4 substrate was presented in [41]. The feed to the antenna was flared just before the patch to improve the current distribution in the patch. The dimensions and positions of the slots were optimized to improve the bandwidth. The antenna covered the entire UWB bandwidth from 2.9 GHz to 11 GHz but exhibited poor efficiencies when in proximity to the human body. The SAR was measured from simulations on a 3-layer lossy and dispersive phantom model at 10 mm from the human body and it was found out to be 1.6 W/kg. A UWB antenna operating over a UWB frequency range of 5.24–8.83 GHz is designed in [42]. The antenna was designed on an FR4 substrate. The antenna was simulated on the 3-layer human phantom model. The antenna exhibited a bandwidth of 3.59 GHz and a gain of 5.37 dBi, an efficiency of 1.11%, and SAR of 0.264 W/kg.

A wideband antenna for Wireless Body Area Network applications is proposed in [43]. The antenna was a semi-circular patch on Rogers RT-Duroid 5880. The antenna was backed by an artificial magnetic conductor structure fabricated on the same substrate as that of the antenna. The artificial magnetic conductor was made of 4 × 6 semi-circular patch unit cells. The antenna was fed by a microstrip line. The return loss bandwidth was observed to be 7 GHz, a gain of 1.84 dB was observed, and SAR was observed to be 2×10^{-5} W/kg. A UWB CPW fed antenna with a wide square slot is presented in [44]. The antennas were fabricated on the non-woven substrate, cotton substrate, and rigid FR4 as substrate. The radiating element was of copper and silver. The antenna covered the frequencies from 3.2 GHz to 16.3 GHz. The antennas were evaluated in free-space and on the chest. When on the body, the overall efficiencies of the antennas decreased. The antenna with a cotton substrate showed lesser variations than the other two. The antenna is aimed at wearable applications.

1.5 ANTENNA DESIGN

1.5.1 Design of UWB Microstrip Antenna

Enhancement of bandwidth and gain is an important aspect of designing a microstrip antenna [45,46]. Even some of the authors have tried to design UWB antenna as in [47–50]. The upper part of truncated ground plane is reshaped in the proposed antenna to increase the bandwidth. The shape is modified by adding a right angled triangular slot for achieving monopole like wideband width, steady radiation pattern, and compact in size. The results are provided for various slot shape of the UWB antenna. In ultra wide band wireless communication radiator is a vital component. The important main reason for designing UWB radiator is to make it simple and small with less distortion over a broad bandwidth. The geometry of the proposed UWB antenna is shown in Figure 1.1. The shortcoming of the antenna is limited bandwidth. The main objective of this design is to make the modifications in the partial ground plane in order to obtain broad bandwidth.

The variables of the antenna can be modified to tune voltage standing wave ratio along with impedance operating bandwidth [51,52]. The upper part of the truncated ground plane is reformed by inserting different slots such as rectangular, square, semi circular, triangular, and right angled triangle shaped slots for further improvement of the proposed antenna and are presented in Figure 1.2.

In Figure 1.3, the return loss of the proposed antenna is provided and the simulated return loss of the proposed radiator is provided in Figure 1.4.

1.5.2 Design of UWB Microstrip Fractal Antenna

Another star shaped fractal antenna for UWB application is also provided in this chapter. FR4-epoxy substrate is considered in this design. The parametric values are shown in Table 1.2.

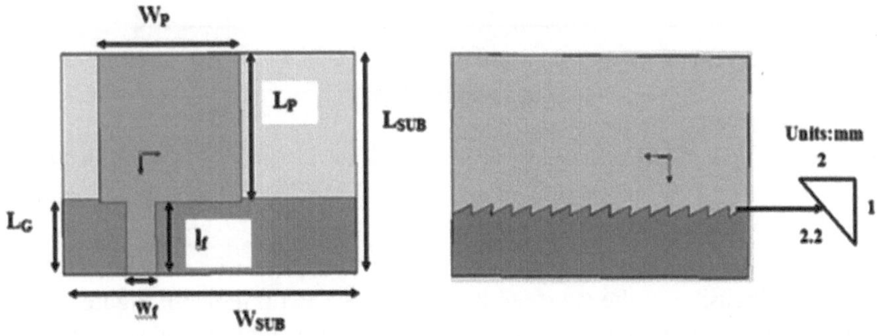

FIGURE 1.1 Structure of proposed antenna.

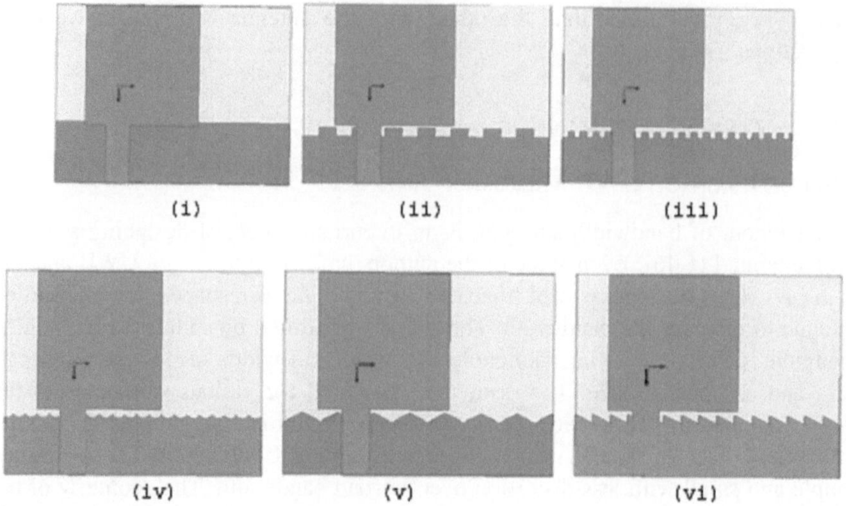

FIGURE 1.2 Antennas with different slots: (i) no slot, (ii) rectangular slot, (iii) square slot, (iv) semi circular slot, (v) triangular, and (vi) right angled triangular slot.

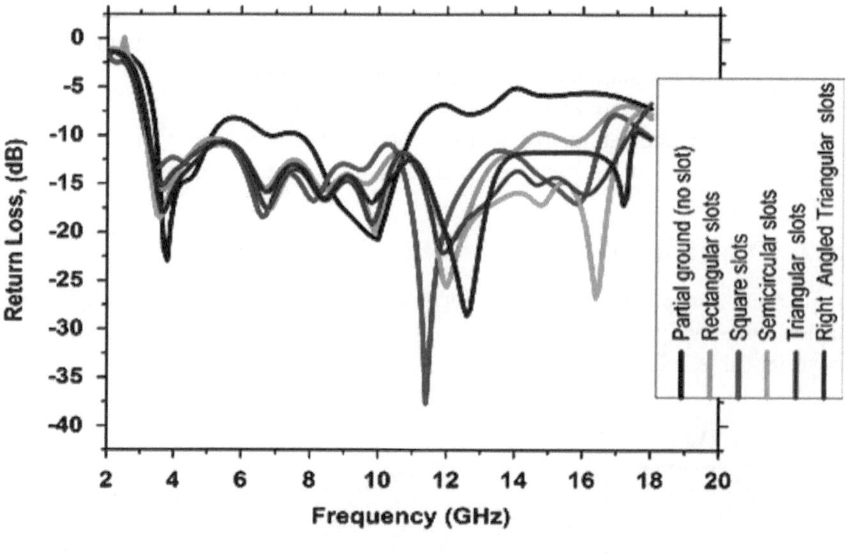

FIGURE 1.3 Return loss (simulation) of proposed antenna without slot and with different slots on the partial ground plane.

For enhancing the bandwidth, a substrate of low dielectric constant is chosen here. The dimension of the substrate is 21.45 × 19.7 × 1.6 mm^3. In Figure 1.5, the geometry of the proposed star shaped antenna is provided. The results obtained from both the antenna structures are provided in Figures 1.6–1.8.

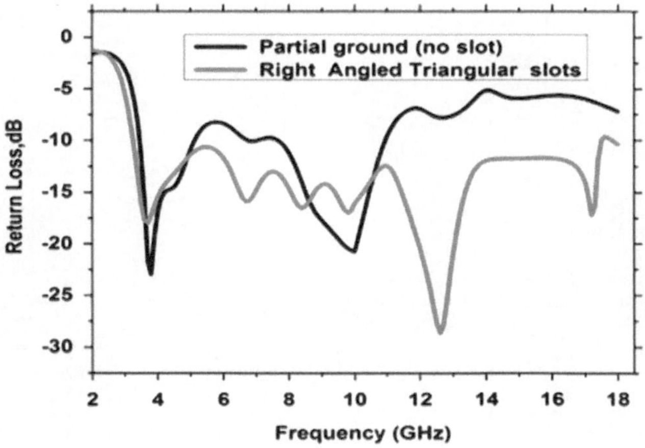

FIGURE 1.4 Return loss (simulation) of proposed antenna with and without slots on the partial ground plane.

TABLE 1.2
Parameters of the Proposed Star-Shaped Antenna

Parameters	Value for Antenna 1 (Unoptimized)	Value for Antenna 2 (Optimized)
Substrate material	FR4-epoxy	FR4-Epoxy
Centre frequency (f_r)	8.6 GHz	8.6 GHz
Substrate height (h)	1.6 mm	1.6 mm
Loss tangent	0.02	0.002
Dielectric constant (ε_r)	4.4	4.4
Sides of equilateral triangle ($A1$)	15.9 mm	15.9 mm
Sides of equilateral triangle ($B1$, $B2$)	5.3 mm	5.3 mm
Feed width	5.52 mm	4.1 mm
Patch thickness	0.1 mm	0.1 mm

Return loss is the important parameter of antenna; it is the difference between forward and reflected power in dB. The acceptable value of return loss is less than −10 dB for the antenna to work efficiently. Figure 1.6 shows the simulated return loss v/s frequency plot of optimized and non-optimized value of proposed antenna. The comparison of the curves shows that the optimized structure has −26.4901 dB return loss at 8.6 GHz frequency, non-optimized structure has −14.3508 dB return

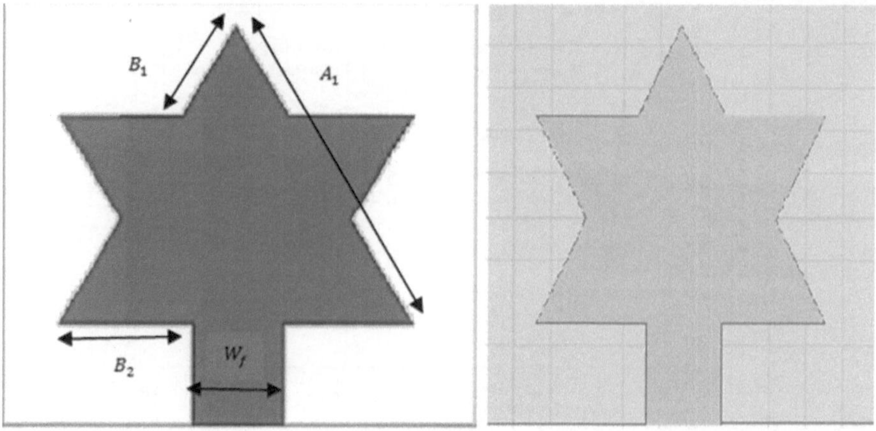

FIGURE 1.5 Geometry of the proposed star shaped antenna.

loss at 8.6 GHz frequency. The return loss was improved by optimizing the star-shape patch using sequential nonlinear programming algorithm in the HFSS environment. Sequential nonlinear programming algorithm is one of the EM optimization techniques integrated with Ansoft HFSS. This has been utilized to reduce the efforts of manual tuning of the star-shape patch dimensions in order to achieve the desired goal.

The VSWR is the measure of impedance mismatch between the feed line and the antenna. The mismatch increases the value of VSWR. The minimum value of VSWR is unity and the maximum VSWR is 2 for a perfect impedance matching. The simulated VSWR v/s frequency plot of optimized and non-optimized value of proposed antenna is shown in Figure 1.7. The optimized of proposed antenna shows that the value of VSWR is 1.0994 at 8.6 GHz frequency and also non-optimized of proposed antenna shows the value of VSWR is 1.4741 at 8.6 GHz frequency.

The gain describes the efficiency and directional capabilities of antenna. The gain plot (Figure 1.8) of optimized and non-optimized antenna shows the gain of 2.2989 dB, 2.1250 dB respectively. In Table 1.3, performance of the unoptimized version (Antenna 1) is compared with the optimized version (Antenna 2).

1.6 CONCLUSION

The proposed chapter has discussed the concepts of microstrip antennas for recent applications and possible advancements. Design of UWB antennas with multiple notch bands is also reported. Recent applications including cognitive radio, wearable and flexible antennas are of main issues in recent times. Antenna design methodologies have also evolved since a long period to meet the design specifications. However, it appears to become more challenging with time as the number of functionalities is required. Two examples have been considered for UWB and

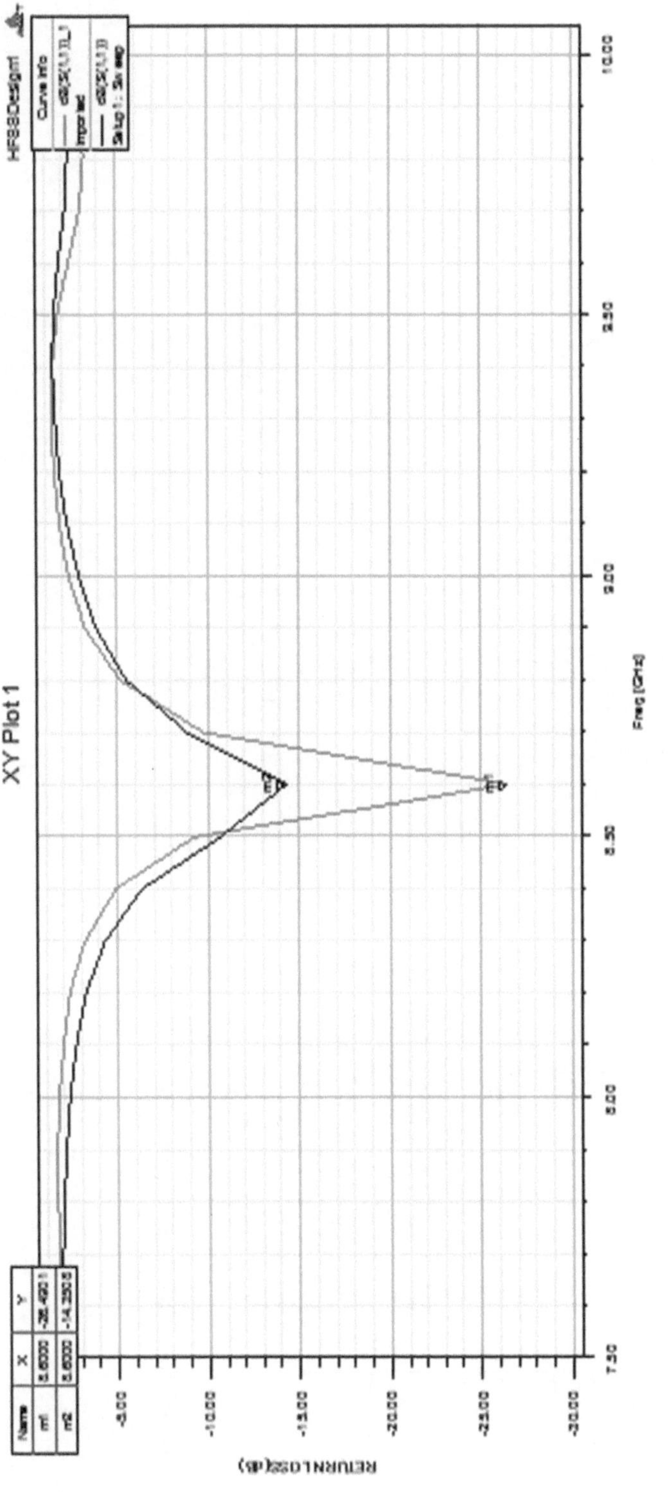

FIGURE 1.6 Return loss vs Frequency plot.

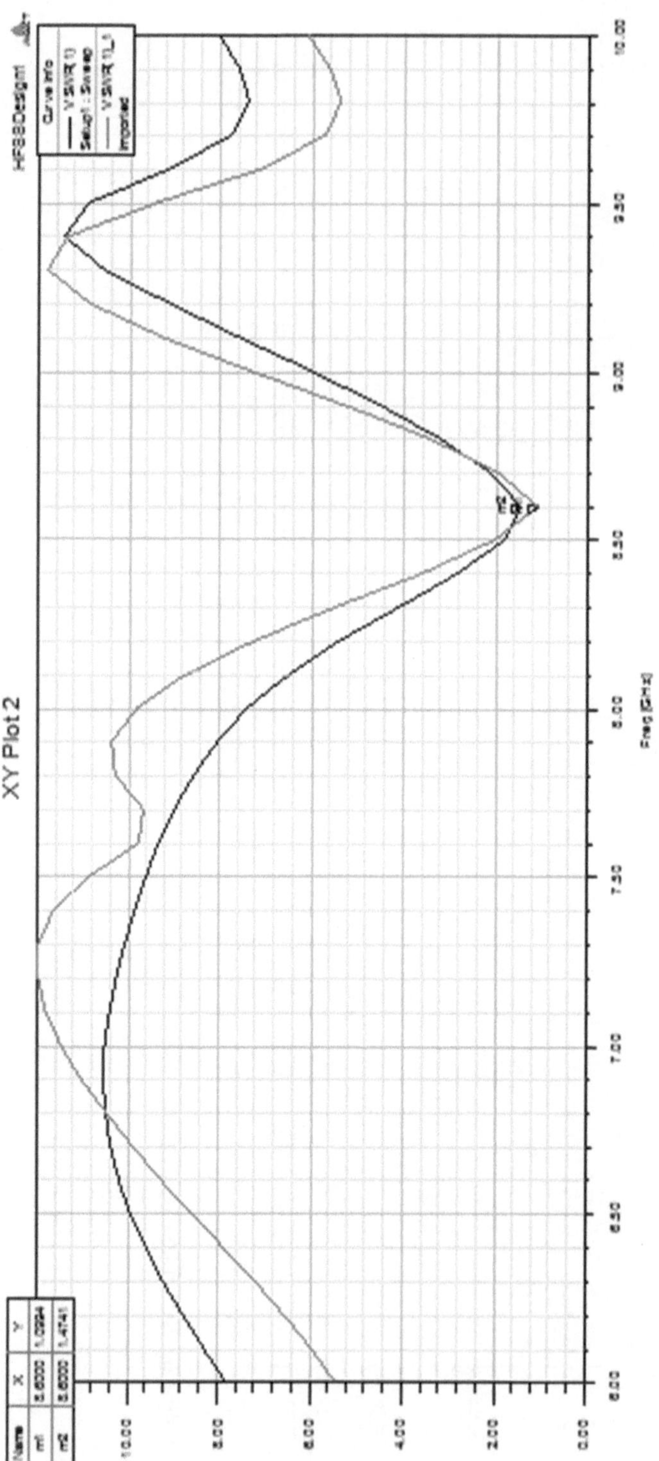

FIGURE 1.7 VSWR vs Frequency plot.

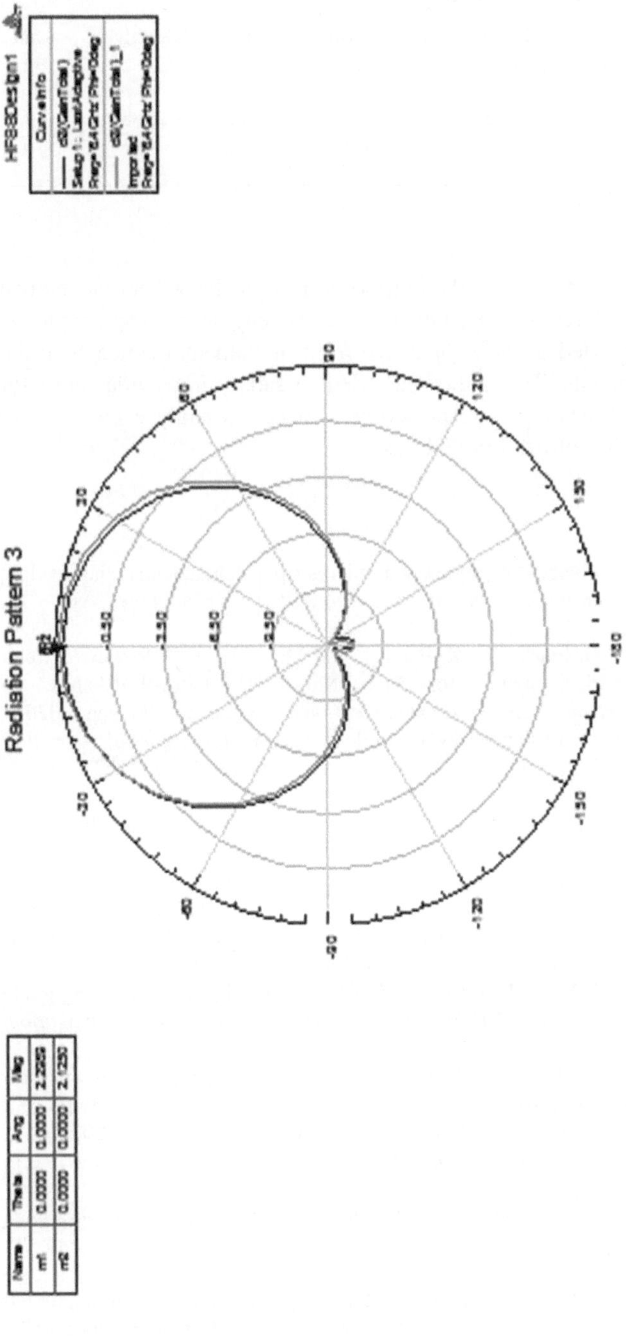

FIGURE 1.8 Plot for Gain in dB.

TABLE 1.3
Performance of the Proposed Antenna

Antenna	Resonating Frequency (GHz)	Return Loss (dB)	Bandwidth (MHz)	VSWR	Directivity	Gain (dB)
Antenna 1	8.6	−14.5	130	1.5	5.38	2.1
Antenna 2	8.6	−26.5	240	1.9	5.40	2.3

fractal type for different applications. The proposed designs have been developed for the new generation of researchers, developers, and engineers. The chapter is prepared to have the knowledge about the range from fundamental concepts to the state-of-the-art developments. The author has tried to satisfy in a wide sense for readers, researchers, and developers respectively. It needs to analyze the strength and weaknesses for further utilization and design.

REFERENCES

[1]. A. Bansal and R. Gupta, "A review on microstrip patch antenna and feeding techniques," *International Journal of Information Technology*, vol. 12, pp. 149–154, 2020.

[2]. A. Singh, K. Shet, D. Prasad, A. K. Pandey, and M. Aneesh, "A Review: Circuit Theory of Microstrip Antennas for Dual-, Multi-, and Ultra-Widebands," in *Modulation in Electronics and Telecommunications*. London: IntechOpen, 2020.

[3]. M. J. M. Saji, "Ultra-wideband (UWB) wireless system," *International Journal of Current Research In Science, Engineering and Technology (IJCRSET)*, vol. 1, pp. 53–59, 2016.

[4]. T. Saeidi, I. Ismail, W. P. Wen, A. R. Alhawari, and A. Mohammadi, "Ultra-wideband antennas for wireless communication applications," *International Journal of Antennas and Propagation*, vol. 10, pp. 1422–1429, 2019.

[5]. S. Elajoumi, A. Tajmouati, J. Zbitou, A. Errkik, A. Sanchez, and M. Latrach, "Bandwidth enhancement of compact microstrip rectangular antennas for UWB applications," *Telkomnika*, vol. 17, pp. 1559–1568, 2019.

[6]. Q. Chen, J.-Y. Li, G. Yang, B. Cao, and Z. Zhang, "A polarization-reconfigurable high-gain microstrip antenna," *IEEE Transactions on Antennas and Propagation*, vol. 67, pp. 3461–3466, 2019.

[7]. N. Nhlengethwa and P. Kumar, "Microstrip Patch Antenna with Enhanced Gain for 2.4 GHz Wireless Local Area Network Applications," in *Micro-Electronics and Telecommunication Engineering*. New York: Springer, pp. 583–591, 2020.

[8]. K.-L. Wong, *Compact and Broadband Microstrip Antennas,* vol. 168. New York: John Wiley & Sons, 2004.

[9]. N. M. Awad and M. K. Abdelazeez, "Multislot microstrip antenna for ultra-wide band applications," *Journal of King Saud University-Engineering Sciences*, vol. 30, pp. 38–45, 2018.

[10]. G. Varamini, A. Keshtkar, and M. Naser-Moghadasi, "Compact and miniaturized microstrip antenna based on fractal and metamaterial loads with reconfigurable

qualification," *AEU-International Journal of Electronics and Communications*, vol. 83, pp. 213–221, 2018.
[11]. K. Wei, J. Li, L. Wang, R. Xu, and Z. Xing, "A new technique to design circularly polarized microstrip antenna by fractal defected ground structure," *IEEE Transactions on Antennas and Propagation*, vol. 65, pp. 3721–3725, 2017.
[12]. P. Lopato and M. Herbko, "A circular microstrip antenna sensor for direction sensitive strain evaluation," *Sensors*, vol. 18, pp. 310, 2018.
[13]. E. Docket, "Docket 98-153, FCC, Revision of part 15 of the Commissions Rules Regarding Ultra-Wideband Transmission Systems," Technical Report, 2002.
[14]. A. Kumar, S. Dwari, G. P. Pandey, B. K. Kanaujia, and D. K. Singh, "A high gain wideband circularly polarized microstrip antenna," *International Journal of Microwave and Wireless Technologies*, vol. 12, pp. 678–687, 2020.
[15]. K. Da Xu, H. Xu, Y. Liu, J. Li, and Q. H. Liu, "Microstrip patch antennas with multiple parasitic patches and shorting vias for bandwidth enhancement," *IEEE Access*, vol. 6, pp. 11624–11633, 2018.
[16]. G. Liu, D. Wang, F. Jin, and S. Chang, "Compact Broad Dual-Band Antenna Using a Shorted Patch With a Thick Air Substrate for Wireless Body Area Network Application," in *2011 4th IEEE International Symposium on Microwave, Antenna, Propagation and EMC Technologies for Wireless Communications*, 2011, pp. 18–21.
[17]. N. Barroca, H. M. Saraiva, P. T. Gouveia, J. Tavares, L. M. Borges, F. J. Velez, *et al.*, "Antennas and Circuits for Ambient RF Energy Harvesting in Wireless Body Area Networks," in *2013 IEEE 24th annual International Symposium on Personal, Indoor, and Mobile Radio Communications (PIMRC)*, 2013, pp. 532–537.
[18]. S. Chamaani and A. Akbarpour, "Miniaturized dual-band omnidirectional antenna for body area network basestations," *IEEE Antennas Wireless Propagation Letters*, vol. 14, pp. 1722–1725, 2015.
[19]. D. Gaetano, P. McEvoy, M. J. Ammann, M. John, C. Brannigan, L. Keating, *et al.*, "Insole antenna for on-body telemetry," *IEEE Transactions on Antennas Propagation*, vol. 63, pp. 3354–3361, 2015.
[20]. X.-Q. Zhu, Y.-X. Guo, and W. Wu, "A compact dual-band antenna for wireless body-area network applications," *IEEE Antennas Wireless Propagation Letters*, vol. 15, pp. 98–101, 2015.
[21]. P. Soontornpipit, "A dual-band compact microstrip patch antenna for 403.5 MHz and 2.45 GHz on-body communications," *Procedia Computer Science*, vol. 86, pp. 232–235, 2016.
[22]. Y. Hong, J. Tak, and J. Choi, "An all-textile SIW cavity-backed circular ring-slot antenna for WBAN applications," *IEEE Antennas and Wireless Propagation Letters*, vol. 15, pp. 1995–1999, 2016.
[23]. G. P. Venkatesan, M. Pachiyaannan, S. Karthik, T. S. Ananth, and J. Kirubakaran, "A Low Profile UWB Antenna for Wireless Body Area Network: Design and Analysis," in *Proceedings of the 2nd International Conference on Biomedical Signal and Image Processing*, 2017, pp. 77–80.
[24]. M. N. Shakib, M. Moghavvemi, and W. N. L. B. W. Mahadi, "Design of a tri-band off-body antenna for WBAN communication," *IEEE Antennas Wireless Propagation Letters*, vol. 16, pp. 210–213, 2016.
[25]. H. Xiaomu, S. Yan, and G. A. Vandenbosch, "Wearable button antenna for dual-band WLAN applications with combined on and off-body radiation patterns," *IEEE Transactions on Antennas Propagation*, vol. 65, pp. 1384–1387, 2017.
[26]. X. Y. Zhang, H. Wong, T. Mo, and Y. F. Cao, "Dual-band dual-mode button antenna for on-body and off-body communications," *IEEE Transactions on Biomedical Circuits Systems*, vol. 11, pp. 933–941, 2017.

[27]. A. Kumar and R. K. Badhai, "A Dual-Band on-Body Printed Monopole Antenna for Body Area Network," in *2017 International Conference on Inventive Systems and Control (ICISC)*, 2017, pp. 1–5.
[28]. M. Wang, Z. Yang, J. Wu, J. Bao, J. Liu, L. Cai, *et al.*, "Investigation of SAR reduction using flexible antenna with metamaterial structure in wireless body area network," *IEEE Transactions on Antennas Propagation*, vol. 66, pp. 3076–3086, 2018.
[29]. B. Hazarika, B. Basu, and J. Kumar, "A multi-layered dual-band on-body conformal integrated antenna for WBAN communication," *AEU-International Journal of Electronics Communications*, vol. 95, pp. 226–235, 2018.
[30]. Roopan, D. Samantaray, and S. Bhattacharyya, "A Multiband Wearable Antenna with Defected Ground Structure," presented at the *2019 URSI Asia-Pacific Radio Science Conference (AP-RASC)*, 2019, New Delhi, India.
[31]. M. M. Bait-Suwailam and A. Alomainy, "Flexible analytical curve-based dual-band antenna for wireless body area networks," *Progress In Electromagnetics Research*, vol. 84, pp. 73–84, 2019.
[32]. I. Suriya and R. Anbazhagan, "Inverted-A based UWB MIMO antenna with triple-band notch and improved isolation for WBAN applications," *AEU-International Journal of Electronics Communications*, vol. 99, pp. 25–33, 2019.
[33]. G. Kumar and K. P. Ray, *Broadband Microstrip Antennas*. Noorwood: Artech House, 2003.
[34]. M. Klemm and G. Troester, "Textile UWB antennas for wireless body area networks," *IEEE Transactions on Antennas Propagation*, vol. 54, pp. 3192–3197, 2006.
[35]. G. Almpanis, C. Fumeaux, J. Frohlich, and R. Vahldieck, "A truncated conical dielectric resonator antenna for body-area network applications," *IEEE Antennas Wireless Propagation Letters*, vol. 8, pp. 279–282, 2008.
[36]. T. S. See and Z. N. Chen, "Experimental characterization of UWB antennas for on-body communications," *IEEE Transactions on Antennas Propagation*, vol. 57, pp. 866–874, 2009.
[37]. H. Shin, J. Kim, and J. Choi, "A Stair-Shaped CPW-Fed Printed UWB Antenna for Wireless Body Area Network," in *2009 Asia Pacific Microwave Conference*, 2009, pp. 1965–1968.
[38]. C.-H. Kang, S.-J. Wu, and J.-H. Tarng, "A novel folded UWB antenna for wireless body area network," *IEEE Transactions on Antennas Propagation*, vol. 60, pp. 1139–1142, 2011.
[39]. S. Dhupkariya and V. K. Singh, "Textile antenna for C-band satellite communication application," *Journal of Telecommunication, Switching Systems Networks*, vol. 2, 2015.
[40]. M. Pachiyaannan and G. Venkatesan, "Optimal design of 6.1 GHz UWB antenna for off body communication," *Asian Journal of Information Technology*, vol. 15, pp. 4229–4235, 2016.
[41]. S. Doddipalli, A. Kothari, and P. Peshwe, "A low profile ultrawide band monopole antenna for wearable applications," *International Journal of Antennas Propagation*, vol. 2017, pp. 1–9, 2017.
[42]. A. Biswas, A. J. Islam, A. Al-Faruk, and S. S. Alam, "Design and Performance Analysis of a Microstrip Line-Fed on-Body Matched Flexible UWB Antenna for Biomedical Applications," in *2017 International Conference on Electrical, Computer and Communication Engineering (ECCE)*, 2017, pp. 181–185.
[43]. T. M. Das, S. N. Islam, G. Sen, and S. Das, "A Novel AMC Backed Wide Band Wearable Antenna," in *2019 IEEE Region 10 Symposium (TENSYMP)*, 2019, pp. 628–630.
[44]. S. M. H. Varkiani and M. Afsahi, "Compact and ultra-wideband CPW-fed square slot antenna for wearable applications," *AEU-International Journal of Electronics Communications*, vol. 106, pp. 108–115, 2019.

[45]. B. Roy, S. K. Chowdhury, and A. K. Bhattacharjee, "Symmetrical hexagonal monopole antenna with bandwidth enhancement under UWB operations," *Wireless Personal Communications*, vol. 108, no. 2, pp. 853–863, 2019.

[46]. T. Parveen, Q. U. Khan, D. Fazal, U. Ali, and N. Akhtar, "Design and analysis of triple band circular patch antenna," *AEU-International Journal of Electronics and Communications*, vol. 112, p. 152960, 2019.

[47]. A. Abbas, N. Hussain, M. J. Jeong, J. Park, K. S. Shin, T. Kim, and N. Kim, "A rectangular notch-band UWB antenna with controllable notched bandwidth and centre frequency," *Sensors*, vol. 20, no. 3, p. 777, 2020.

[48]. A. Iqbal, A. Smida, N. K. Mallat, M. T. Islam, and S. Kim, "A compact UWB antenna with independently controllable notch bands," *Sensors*, vol. 19, no. 6, p. 1411, 2019.

[49]. Lakrit, S., Das, S., El Alami, A., Barad, D., and Mohapatra, S., "A compact UWB monopole patch antenna with reconfigurable Band-notched characteristics for Wi-MAX and WLAN applications," *AEU-International Journal of Electronics and Communications*, vol. 105, pp. 106–115, 2019.

[50]. F. A. Shaikh, S. Khan, A. H. Z. Alam, D. Baillargeat, M. H. Habaebi, M. B. Yaacob, … and Z. Shahid, "Design and parametric evaluation of UWB antenna for array arrangement," *Bulletin of Electrical Engineering and Informatics*, vol. 8, no. 2, pp. 644–652, 2019.

[51]. Arora V., Malik P. K. (2020) Analysis and Synthesis of Performance Parameter of Rectangular Patch Antenna. In: Singh P., Pawłowski W., Tanwar S., Kumar N., Rodrigues J., Obaidat M. (eds) *Proceedings of First International Conference on Computing, Communications, and Cyber-Security (IC4S 2019)*. Lecture Notes in Networks and Systems, vol 121. Springer, Singapore https://doi.org/10.1007/978-981-15-3369-3_12

[52]. A. Kaur and P. K. Malik, "Tri State, T Shaped Circular Cut Ground Antenna for Higher 'X' Band Frequencies," *2020 International Conference on Computation, Automation and Knowledge Management (ICCAKM)*, 2020, pp. 90–94, doi: 10.1109/ICCAKM46823.2020.9051501.

Part II

Performance Analysis of Microstrip Antenna

2 Design and Development of a Printed Circuit Microstrip Patch Antenna at C-Band for Wireless Applications with Coaxial Coupled Feed Method

P. Upender[1] *and Dr. P.A. Harsha Vardhini*[2]
[1]Assistant Professor, VITS, Deshmukhi, Telangana, India
[2]Professor, VITS, Deshmukhi, Telangana, India

2.1 INTRODUCTION

This work aims at design and development of a printed circuit microstrip patch antenna at 4.1 to 4.6 GHz (C-band) having linear polarization for wireless applications with coaxial coupled feed method. Radar and satellite wireless applications include a radar altimeter where they prefer low weight and high-grade antennas. This application is feasible with the mentioned microstrip antenna, also known as a planar antenna. The microstrip antenna is equipped with a coaxial feed at 4.3 GHz (C-band) in a rectangular shape. The performance of this shape is analyzed in terms of radiation pattern, half power points, and gain and impedance bandwidth. This work is extended with fabrication using a photo etching process similar to printed circuit board technology and, further, these parameters are simulated in ANSOFT HFSS. The C-band micro-trip antenna is used to build, optimize, and test high-performance wireless applications. Because of their low cost and comfortable and easy architecture, patch antenna is an attractive way to create wireless communications, particularly in radar altimeters. Patch antenna is a resonant style radiator having narrow bandwidth. The simulated results are compared with the obtained fabrication results. Proposed coaxial couple feed

DOI: 10.1201/9781003187325-2

antenna performs for a band width of 100 MHz and providing a gain of 5 dB at 4.3 GHz resonant frequency with a return loss of −12.3 dB. With a resonant frequency centered at 4.3 GHz, the proposed antenna has greater advantage and is suitable for altimeter application. Radar altimeter (RA) is a system used to measure a low altitude or distant distance to the ground or sea level from an aircraft or spacecraft. A part of the radar is radio altimeter. The fundamental theory of radar is that radio waves are transmitted to land or sea level and receive an echo signal for a long time [1].

For radar altimeter applications MSA in C- band is carried out [2]. The basic antenna needs are increased, bandwidth, polarization, small size, low weight, and easy manufacturing using modern wireless communication methods. All such criteria could be made using circuit board antennas. It is an attractive solution to many wireless communication scenarios because of their low cost, conformable, and easy-to-manufacture architecture. It is comprised of ground plane at the last, a dielectric substrate of stature h, and a fix based on the substrate. Measurements of ground plane are more prominent than substrate and fix [3]. The range of the transmitting frequency of altimeter is from 4.2 GHz to 4.4 GHz and the transmitted signal at a frequency of 40 Hz per foot varies. Parameters involved in the design are given.

2.1.1 Radiation Pattern and HPBW

An antenna radiation is characterized as a realistic portrayal of the receiving wire's radiation properties as space coordinates work. It gives data on the radio wire shaft width, receiving wire side flaps. The radio wire examples of most viable reception apparatus contain a principle flap and a few assistant projections named as side flaps. The half power beam width (HPBW) is the exact division wherein the size of the radiation configuration reduces considerably (or −3 dB) from the zenith of the fundamental pillar [4,5].

2.1.2 Voltage Standing Wave Ratio (VSWR)

The standing wave is characterized as "The proportion of most extreme to least current or voltage on a line having Voltage Standing Waves and this Ratio is abridged (VSWR)" [6,7]. Thus,

$$VSWR = \left| \frac{V_{\max}}{V_{\min}} \right| \tag{2.1}$$

Relation between (S) and (Γ):

$$VSWR = \left| \frac{V_{\max}}{V_{\min}} \right| = \frac{1+\rho}{1-\rho} \tag{2.2}$$

2.1.3 Return Loss

When the other end is mismatched the returned power is return loss [8]. In dB it is given as

$$RL = -20 \log|\Gamma| \text{ dB}, \tag{2.3}$$

$$\Gamma = \frac{V_o^+}{V_o^-} = \frac{Z_L - Z_o}{Z_L - Z_o}. \tag{2.4}$$

2.1.4 Gain

Gain is a primary output number integrating direction and strength of the antenna [8].

2.2 METHODOLOGY

MSA is comprised of a conductive layer of planar arrangement, one hand of the dielectric base inverse leg upheld by a ground line [9]. Various focal points include lightweight, low volume, minimal effort, planar arrangement, and coordinated circuit similarity. Low-profile receiving wires are required for remote applications. A feed line is used to excite the patch [10,11]. In the proposed design coaxial feed is utilized.

2.2.1 Coaxial Feeding

Coaxial is a technique during which the coaxial inward conductor is associated with the radio wire's radiation field where the external conduct is connected to the basement [12,13].

2.2.2 MSPA Design Calculations

To plan any MSPA, barely any boundaries should have been picked earlier. The plan boundaries of proposed radio wire are operational recurrence fr = 4.3 GHz, RT/Duroid 5880 is taken as dielectric substrate whose dielectric steady is 2.20 ± 0.02, and scattering factor is 0.0009 [14,15].

2.2.3 Patch Design Procedure

Width (W): With the details including data about the dielectric consistent of the substrate ($\in_r$), the thunderous recurrence (fr), and the stature of the substrate (h). The procedure is as follows:

$$W = \frac{c}{2f_r}\left(\frac{\varepsilon_r + 1}{2}\right)^{\frac{-1}{2}}. \tag{2.5}$$

For $C = 3 * 10^8$ m/sec, $f = 4.3$ GHz, $\varepsilon_r = 2.2$.
$W = 27.5$ mm.

Effective dielectric constant (ε_{eff}):
Find Eeff for MSA using Equation (2.6):

$$\varepsilon_{eff} = \frac{\varepsilon_r + 1}{2} + \frac{\varepsilon_r - 1}{2}\left(1 + \frac{12h}{W}\right)^{\frac{-1}{2}} \tag{2.6}$$

$W = 27.5$ mm; $\varepsilon_r = 2.2$; $h = 0.8$ mmε_r= 2.116

Extension length (ΔL):

$$\frac{\Delta l}{h} = 0.412\frac{(\varepsilon_{eff} + 0.3)\left(\frac{W}{h} + 0.264\right)}{(\varepsilon_{eff} - 0.258)\left(\frac{W}{h} + 0.8\right)} \tag{2.7}$$

where $\varepsilon_{\text{eff}} = 2.116$, $h = 0.8$ mm, $W = 27.5$ mm.
Length extension calculated using above values (**Δ**L) = 0.422 mm

Actual length (L) is given by:

$$L = \frac{c}{2f_r\sqrt{\varepsilon_{eff}}} - 2\Delta l \tag{2.8}$$

Actual length calculated using above values is $L = 23.131$ mm.

Effective length (L_{eff}):
The L_{eff} is found by using Equation (2.2).

$$L_{eff} = L + \Delta L \tag{2.9}$$

By inserting values of L and Δl we get $L_{\text{eff}} = 23.975$ mm.

Ground Plane (GP) parameters:
Parameters are its length Ls and width Wg, which are twice the patch dimensions. This will allow the reflected waves from the edge of the ground plane to have a complete zero shifts from the point of origin (center of the patch).
Ls = 46 mm and Wg = 56 mm.

Microstrip line width (Wo):
Using Equations (2.10) and (2.11), line width is depicted as

$$Z_o = \frac{120\pi}{\sqrt{\varepsilon_{eff}}\left[\frac{W_0}{h} + 1.393 + 0.667\ln\left\{\frac{W_0}{h} + 1.414\right\}\right]} \quad \text{for } \frac{W_0}{h} > 1 \tag{2.10}$$

$$Z_o = \frac{60}{\sqrt{\varepsilon_{eff}}} \ln\left[\frac{8h}{W_o} + \frac{W_o}{4h}\right] \text{ for } \frac{W_0}{h} < 1 \tag{2.11}$$

By iterations, Wo = 2.95 mm.

Formula for wavelength is depicted as

$$\lambda = cf. \tag{2.12}$$

Wavelength λ = 62.76 m.

Microstrip length yo calculation:
Radiating Conductance, Gr = 1.7 * 10^{-3}

$$Z_c = \frac{1}{2(G_T)}\left[\cos^2\left(\frac{\pi}{2}y_0\right)\right] \tag{2.13}$$

Microstrip length, Yo = 8.77 mm. Table 2.1 depicts the designed values.

Dimension of MSPA:
Figure 2.1 depicts the pictorial representation of MSPA with all the dimensions.

2.3 SIMULATION OF MSA USING COAXIAL COUPLED FEED METHOD USING HFSS

For good impedance coordinating the coaxial procedure is being chosen. A square miniature strip fix utilizing 50-ohm test feed is utilized for this situation. Coaxial is likewise best appropriate for simple mounting utilization of altimeter, and it offers ease and simplicity of creation favorable circumstances. HFSS is a vivid test system

TABLE 2.1
Final MSA Designed Values

Parameter	Units
Length, L	23.13 mm
Width, W	27.5 mm
Microstrip Width, W_o	2.95 mm
Microstrip Length, Y_o	8.77 mm
Length of the ground plane, L_g	46 mm
Width of the ground plane, W_g	56 mm
Effective length, L_{eff}	23.975 mm
Effective dielectric constant, ε_{eff}	2.116
Wavelength, λ	69.76

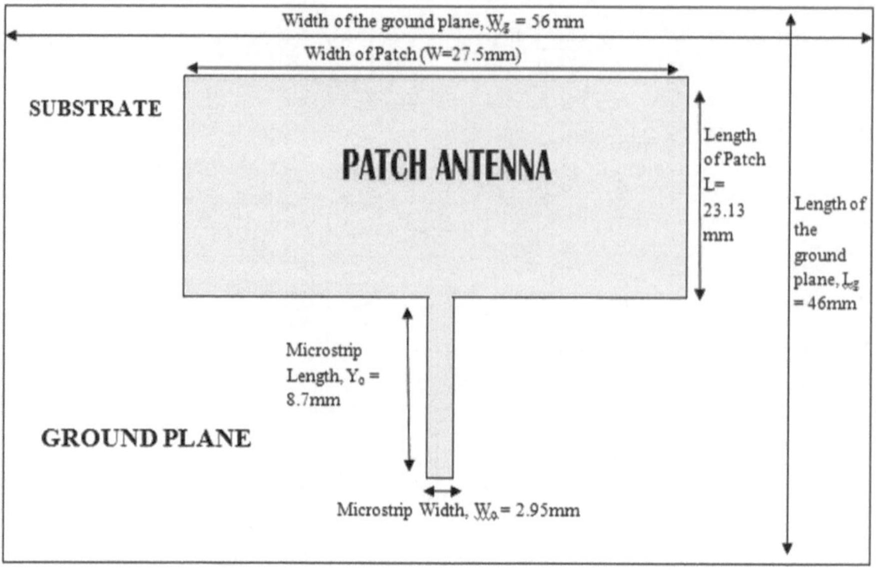

FIGURE 2.1 Dimensions of MSPA.

whose base work component is tetrahedron. This lets us beat certain discretionary 3D math, especially those that utilize certain methods inside a small amount of the time it would take for muddled bends and shapes. Following advances are to be acted in the plan of MSA.

2.3.1 Design of Patch Antenna Using HFSS:

Plan of antenna measurements utilizing HFSS includes the following advances:

- The coordinates for the HFSS is planned or determined utilizing the element of the fix radio wire.
- From the get facilitates at 3D coaxial took care of fix is drawn.
- Patch, ground plane, and coaxial feed are allotted as flawless electric channel material and substrate is relegated RT/5880.
- An air box (radiation box) is made around the fix and the radiation limit is allotted to all sides of the container.
- Wave port is relegated to the fix radio wire for excitation reason.

2.3.2 Optimization Using HFSS

When characteristics are portrayed for the development, a 3D setup is built. On the off chance that the trademark doesn't coordinate with the normal, the component of fix takes care of or fluctuated in little advances and re-reproduces until required attributes are acquired. In this way, streamlining of boundary like feed length underneath the fix, feed width, and fix length is accomplished straightforwardly without making a high quality fix and afterward testing it with functional estimation methods.

2.3.3 Simulation

In the planning of the microstrip fix radio wire the reenactment is performed. Right off the bat, validation check is performed to watch if there are any mistakes in the plan. Next, the setup is allotted to the receiving wire, which incorporates working recurrence 4.1 GHz to 4.6 GHz with step size of 0.1 GHz. Investigate everything that is chosen to play out the reenactment. At last, the necessary reception apparatus boundaries charts are seen in HFSS.

2.3.4 Return Loss Plot

The return loss is acquired from the chart that appeared in Figure 2.2. As the return is exceptionally less, acceptable impedance coordinating is acquired between the feed and the fix.

2.3.5 VSWR Plot

As appeared in Figure 2.3, the VSWR is least at 4.3 GHz. As VSWR is roughly more like 1, no signal is reflected back toward the feed, i.e., the whole signal will be communicated to the fix.

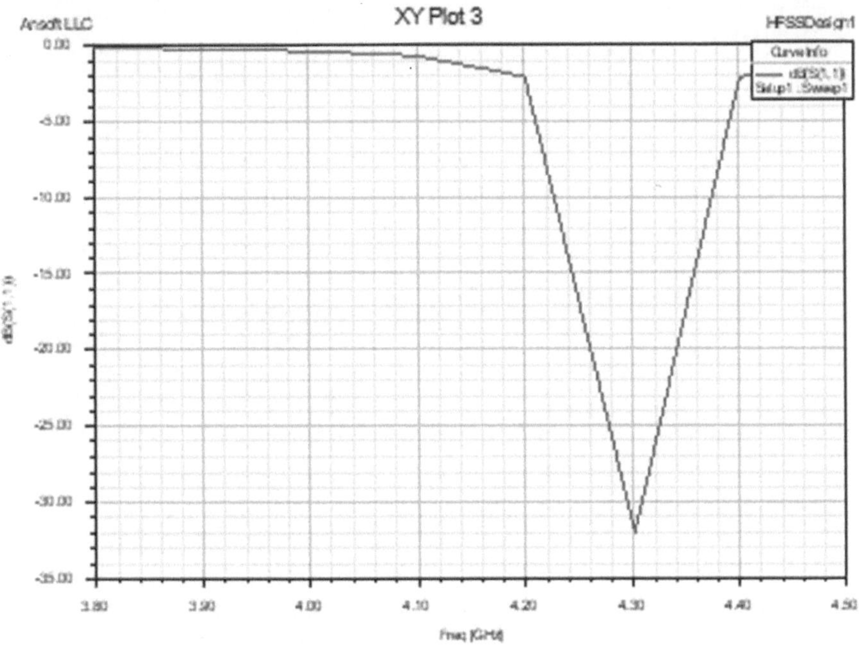

FIGURE 2.2 Return loss (dB) plot.

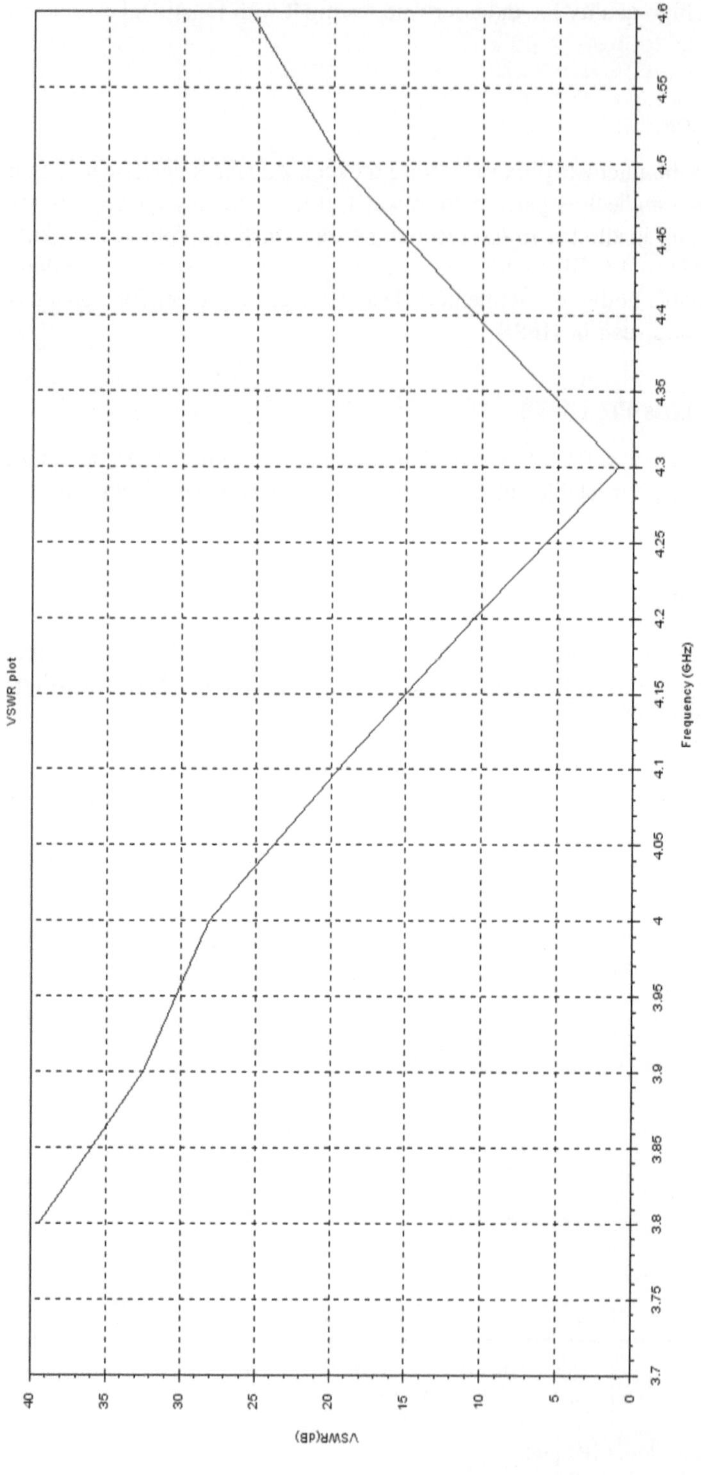

FIGURE 2.3 VSWR graph.

2.3.6 Graph of 2D Radiation Pattern

A 2D rectangular radiation plot in the far-field regions is shown in Figure 2.4.

2.3.7 3D Polar Radiation Pattern Graph

Figure 2.5 is a 3D polar radiation plot where radiation intensity with respect to spherical coordinates is observed.

2.3.8 Graph of 2D Radiation Pattern

Figure 2.6 depicts 2D radiation pattern resonates at 4.3 GHz. From the simulation results it is clear that return loss is a convenient way to characterize the input and output of the signal sources or when the load is mismatched, not all the available power from the generator is delivered to the load. Since RL is close to −32 db, which is less, acceptable impedance coordinating is received between the feed and the fix. From the VSWR plot as appeared in Figure 2.8, the VSWR is least at 4.3 GHz. As, VSWR is around 1, no signal is reflected back toward the feed. From the radiation plots it is clear that at 4.3 GHz gain of 9 dB is achieved, which is good and has a great advantage to be used for wireless applications.

2.4 ANTENNA TESTING AND FABRICATION

The plan boundaries decide the entire components of the fix receiving wire. Directions are gotten dependent on reception measurements. The outcomes get

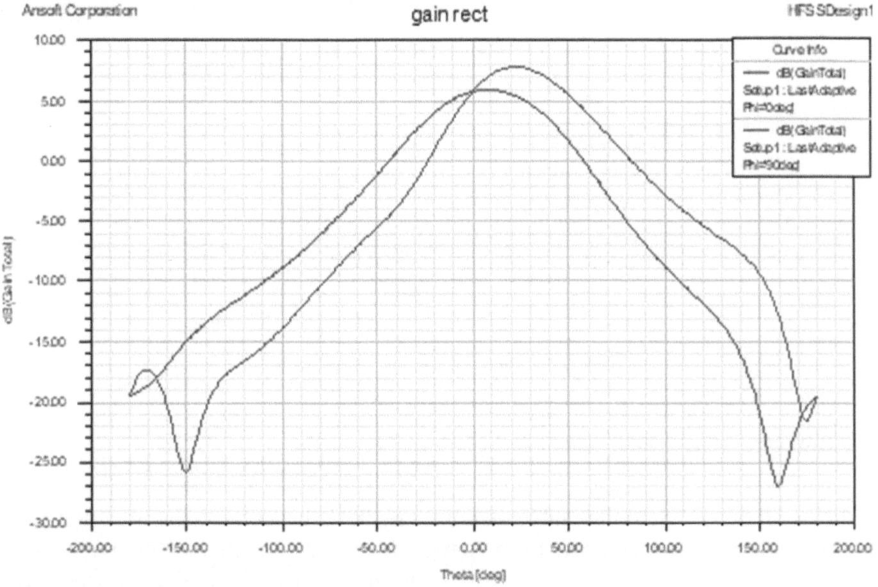

FIGURE 2.4 2D radiation pattern graph.

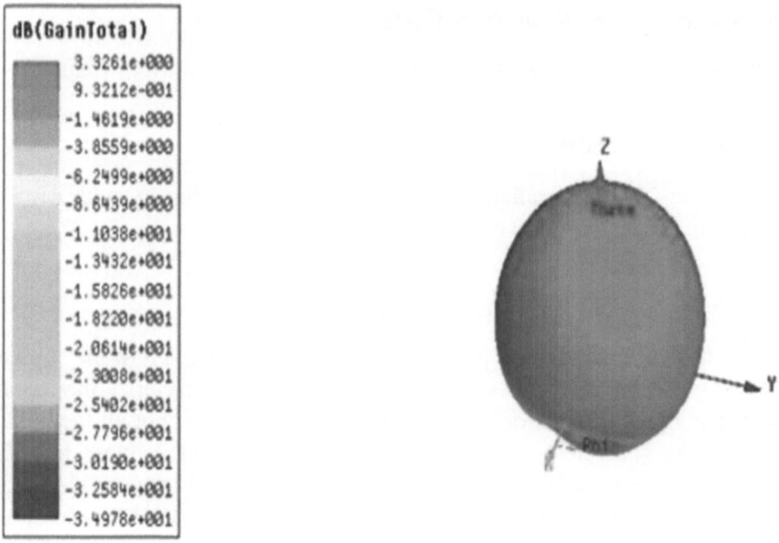

FIGURE 2.5 3D polar radiation pattern plot.

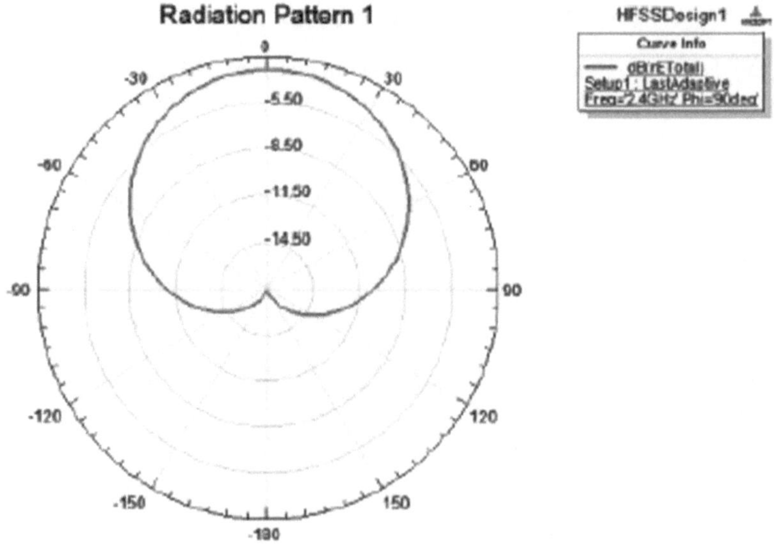

FIGURE 2.6 2D radiation pattern plot (gain vs frequency).

utilizing HFSS match to the viable outcomes by 95%; henceforth, the directions can be straightforwardly taken care of in AutoCAD for creating the reception apparatus design. MSA qualities enormously rely on the sort of material utilized as dielectric substrate.

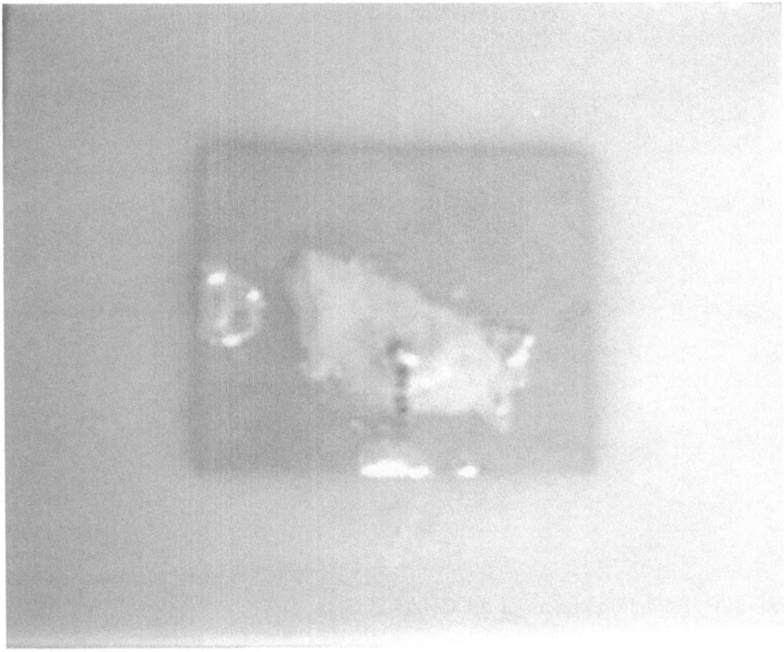

FIGURE 2.7 Fabricated coaxial coupled MPA.

2.4.1 Dielectric Substrate Choice

RT/Duroid 5880 is chosen as substrate whose $\mathcal{E}$ is 2.20 ± 0.02.

Metallization

In this proposed design, PEC "perfect electrical conductor" for metallization design is chosen.

2.4.2 Fabrication

Creation of the reception antenna is completed and the substrate material metalized on utilization of photolithographic measure on the two sides.

2.4.3 Photolithographic Process

The manufacture of the MSA depends on the photolithographic cycle in which a photosensitive resistive layer is presented to bright radiation through a veil. A few methodologies can be utilized to configuration veil.

2.4.4 Assembly of Antenna

L shape strip is mounted the tab on the line, which is soldered for matching the impedance.

The fabricated coaxial coupled microstrip antenna is shown in Figure 2.7.

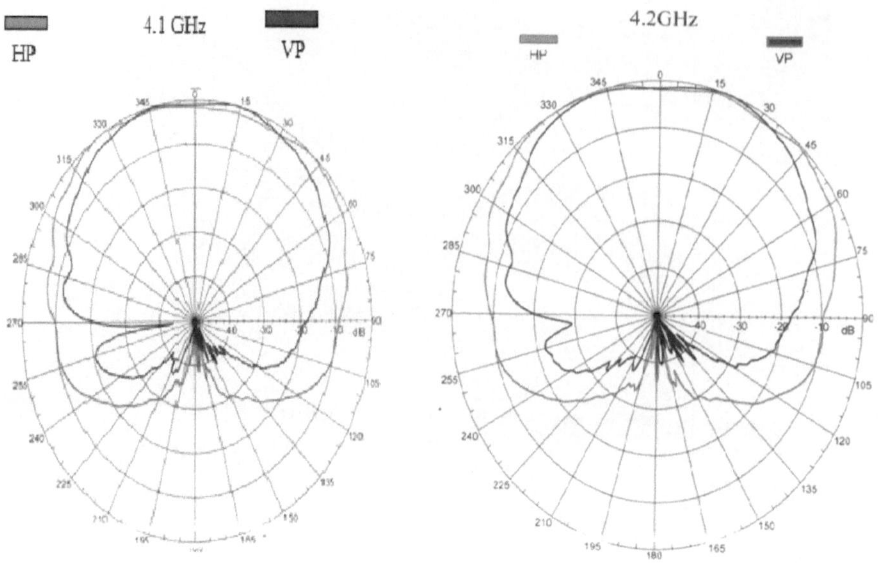

FIGURE 2.8 Radiation pattern at 4.1 GHz, 4.2 GHz.

2.5 FABRICATION RESULTS

For the manufactured receiving wire radiation design, RL, VSWR diagrams are drawn and are portrayed in Figures 2.16–2.23. In the radiation designs in the figures, smooth radiation design in blue tone are E and in red tone is in H-plane.

2.5.1 Radiation Pattern Plot

Following Figures 2.8–2.10 portrays the radiation design plots of MSA. Radiation design at 4.1, 4.2, 4.3, 4.4, 4.5, and 4.6 GHz appeared beneath in Figures 2.8–2.10. It is seen that at 4.3 GHz radiation example of microstrip fix radio wire is wide and it will have low radiation and restricted recurrence data transfer capacity. MSA has lesser directivity.

2.5.2 Return Loss plot

The RL is acquired from the chart which appears in the figure underneath. The RL acquired at 4.215 GHz is −12.391 dB as appeared in Figure 2.11. As the RL is less acceptable, impedance coordinating is acquired.

2.5.3 VSWR Plot

For Patch antenna VSWR calculation an HP-8722D Vector Network analyzer was used.

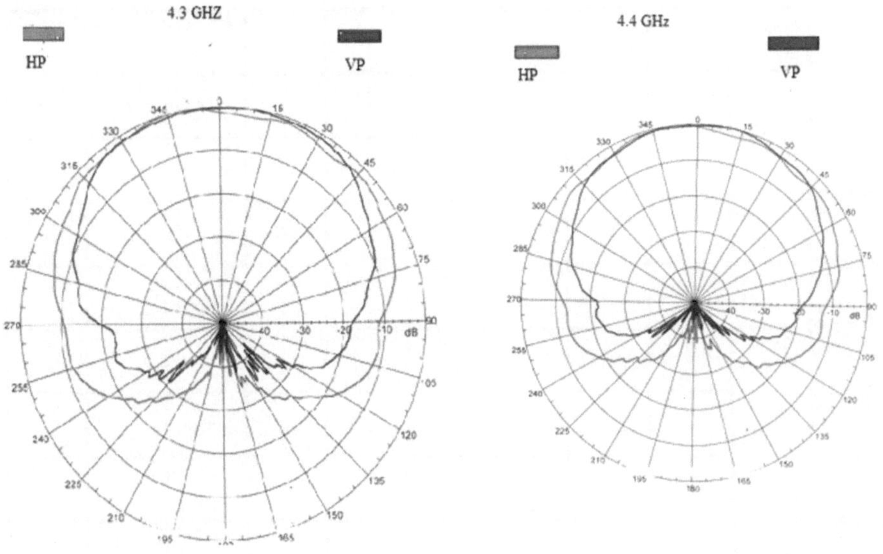

FIGURE 2.9 Radiation pattern at 4.3 GHz, 4.4 GHz.

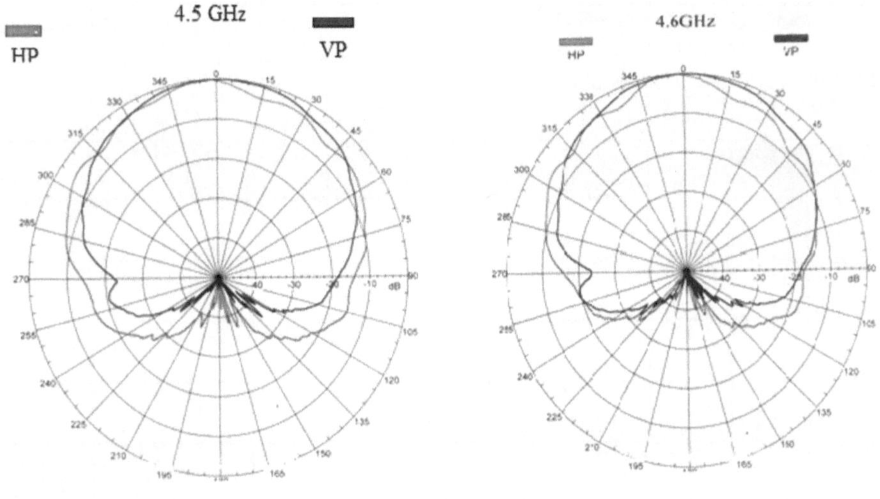

FIGURE 2.10 Radiation pattern at 4.5 GHz, 4.6 GHz.

VSWR is a minimum resonant at 4.3 GHz from the VSWR plot as shown in Figure 2.12. Since the VSWR is nearer to 1 roughly, there are no signals back to the feed.

2.5.4 Gain Calculation

In Anechoic chamber pattern and gain are measured. The gain calculation formula is given below:

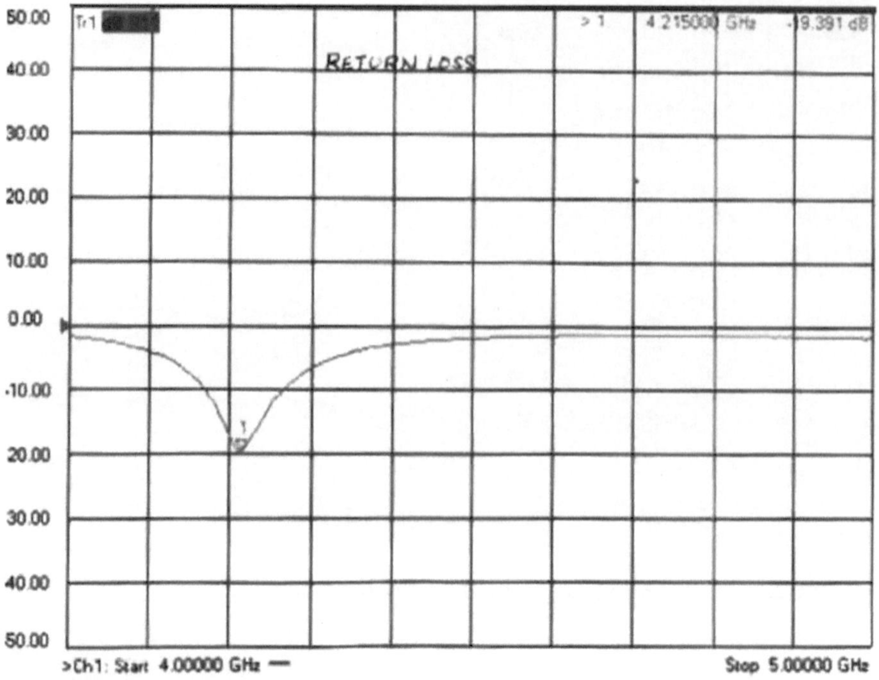

FIGURE 2.11 Return loss plot.

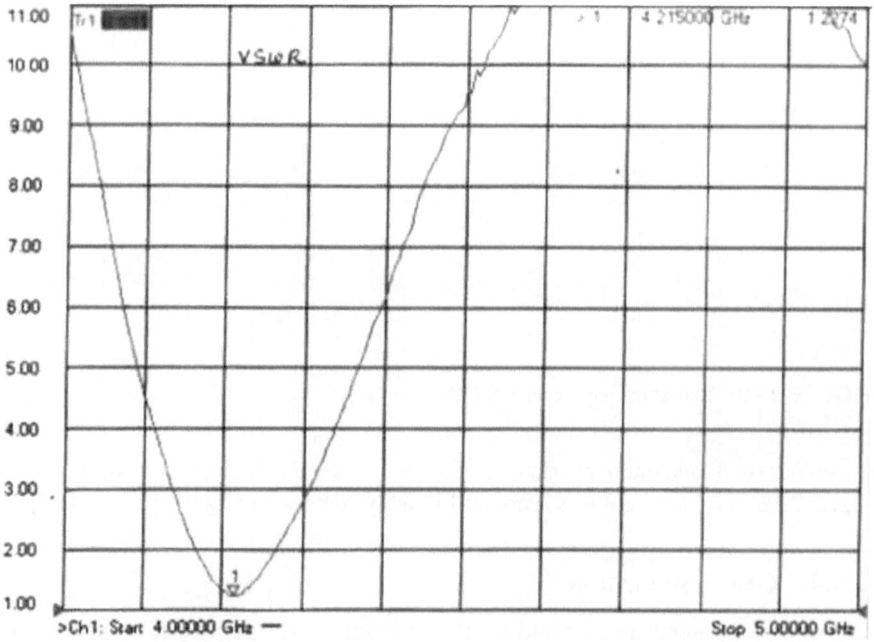

FIGURE 2.12 VSWR plot.

$$G = \left[\frac{P_{max}(AUT)}{P_{max}(REF)}\right] XGain(REF). \tag{2.14}$$

Using Equation (2.14), gain is calculated and depicted in Table 2.2.

2.5.5 Comparison of Simulated and Measured

Table 2.3 depicts results comparison of simulated and measured with frequency from 4.1 GHz to 4.6 GHz.

2.5.6 Comparison Between Simulated and Fabricated VSWR and Return Loss Graphs

Figures 2.13 and 2.14 depicts comparison graphs between simulated and fabricated patch antenna design between VSWR and return loss.

TABLE 2.2
Calculated Gain Values with AUT

S.NO	Frequency (GHz)	AUT Gain (dB)
1	4.1	4.89
2	4.2	4.94
3	4.3	5
4	4.4	5.56
5	4.5	5.6
6	4.6	6.02

TABLE 2.3
Comparison between Measured and Simulated Results

Simulated				Measured			
Frequency (GHz)	Return Loss (dB)	VSWR	Gain (dB)	Frequency (GHz)	Return Loss (dB)	VSWR	Gain (dB)
4.1	−1	19.4	1.3	4.1	−4.2	4.5	4.89
4.2	−2	11.2	8.2	4.2	−19.2	1.29	4.94
4.3	−32	1.1	9.01	4.3	−10.3	2.9	5
4.4	−2	11.2	7.1	4.4	−4.3	6.2	5.56
4.5	−2	19.3	3.6	4.5	−4.2	9.2	5.6
4.6	−2	25.1	2.2	4.6	−4.1	11.3	6.02

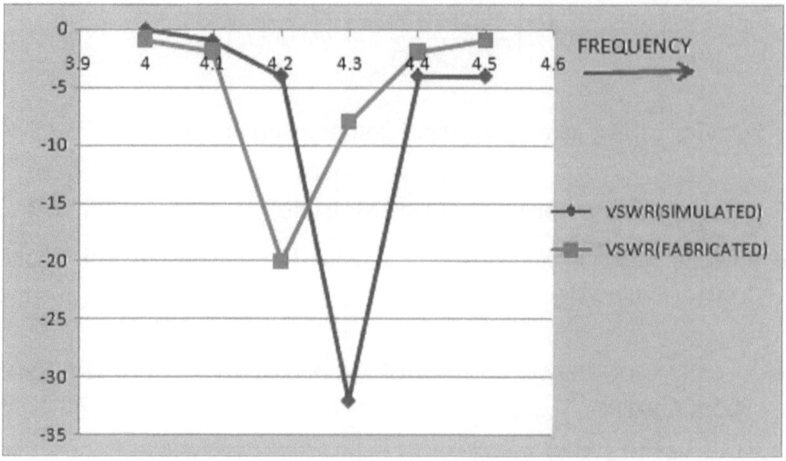

FIGURE 2.13 Comparison of VSWR between simulated and fabricated.

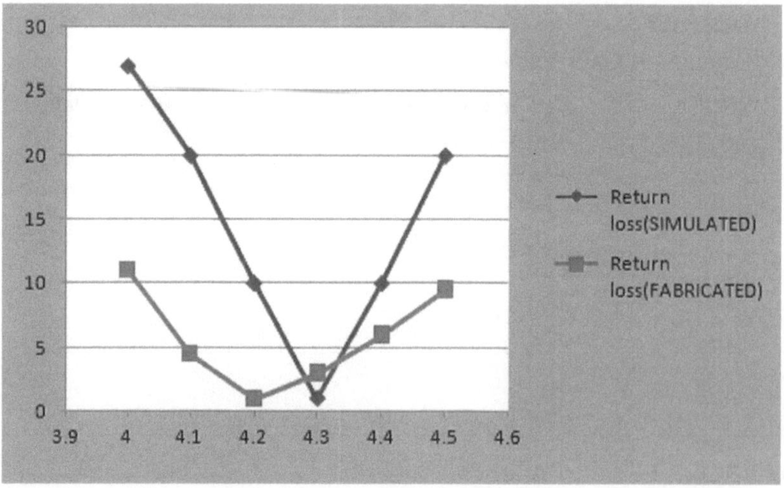

FIGURE 2.14 Comparison of return loss between simulated and fabricated.

CONCLUSION

Along these lines the plan of MSA is effectively reproduced and created at 4.3 GHz with direct polarization for radar (altimeter) applications. With the current remote technique for contact, the essential reception apparatus prerequisites are gain, bandwidth, polarization, size must be low, weight must be low, and simplicity of manufacture. All the necessities referenced above should be possible with the assistance of MSA. MSA is an answer for some remote situations on account of their lowness cost, and if they are comparable and simple to fabricate. MSA is being created in both programming and manufactured. MSPA planned at 4.3 GHz for band width of 100 MHz can be utilized for different C-band applications where low deceptive radiations, similarity with measured plans and smallness, are of essential significance. It very well may be seen that the above coaxial coupled feed method have the band width 90 MHz to 100 MHz, give an increase of 5 dB and resounding recurrence of 4.215 GHz, and return loss of −12.3 dB.

REFERENCES

[1]. Girish Kumar, Ray K.P., *Microstrip Antennas Broadband.* Boston: Aretch House, 2003.

[2]. Vidya V. Deshmukh, Suvarna S. Choragea, "Microstrip Antennas used for Noninvasive Determination of Blood Glucose Level," in *2020 4th International Conference on Intelligent Computing and Control Systems (ICICCS)*, 2020, Madurai, India, pp. 720–725.

[3]. Prakasam, Anudeep L and P. Srinivasu, "Design and Simulation of Circular Microstrip Patch Antenna with Line Feed Wireless Communication Application," in *2020 4th International Conference on Intelligent Computing and Control Systems (ICICCS)*, 2020, Madurai, India, pp. 279–284, doi: 10.1109/ICICCS48265.2020.9121162.

[4]. Prakasam, V. and Sandeep, P., "Design and Analysis of 2 × 2 Circular Micro-Strip Patch Antenna Array for 2.4 GHZ Wireless Communication Application," *International Journal for Innovative Engineering & Management Research*, vol. 7, no. 12, 9 pp. 2018.

[5]. Pradeep Kumar, Neha Thakur. "Micro Strip Antenna for 2.4 GHz Wireless Applications," *International journal of engineering Trends and Technology*, vol. 4, no. 8, pp. 3544–3547, 2013.

[6]. Prasad K. D., *Antenna and Wave Propagation.* New Delhi: Satya Prakashan, 2003.

[7]. Ramirez and Santos, "Design, Simulation of an Irregularly Shaped MS for Air-to-Ground Communications," *IJAP*, 2017.

[8]. Jaume Anguera, Aurora Andújar, Jeevani Jayasinghe, "High-Directivity Microstrip Patch Antennas Based on TModd-0 Modes," *IEEE Antennas and Wireless Propagation Letters*, vol. 19, no 1, pp. 39–43, 2020, doi: 10.1109/LAWP.2012.2952260.

[9]. A. D. Novella, H. Wijanto and A. D. Prasetyo, "Dual-Feed Circularly Polarized microstrip antenna for S-Band transmitter of Synthetic Aperture Radar (SAR) system," in *2015 International Conference on Quality in Research (QiR)*, 2015, Lombok, pp. 4–7, doi:10.1109/QiR.2015.7374882.

[10]. P. A. H. Vardhini and N. Koteswaramma, "Patch Antenna Design With FR-4 Epoxy Substrate Formultiband Wireless Communications Using CST Microwave Studio," in *2016 International Conference on Electrical Electronics and Optimization Techniques (ICEEOT)*, 2016, pp. 1811–1815.

[11]. N. Koteswaramma, P.A. Harsha Vardhini and K. Murali Chandra Babu, "Realization of Minkowski Fractal Antenna for Multiband Wireless Communication," *International Journal of Engineering and Advanced Technology (IJEAT)*, vol. 9, no. 1, pp. 5415–5418, 2012.

[12]. Syihabuddin, Wijanto, Heroe, Prasetyo, AgusOwi, *Design of Ground Segment System for Tel-Usat 1 Nano-Satellite, in Bahasa, Seminar Nasional dan Expo TeknikElektro (SNETE) ke-4.* Kuala: UniversitasSyiah Kuala, 2011.

[13]. P. Upender, P. A. Harsha Vardhini, "Design of GPR for Buried Object Detection Using UWB Antenna," *IJEAT*, vol. 9, no 1, pp. 5419–5423, 2019.
[14]. Balanis CA, *Analysis and Design Antenna Theory,* 2nd ed. New York: John Wiley and Sons, pp. 28–30.
[15]. P. Upender, R. Tanisha, G. Priya, N. Bhargavi, "Rectangular Microstrip Patch antenna using HFSS," *International Journal of Research in Electronics and Computer Engineering,* vol. 7, pp. 349–351, 2019.

3 Study of Performance Parameters of Stub Loaded Oval-Shaped Patch Antenna Using Metamaterials, Electromagnetic Bandgap Structures, and DGS of Dumbbell Shape

Karteek Viswanadha[1] *and N.S. Raghava*[2]
[1]Delhi Technological University, New Delhi, India
[2]Delhi Technological University, New Delhi, India

3.1 INTRODUCTION

Recent developments in the microstrip patch antenna are based on miniaturization, gain, impedance bandwidth, and efficiency improvement [1]. Many high impedance surfaces like metamaterials (MTMs), electromagnetic bandgap structures (EBG), and defective ground structures (DGS) are used to improve the performance parameters of the microstrip patch antennas.

Different types of patch antennas utilizing different high impedance structures and their role in miniaturizing as well as improving performance parameters of the later are discussed in [2–17]. 2-segment split ring resonator (SRR) loaded patch antenna resonating at 4 GHz is proposed in [2]. The antenna is miniaturized by 400% and bandwidth is increased from 0.4 GHz to 2.4 GHz, i.e., 600%. A strip line metamaterial loaded patch antenna is proposed in [3]. Miniaturization of 400% and bandwidth improvement by 600% is achieved in [3]. A peak gain of 3 dBi is achieved at center frequency 4 GHz. An S-shaped metamaterial loaded patch antenna is proposed in [4]. The proposed antenna in [4] resonates at 43 THz with a bandwidth of 770 GHz and a gain of 7.16 dBi. An S-shaped metamaterial loaded patch antenna is proposed in [5]. The proposed patch antenna resonates at 4.3 GHz

DOI: 10.1201/9781003187325-3

with a 74% improvement in bandwidth and an 11% improvement in the directivity. A double negative (DNG) U-T shaped metamaterial loaded fractal patch antenna resonating at 7 GHz is proposed in [6]. The proposed antenna is miniaturized by 58% and 75% along with improvement in bandwidth. The proposed antenna in [6] possesses bandwidth of 761 MHz, and a gain of 9.1 dBi. A C-shaped pair metamaterial loaded monopole patch antenna working at 7.5 GHz is proposed in [7]. By loading the patch antenna in [7] with C-shaped metamaterials, directivity is improved by 2% (from 0.657 dBi to 0.67 dBi). Patch antenna loaded with labyrinth SRR MTM is proposed in [8]. The patch antenna in [8] resonates at 30 GHz and secondary resonances have been created around 8.5, 17.7, 20, and 23.7 GHz. The whole structure is miniaturized by 72%. Patch antenna loaded with swastika metamaterial is proposed in [9]. The proposed patch antenna in [9] resonates at 3.8 THz and its bandwidth is improved to 0.92 THz. A peak gain and directivity of 6.4 dBi and 6.72 dBi are observed in [9]. Patch antenna with shorted posts mounted on the photonic bandgap structure (PBG) is proposed in [10]. The antenna in [10] resonates at 2.64 GHz with improved efficiency of 96.53%, a gain of 4.63 dBi, and a directivity of 4.78 dBi. Patch antenna mounted on the cross slotted EBG ground plane is proposed in [11]. The design in [11] resonates at 2.5 GHz is constructed by using the stacking of patches, shorting pin, and cross slotted EBG to form an optimized antenna design with antenna efficiency of approximately 99.06%. The gain and directivity in [11] are 4.74 dBi and 4.78 dBi. In [12], a flower-shaped patch antenna mounted on the double substrate is proposed. Antenna in [12] is loaded with complementary split ring resonator (CSRR) and mounted on the Jerusalem cross-shaped defective ground structure (DGS). Antenna resonates at 5.2 GHz and 8.25 GHz with the bandwidths of 1.2 GHz (4.95 GHz–6.15 GHz) and 2.2 GHz (7.1 7GHz–9.3 GHz) respectively. The gains in these bands are observed to be 3.93 dBi and 5.02 dBi. Dumbbell shaped patch antenna mounted on splatteredring EBG is proposed in [13]. The proposed antenna in [13] possesses bandwidth of 20.2 GHz (171.4%) with the gain varying from 6.21 dB to 9.03 dB in the frequency range of 10.8 GHz–31 GHz. This further possesses efficiencies varying from 88.34% to 91.06% in the said frequency range. A double broken ring slotted patch antenna mounted on the swastika EBG structure is proposed in [14]. The proposed patch antenna in [14] resonates at 21.29 GHz with a bandwidth of 3.2 GHz and a gain of 11.92 dBi. Patch antenna mounted on the hexagonal slotted EBG ground plane is proposed in [15]. The whole structure in [15] is backed by a plane ground with an air gap of 10.5 mm. When compared to an RMSA of the same dimensions the antenna efficiency increased by 74%, the gain by 10 dBi, and the directivity showed an enhancement of 2 dBi. A two-layer highly efficient directive E-shaped patch radiator is proposed in [16]. By modifying the geometry of a rectangular patch and by introducing two slits, the size of the original rectangular patch is reduced. Further reduction in the size is achieved by stacking E-shaped patches. Both the gain and efficiency of this modified antenna is increased by 16% in [16]. A stacked square patch antenna mounted on EBG ground plane is proposed in [17]. A reflector is placed at 8.5 mm below the ground and the performance parameters of antenna in [17] are studied by varying the airgap. The proposed antenna in [17] resonates at 2.57 GHz with a peak gain of 3.64 dBi and directivity of 4.5 dBi.

In this chapter, the study of performance parameters of the stubs loaded oval ring patch is presented. Various performance parameters like gain, bandwidth, and efficiency are studied in this chapter. Section 3.2 presents the design procedure of a basic stub loaded patch antenna. Section 3.3 represents the parametric analysis on variation in the return loss of the patch antenna with the change in the number of metamaterial cells, EBG slots, and dimensions of the DGS. Section 3.4 presents the results of the performance parameters.

3.2 OVAL SHAPED PATCH ANTENNA LOADED WITH STUBS

The structure of the proposed oval ring patch antenna [18] with λ/4 folded meander lines and L-shaped stub-lines is shown in Figure 3.1. The antenna is printed on one side of an RT/duroid-6202 with a thickness of 1.524 mm, the relative permittivity of 2.94, and loss tangent of 0.0015. Initially, an oval patch is excited by a 50 ohms line having a length (*l*) and width (*w*). The patch possesses low impedance bandwidth. To further increase the impedance bandwidth, the proposed patch is connected to a folded meander line strip of length equal to λ/4. Finally, two L-stub lines of lengths equal to $\lambda/4$ are connected outside the oval ring patch that makes the proposed antenna. When the L-stubs and folded meander lines are connected to the proposed patch, the required multiband feature is achieved. The proposed patch antenna is mounted on the ground plane having the dimensions ($W_g \times L_g \times t$) of 10 × 12 × 0.035 mm^3. The detailed dimensions of the proposed patch antenna are given in Table 3.1.

Table 3.1 shows the detailed dimensions of the proposed patch antenna.

The resonant frequency of the right stub-line portion (f_1) is given by

$$f_1 = \frac{c}{4(l_{rs})\sqrt{\varepsilon_{eff}}}, \tag{1.1}$$

where 'c' is the velocity of the light, 'l_{rs}' is the length of the right stub, and 'ε_{eff}' is the effective dielectric constant given by

$$\varepsilon_{eff} = \frac{\varepsilon_r + 1}{2} + \left(\frac{\varepsilon_r - 1}{2}\right)\left[\frac{1}{\sqrt{1 + \frac{12h}{w}}}\right], \tag{1.2}$$

where 'ε_r' is the relative dielectric constant, 'h' is the thickness of the dielectric, and 'w' is the width of the feed line.

Mathematically, the frequency (f_1) at which the right stub-line resonates is obtained as 6.26 GHz and a dip is observed at 6.12 GHz in the simulated result of the return loss of the proposed patch antenna. The difference in the mathematical and simulated results is due to the non-consideration of the dielectric and conductor losses.

Similarly, the resonant frequency (f_2) of the left stub section is given by

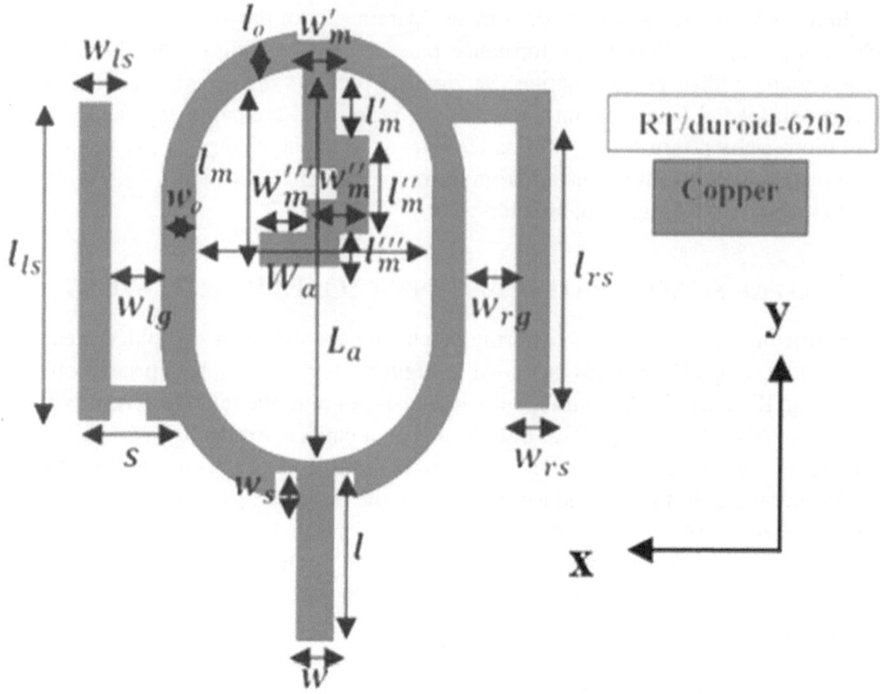

FIGURE 3.1 The geometry of the proposed patch antenna.

TABLE 3.1
Detailed Dimensions of the Proposed Antenna

Parameter	L_a	W_a	l_m	l'_m	l''_m
Value(mm)	6	4	3.25	2	0.75
Parameter	l'''_m	w'_m	w''_m	w_{rg}	w_{lg}
Value(mm)	0.5	0.5	1	1	1
Parameter	L_g	W_g	w_{ls}	w_{rs}	l_{ls}
Value(mm)	12	10	1	1	5.5
Parameter	l_{rs}	s	l	w	w_s
Value(mm)	7.75	2.5	2.5	1	0.5
Parameter	w_o	l_o	w''_m		
Value(mm)	1	1	1.5		

$$f_2 = \frac{c}{4(l_{ls})\sqrt{\varepsilon_{eff}}}. \tag{1.3}$$

Mathematically, the frequency at which the left stub portion resonates is obtained as 9.4 GHz and a dip is observed at 9.42 GHz in the simulated result of the return loss of the proposed patch antenna. The difference in the mathematical and simulated results is due to the non-consideration of the dielectric and conductor losses.

The frequency of the meander strip line is given by:

$$f_3 = \frac{c}{4(l_m)\sqrt{\varepsilon_{eff}}}. \tag{1.4}$$

Mathematically, the frequency at which strip line portion resonates is obtained as 15.91 GHz and a dip is observed at 15.6 GHz in the simulated result of the return loss of the proposed patch antenna. The difference in the mathematical and simulated results is due to the non-consideration of the dielectric and conductor losses.

The frequency of the oval ring is given by

$$f_4 = \frac{c}{\pi (l_o)_{eff}\sqrt{\varepsilon_{eff}}}, \tag{1.5}$$

where 'l_o' is the perimeter of the oval ring

$$l_{o_{eff}} = l_o\left[1 + \left(\frac{2h}{l_o\pi\varepsilon_r}\right)\left[\ln\left(\frac{a}{2h}\right) + (1.41\varepsilon_r + 1.77) + \frac{h}{l_o}(0.265\varepsilon_r + 1.65)\right]\right]^{\frac{1}{2}} \tag{1.6}$$

3.2.1 Proposed High Impedance Surfaces

Figure 3.2 shows the proposed structures. The proposed patch antenna is mounted on a partial ground plane of dimensions 10 × 6 × 0.035 mm^3. The proposed oval patch antenna is loaded with 24 metamaterial cells. The reason for choosing 24 metamaterial cells is discussed in the parametric analysis section. Table 3.2 shows the dimensions of the proposed unit metamaterial cell along with the vertical and horizontal spacings between two cells. The introduction of metamaterials not only miniaturizes the patch antenna but also introduces the multi-band feature.

An oval patch antenna is mounted on the dumbbell shaped EBG (full ground plane with the dimensions of 10 × 12 × 0.035 mm^3) to study its performance characteristics. Fourteen EBG slots are etched on the ground plane to analyze the performance parameters of the patch antenna. The reason for etching 14 EBG slots on the ground plane is discussed in the parametric analysis section. To further study the change in the performance parameters, stub loaded patch antenna is mounted on a dumbbell-shaped defective ground structure (DGS) (full ground plane with the dimensions of 10 × 12 × 0.035 mm^3). The change in the return loss is studied by varying the total area of a DGS slot.

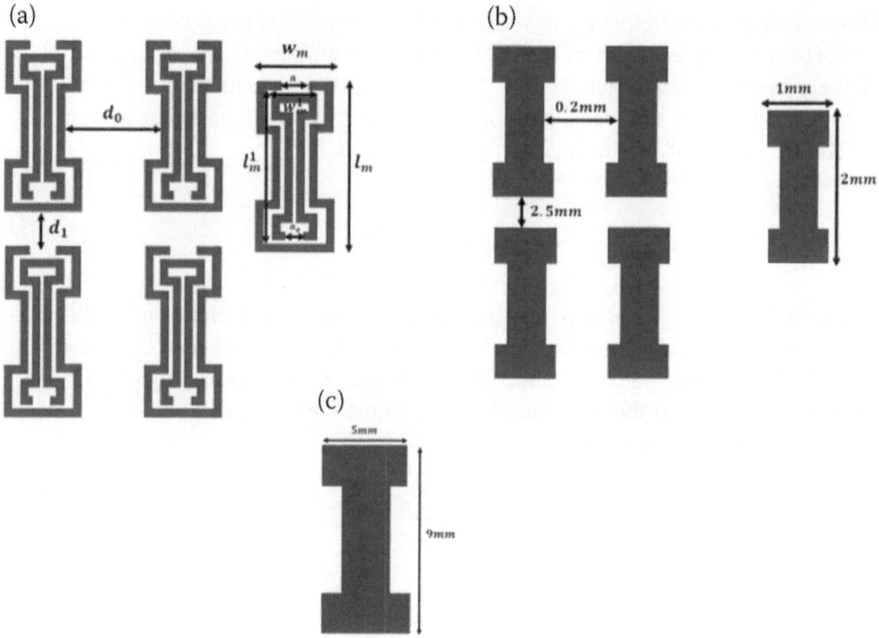

FIGURE 3.2 Proposed a. Metamaterials. b. EBG slots. c. DGS.

TABLE 3.2
Dimensions of the Proposed Unit Metamaterial Cell

Parameter	d_o	d_1	l_m	w_m	a	a_o	l'_m	w'_m
Value (mm)	0.2	1.2	2	1	0.75	0.2	1.5	0.5

3.2.2 Parametric Analysis of the Proposed High Impedance Structures

This section deals with the variation of impedance bandwidth and resonant frequencies of the oval shaped patch antenna with the change in the number of metamaterial cells, EBG slots, and the dimensions of a DGS slot on the ground plane. Figure 3.3a shows the variation of return loss with the number of metamaterials. A decrease in the number of metamaterials decreases bandwidths in two bands circled with brown and black dotted ovals. The decrease in the number of metamaterial cells not only reduces bandwidth in the desired frequency bands but also shifts operational frequencies. Additional resonant frequencies are introduced if 8 metamaterial cells are introduced (as circled with a green dotted oval). Therefore, the number of metamaterial cells are optimized to 24 to achieve better results.

Figure 3.3b shows the variation of return loss with the number of EBG slots on the ground plane. With the decrease in the number of EBG slots on the ground

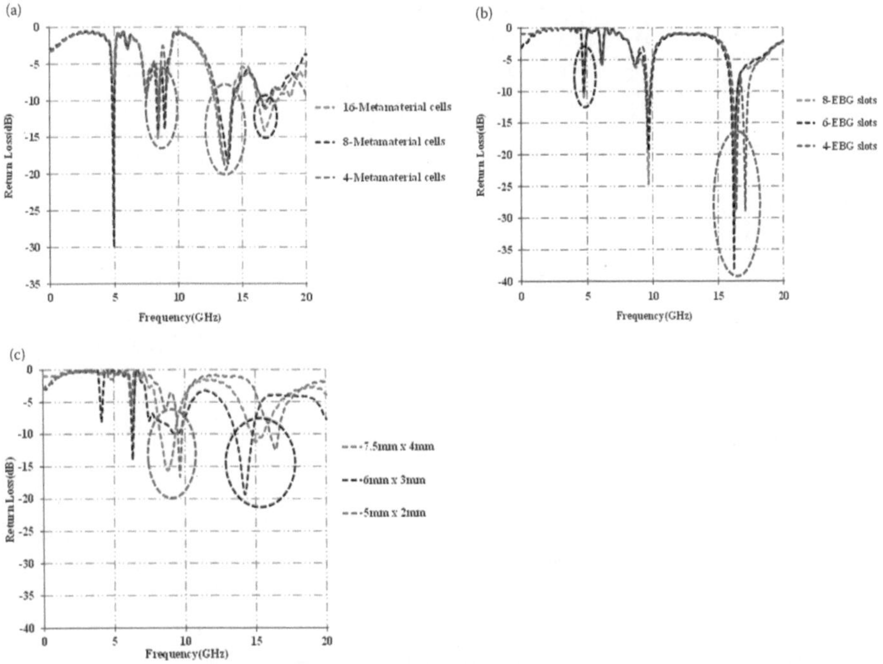

FIGURE 3.3 a. Variation of return loss with the number of metamaterials. b. Variation of return loss with the number of EBG slots. Variation of return loss with the dimensions of the DGS slot.

plane, the lower resonant frequency gets shifted towards higher frequencies. Hence, the number of EBG slots are optimized to 14.

Figure 3.3c shows the variation of a variation of return loss with the dimensions of the DGS slot. Reducing the dimensions of the DGS slot not only shifts the resonant frequency to a higher frequency range but also diminishes the multiband feature of the oval patch antenna. Hence, the dimensions of the DGS slot are optimized to 9 mm × 5 mm.

3.2.3 Performance Analysis of the Proposed Antenna Structures

Figure 3.4 shows the return losses of the proposed patch antenna loaded with metamaterials, mounted on EBG and DGS ground planes. When the oval patch antenna is loaded with metamaterials (inside the substrate), the patch antenna resonates at 4.92 GHz, 7.7 GHz, 13.88 GHz, 16.78 GHz and 18.18 GHz. The patch antenna not only retained its multiband characteristics but also enhanced its bandwidth in the frequency ranges of 12.72 GHz–14.23 GHz (1.51 GHz) and 16.52 GHz–18.71 GHz (2.19 GHz). When the oval patch antenna is mounted on the EBG ground plane, the patch antenna resonates at 4.74 GHz, 9.62 GHz, and 16.59 GHz. Miniaturization of the patch antenna is improved. A simple stub loaded oval patch antenna with the dimensions of 7.5 × 9 mm^2 has a lower resonant frequency of

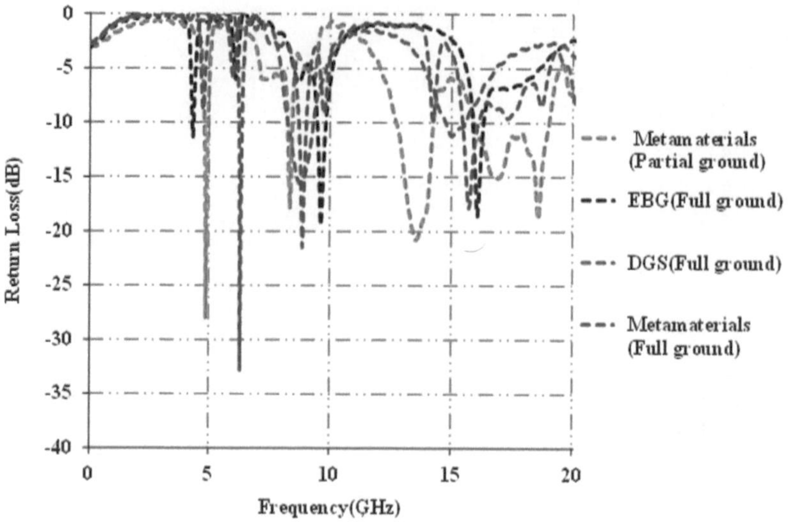

FIGURE 3.4 Return losses of the oval patch antenna with the proposed high impedance surfaces.

6.12 GHz and the same patch antenna when mounted on the proposed EBG structure possesses the lower resonant frequency of 4.74 GHz. Miniaturization of 74% is achieved with the simple patch antenna. Miniaturization is further improved to 84% when the stub loaded patch antenna is mounted on the proposed dumbbell shaped EBG structure. When the patch antenna is mounted on the DGS ground plane, the patch antenna resonates at 6.83 GHz, 8.85 GHz, and 15.01 GHz. Bandwidth is highly reduced in the corresponding frequency bands. When the metamaterial loaded patch antenna is mounted on a full ground plane, it resonates at 9.1 GHz and 16.23 GHz. The full ground plane shifts the resonant frequencies to a higher frequency range and reduces the gain. Hence, those results are not reported in this chapter. Figure 3.5 shows the surface currents of metamaterial loaded, DGS, and EBG mounted patch antenna. For metamaterial loaded patch antenna, the current is concentrated in the oval patch antenna, partial ground, and a part of the current is diverted towards metamaterials. As the frequency increases, the current is fully concentrated in metamaterials. When the patch antenna is mounted on the EBG ground, the current is highly concentrated on the patch and the ground plane at a lower frequency. As the frequency increases, the current is concentrated on the stubs of the patch and center of the ground plane. When the patch is mounted on the DGS ground plane, the current is concentrated on the stubs. As the frequency increases, the current is maximum of the ground plane and on the oval ring.

Figure 3.6 shows the real and imaginary parts of permittivity and permeability. For calculating the permittivity and permeability, Nicolson-Ross-Weir is used. The above method is carried out by extracting s-parameters from the CST-MWS package and the subsequent calculations are done using MATLAB [19]. Permittivity is negative 4.91 GHz, 7.61 GHz, 13.1 GHz, 16.43 GHz, and 19.1 GHz.

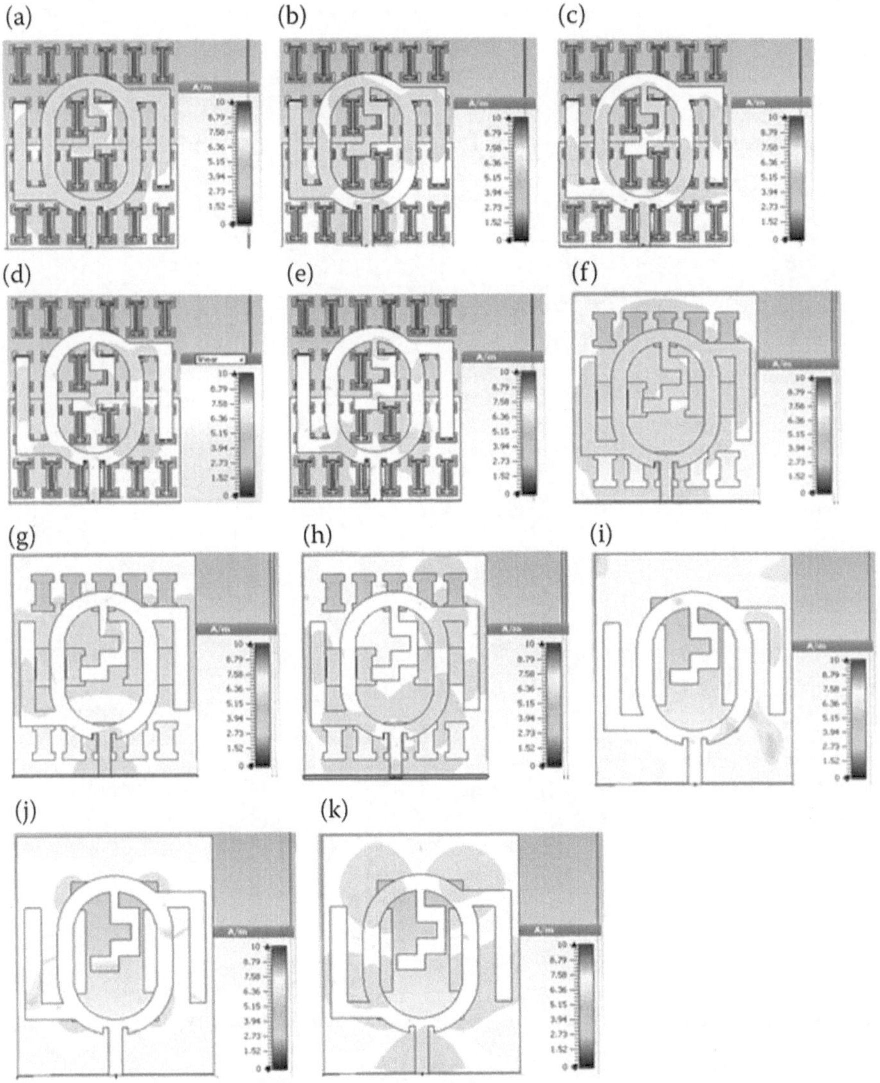

FIGURE 3.5 Surface currents with metamaterials at a. 4.92 GHz. b. 7.7 GHz. c. 13.88 GHz. d. 16.78 GHz. e. 18.18 GHz. Surface currents with EBG ground plane at f. 4.74 GHz. g. 9.62 GHz. h. 16.59 GHz. Surface currents with DGS ground plane at i. 6.83 GHz. j. 8.85 GHz. k. 15.01 GHz.

Similarly, permeability is negative at 4.81 GHz, 7.71 GHz, 13.1 GHz, and 16.38 GHz. The proposed dumbbell metamaterial converts from Double Negative (DNG) to Epsilon Negative (ENG) at higher frequencies.

Figure 3.7 shows the axial ratio plots of different structures. Metamaterial loaded oval patch antenna is right-handed circularly polarized (Axial Ratio < 2 dB) at 4.36 GHz and 13.88 GHz. It is elliptically polarized at 7.87 GHz and 17.21 GHz

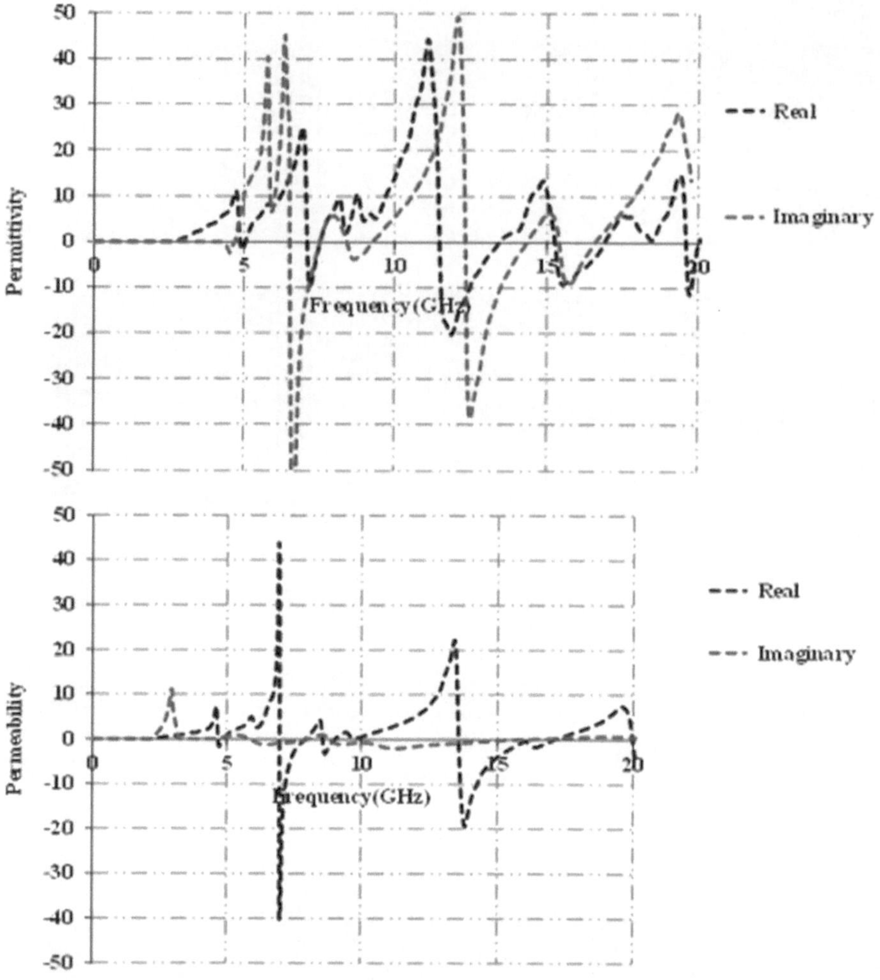

FIGURE 3.6 Real and imaginary parts of permittivity and permeability.

and linearly polarized at remaining frequencies. When the oval patch antenna is mounted on the EBG ground plane, it possesses circular polarization at 16.43 GHz. When the oval patch antenna is mounted on the DGS ground plane, it possesses elliptical polarization.

Figure 3.8 shows the radiation patterns of Metamaterial-, EBG-, and DGS-oval shaped patch antennas at their corresponding resonant frequencies. Metamaterial antenna resonates at 4.92 GHz, 7.7 GHz, 13.88 GHz, 16.78 GHz, and 18.18 GHz with the gains of 1.4 dBic, 1.42 dBi, 2.65 dBic, 2.31 dBi, and 2.89 dBi respectively. EBG-antenna resonates at 4.74 GHz, 9.62 GHz, and 16.59 GHz with gains of 1.45 dBi, 2.6 dBi, and 2.85 dBic respectively. DGS-antenna resonates at 6.83 GHz, 8.85 GHz, and 15.01 GHz with the gains of 3.21 dBi, 3.3.1 dBi, and 4.38 dBi respectively. Radiation patterns are unidirectional in all the planes.

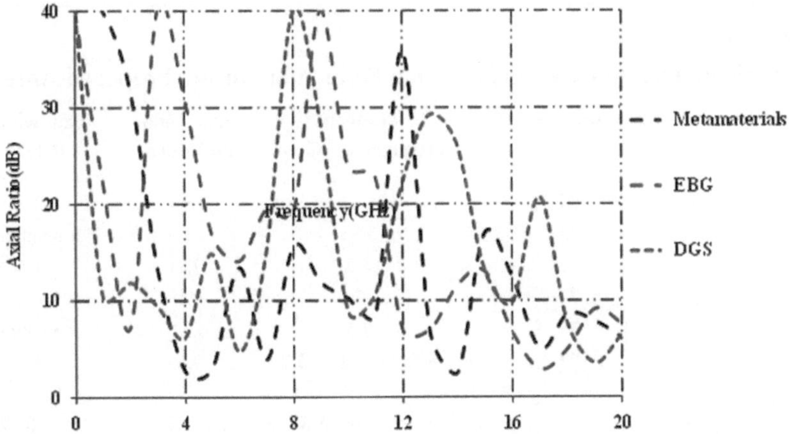

FIGURE 3.7 Axial ratios of oval patch antenna loaded/mounted with/on different structures.

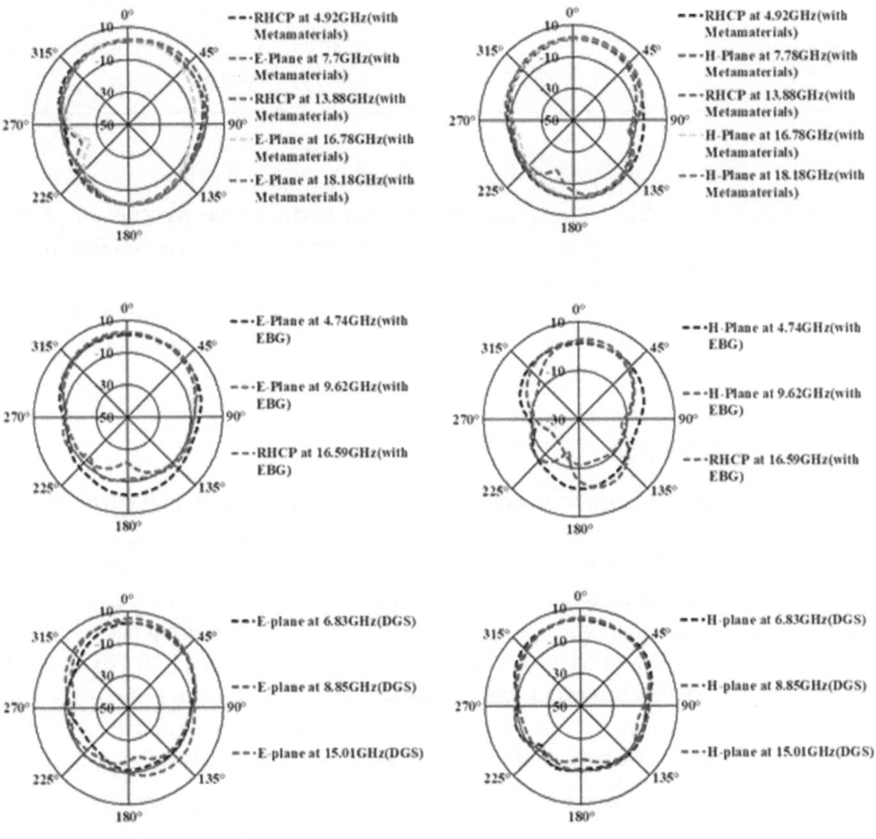

FIGURE 3.8 Radiation patterns of proposed antenna structures.

TABLE 3.3
Comparison between the Performance Parameters of the Proposed Antenna

Type of Antenna	Miniaturization (%)	Resonant Frequencies (GHz)	Peak Gain (dBi/dBic)	Bandwidth (GHz)
[2]	25	4	4.99	2
[3]	25	3.85	3	Not mentioned
[5]	81	4.3	7.2	1
[6]	75	7	9.1	0.761
[7]	35	7.5	0.67	Not mentioned
[8]	72	8.5, 17.7, 20, 23.7 & 30	9.6	3
[18]	74	6.12, 9.4, 16.75 & 25	1.1, 1.34, 5.72 7 8.1	0.315, 0.671, 0.441 & 3.21
Metamaterial loaded	81.2	4.92, 7.7,13.88,16.78 & 18.18	1.42, 1.48, 2.8, 2.45 & 3	0.2, 0.31, 1.51 & 2.19
EBG-mounted	84	4.74, 9.62 & 16.59	1.5, 2.7 & 2.9	0.12, 0.29 & 0.36
DGS-mounted	71.6	6.83, 8.85 & 15.01	3.3, 3.45 & 4.5	0.19, 1.2 & 0.43

Table 3.3 shows the comparison between the performance parameters of the proposed antenna. A suitable antenna is chosen based on the requirement of an application. Metamaterial antenna is chosen if high bandwidth and the multi-band is required for an application. EBG-antenna is chosen if any further miniaturization of the oval patch antenna is required. DGS-antenna is chosen if a high gain is required for an application.

3.3 CONCLUSION

Performance characteristics of various patch antennas are studied. Dumbbell shapes structures (metamaterials, EBG, and DGS) are promising enough to achieve miniaturization, high bandwidth, and gain. These structures are used along with the stub loaded oval-shaped patch antenna to further enhance the performance parameters. Metamaterial antenna possesses high bandwidths in the frequency ranges 12.72 GHz–14.23 GHz (1.51 GHz) and 16.52 GHz–18.71 GHz (2.19 GHz). EBG mounted patch antenna is used to achieve further miniaturization and DGS mounted patch antenna is used to achieve high gain at all its resonant frequencies. These structures are used in various applications of satellite applications ranging from C-band to K-band.

REFERENCES

[1]. D. Chatterjee and A. K. Kundu, "Study of Miniaturized Patch Microstrip Antenna with Circular Slot and SRR Optimization," in *2019 URSI Asia-Pacific Radio Science Conference (AP-RASC)*, 2019, pp. 1–4. doi:10.23919/URSIAP-RASC.2019.8738443

[2]. P. Dawar, N. S. Raghava and A. De, "UWB Metamaterial-Loaded Antenna for C-Band Applications," *International Journal of Antennas and Propagation*, vol. 2019, Article ID 6087039, 2019, pp. 1–14. doi:10.1155/2019/6087039.

[3]. P. Dawar, N. S. Raghava and A. De, "UWB and Miniaturized Meandered Stripline Fed Metamaterial Loaded Antenna for Satellite Applications," *IOP Conference Series Materials Science and Engineering*, vol. 377, no. 1, 2018, pp. 1–11. doi:10.1088/1757-899X/377/1/012080.

[4]. P. Dawar, N. S. Raghava and A. De, "UWB, Miniaturized and Directive Metamaterial Loaded Antenna for Satellite Applications," *International Journal of Networked and Distributed Computing*, vol. 6, no. 1, 2018, pp. 24–34. doi:10.2991/ijndc.2018.6.1.3.

[5]. P. Dawar, N. S. Raghava and A. De, "S-Shaped Metamaterial Ultra-Wideband Directive Patch Antenna," *Radioelectronics and Communications Systems*, vol. 61, 2018, pp. 394–405. doi:10.3103/S0735272718090029.

[6]. P. Dawar, N. S. Raghava and A. De, "High Gain, Directive and Miniaturized Metamaterial C-Band Antenna," *Cogent Physics*, vol. 3, no. 1, 2016, pp. 1–12. doi:10.1080/23311940.2016.1236510.

[7]. P. Dawar, N. S. Raghava and A. De, "Tunable and Directive Metamaterial-Inspired Antennas for 'C' Band Applications," *International Journal of Microwave and Optical Technology*, vol. 10, no. 3, 2015, pp. 168–175. IJMOT 2014-11-645.

[8]. P. Dawar, N. S. Raghava and A. De, "A Novel Metamaterial for Miniaturization and Multi-Resonance in Antenna," *Cogent Physics*, vol. 2, no. 1, 2015, pp. 1–13. doi:1080/23311940.2015.1123595.

[9]. P. Dawar, N. S. Raghava and A. De, "Ultra Wide Band, Multi-resonance Antenna Using Swastika Metamaterial," *International Journal of Microwave and Optical Technology*, vol. 11, no. 6, 2016, pp. 413–420. IJMOT 2014-11-645.IJMOT-2016-6-1035.

[10]. N. S. Raghava and A. De, "Photonic Bandgap Stacked Rectangular Microstrip Antenna for Road Vehicle Communication," *IEEE Antennas and Wireless Propagation Letters*, vol. 5, 2006, pp. 421–423. doi:10.1109/LAWP.2006.883950.

[11]. N. S. Raghava, A. De, N. Kataria and S. Chatterjee, "Stacked Patch Antenna With Cross Slot Electronic Band Gap Structure," *International Journal of Information and Computation Technology*, vol. 3, no. 5, 2013, pp. 1–4, arXiv:1310.6259.

[12]. K. Viswanadha and N. S. Raghava, "Design and Analysis of a Multi-band Flower Shaped Patch Antenna for WLAN/WiMAX/ISM Band Applications," *Wireless Personal Communications*, vol. 112,2020, pp. 863–887. doi:1007/s11277-020-07078-8.

[13]. K. Viswanadha and N. S. Raghava, "Design of a Dual Polarized Ultra Wideband Dual Feed Dumb-Bell Patch Antenna With Splattered Ring EBG for Ultra High Speed Communications," in *IEEE Indian Conference on Antennas and Propogation (InCAP)*, 2019, pp. 1–5. doi:10.1109/InCAP47789.2019.9134652.

[14]. K. Viswanadha and N. S. Raghava, "Design of High Gain, Bandwidth and Efficient Double Split Ring Slotted Antenna with Swastika Shape EBG Structures at 21.29 GHz for High Data Rate Communications," *International Journal of Computer Applications*, vol. 180, no. 11, 2018, pp. 35–38. doi:10.5120/ijca2018916232.

[15]. A. Choudhary, N. S. Raghava and A. De, "A Highly Efficient Rectangular Microstrip Antenna With Hexagonal Holes as an Electromagnetic Bandgap Structure in the Ground Plane," in *International Conference on Recent Advances in Microwave Theory and Applications*, 2008, pp. 152–153. doi:doi:10.1109/AMTA.2008.4763076.

[16]. N. S. Raghava and A. De, "A Novel High-Performance Patch Radiator," *International Journal of Microwave Science and Technology*, vol. 2008, Article ID 562193, 2008, pp. 1–4. doi:doi:10.1155/2008/562193.

[17]. N. S. Raghava and A. De, "Effect of Air Gap Width on the Performance of a Stacked Square Electronic Band Gap Antenna," *International Journal of Microwave and Optical Technology*, vol. 3, no. 5, 2009, pp. 315–317. IJMOT-2009-6-443.

[18]. K. Viswanadha and N. S. Raghava, "Design and Analysis of a Dual-Polarization Multiband Oval Ring Patch Antenna With L-Stubs and Folded Meander Line for C-band/X-Band/ Ku-band/ K-Band Communications," *International Journal of Electronics*, 2020, pp. 1–22. doi:10.1080/00207217.2020.1793411.

[19]. E. J. Rothwell, J. L. Frasch, S. M. Ellison, P. Chahal and R. O. Ouedraogo, "Analysis of the Nicolson–Ross–Weir Method for Characterizing the Electromagnetic Properties of Engineered Materials," *Progress in Electromagnetics Research*, vol. 157, 2016, pp. 31–47. doi:10.2528/PIER16071706

4 Transparent Dielectric Resonator Antenna for Smart Wireless Applications

Preksha Gandhi[1], *Yesha Patel*[1], *Swapnil Shah*[1], *Arpan Desai*[2,3], *Trushit Upadhyaya*[1], and *Nguyen Truong Khang*[2,3]
[1]Department of Electronics and Communication Engineering, CSPIT, Charotar University of Science and Technology (CHARUSAT), Changa 388421, India
[2]Division of Computational Physics, Institute for Computational Science, Ton DucThang University, Ho Chi Minh City, Vietnam
[3]Faculty of Electrical and Electronics Engineering, Ton DucThang University, Ho Chi Minh City, Vietnam

4.1 INTRODUCTION

The demand for smart applications demands smarter devices that are connected through the internet. The best source of internet is through the routers and repeaters which work on the protocol of the IEEE 802.11 WLAN (Wireless Local Area Network) standard. The WLAN standard demands an antenna that works mainly in the frequency band of 2.4/5 GHz.

The advancement in materials used to design the antenna to increase the gain and efficiency is in great demand. One such material known as DRA (dielectric resonator antenna) could be used for making efficient resonators [1]. The tremendous benefits of DRA include compact size, minimum loss, low cost, negligible weight, and facilitation of excitation and have led to a great extent of research in this direction [2,3]. DR also helps in making the antenna size smaller by a factor of $1/(\varepsilon_r)^{1/2}$, where ε_r represents the dielectric constant of the DR. In this era where smart devices, mobile, and other wireless gadgets have shown a hike, it has directed to high gain antennas [4]. Also, the priority has shifted to compactness, so it is a necessity to minimize the system's size [5–7]. The number of devices that use Wi-Fi is

DOI: 10.1201/9781003187325-4

skyrocketing in the world and for that, seamless connectivity and a steady network are very much necessary. For fulfilling the purpose, efficient antennas have to be mounted either externally or internally on the device. Here the DRA is useful, as it gives wide bandwidth with minimum space necessities.

The DRA can be incorporated with transparent antennas since this type of antennas are visually neutral and can be interfaced anywhere without any constraint of the space [8]. The transparent antennas are majorly designed for their feature of camouflaging. The transparent antenna can be achieved using a transparent conductive material, which can be prepared with different oxides like zinc, cadmium, tin indium, AgHT, fluorine-doped tin, etc. [9,10]. Such antennas have a disadvantage in the form of gain that is very low as the conductivity of oxides which are used for designing this antenna is very low as compared to copper. So, increasing the gain of such antennas can make it useful for its use in smart applications. Many such transparent antennas are proposed in the literature, which uses various techniques for gain increments like metamaterials [11], MIMO structure [12], and thin gold layer [13].

In wideband omnidirectional DRA proposed in [14], the coaxial probe was used for excitation, and acrylic was used for the inner layer and glass material was used for the outer layer of DRA. The Refractive index of the glass is 1.5, which gives a dielectric constant of 2.25, but at microwave frequency, the dielectric constant of glass is 7. The glass's dielectric constant of 6.85 was used for designing of two-layer hemispherical DRA and acrylic's dielectric constant of 2.5 was used for the same purpose. The transparent DRA for optical applications was proposed that used hemispherical DRA made up of Borosilicate Crown glass i.e. Pyrex of radius 28 mm. The measured dielectric constant came about 7.0 round 1.9 Ghz of glass [15]. In another research, a new circular polarized cylindrical DRA, open-ended logarithmic spiral slots were used for excitation in the ground plane. A glass DR was fabricated for the prototype. It covered about 2.4 GHz of the WLAN band [16]. The other used the same cylindrical dielectric resonator antenna (DRA), which covered about 2.4 GHz WLAN application. The prototype used a rectangular cross slot and an axial coaxial probe for its excitation. The antenna was fabricated with two unique transparent materials to cover the frequency [17]. One more prototype was designed in 2005. It was a hybrid resonator antenna that consisted of a microstrip patch that was coupled to a dielectric resonator (DR). A 10 dB return-loss bandwidth of 5.14–6.51 GHz was demonstrated. Simultaneously a radiation pattern that was similar to microstrip patch or DRA was obtained. It was claimed to be used for broadband wireless devices in the areas where the antenna is restricted [18]. The shape of DRA can vary from rectangular, cylindrical, triangular, to hemispherical [19] but the cylindrical-shaped DRA is easy to deal with rather than rectangular and hemispherical. A thin conducting metal strip has been placed on cylindrical DRA for both feeding and parasitic fashion. This increases return loss to 11.5% [20]. On the same hand, hemispherical shaped antenna yields exact analytical solutions. Similarly, several feeding mechanisms such as a coaxial probe, coupling slot, microstrip line, coplanar waveguide, aperture, conformal patch, microstrip line, or metallic waveguide are used for coupling energy for DRA [21–23].

In this paper, a transparent DRA resonating in the WLAN frequency band is proposed. The antenna consisting of AgHT-8 (patch and ground), plexiglass (substrate), and borosilicate glass (DR) achieves transparency more than 75% with wide impedance bandwidth of 31.26% ranging from 1.97–2.70 GHz covering WLAN, ZigBee, and Bluetooth applications.

Table 4.1 shows the comparison of transparent DRA performance with other antennas proposed in the literature where it can be observed that along with transparency, the proposed antenna achieves a gain around 2 dBi and an impedance bandwidth of 31.26% making it commercially viable for various smart applications.

4.2 ANTENNA GEOMETRY AND DESIGN PRINCIPLE

The proposed transparent DRA geometry has been shown in Figures 4.1 and 4.2 where the top view and the side view are included along with antenna geometry parameters.

The optical transparent DRA is designed using AgHT-8 (patch and ground), Plexiglas (substrate), and borosilicate glass (DR) with an overall size of 50 × 50 mm^2. AgHT-8 has sheet impedance of 8 ohm/sq whereas Plexiglas and borosilicate glass has dielectric constant, loss tangent, and height of 2.3,0.0003,1.48 mm, and 7.75,0.102, 1.84 mm, respectively. An impedance bandwidth of 31.26% by using a trapezoid feed line is achieved. A tree-shaped structure is designed where the feed

TABLE 4.1
Comparative Analysis of Transparent DRA with Other DRAs from Literature

References	Frequency of Operation	Impedance Bandwidth	DRA Material (Er)	Gain
14	3.4–3.7 GHz (WiMAX application)	31.9%	(Inner Diameter)Acrylic—2.5(Outer Diameter)Glass—6.85	—
16	Entire 2.4 GHz band(WLAN Application)	~6.2%	Glass DR—6.85	0.4 dB
17	2.45 GHz	23.4%	Glass DR—6.85	< 0.1 dB
18	5.14–6.51 GHz	23.5%	Rogers T MM—9.2	6.0 dB at 5.14 GHz6.3 dB at 5.30 GHz7.5 dB at 6.30 GHz
21	67 GHz	16.4%	Rogers 6010—10.2Rogers 5880—2.2	17.2 dB
22	28 GHz	17.8%	Rogers 3210—10.2	10.5 dB
23	6.2–7.35 GHz	17%	Emerson & Cuming make Eccostock HIK—20	13.6 dB
Proposed Antenna	2.42 GHz	31.26%	Borosilicate Glass—6.59(Transparent)	2.04 dB

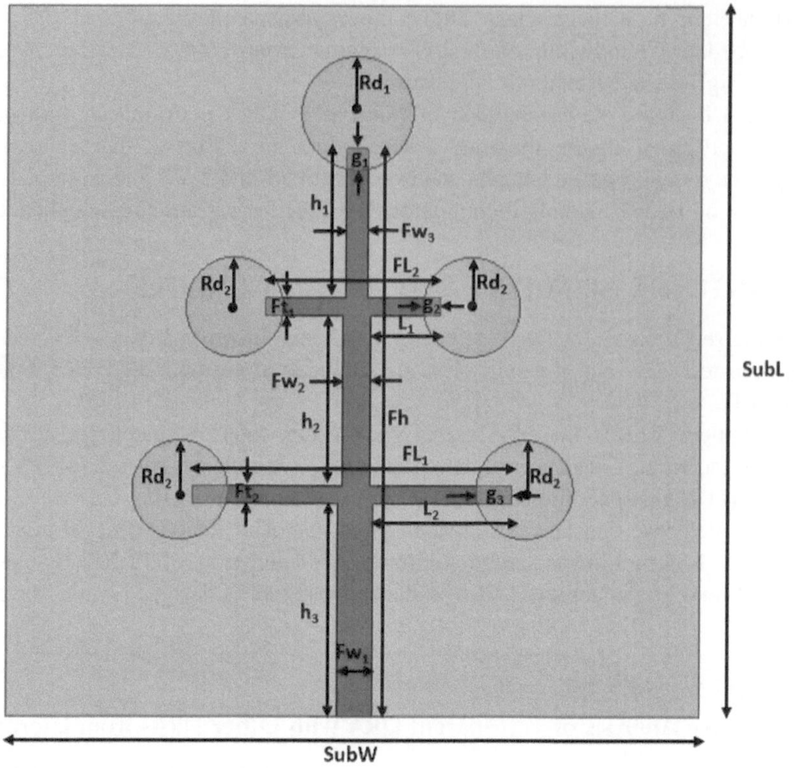

FIGURE 4.1 Proposed DRA based on transparent antenna geometry (top view).

SubL = 50 mm; Subw = 50 mm; Rd1 = 4 mm; Rd2 = 3.5 mm g1 = 1.5 mm; g2 = 1.25 mm; ġ3 = 2.5 mm; h1 = 10.5 mm; h2 = 12 mm; h3 = 15 mm; Ft1 = 1.25 mm; Ft2 = 1.25 mm; Ld = 8 mm; Fl1 = 1.5 mm; Fl2 = 12.5 mm; Fw1 = 2.5 mm; Fw2 = 2 mm; Fw3 = 1.5 mm; Fh = 40 mm; L1 = 5 mm; L2 = 10.25 mm

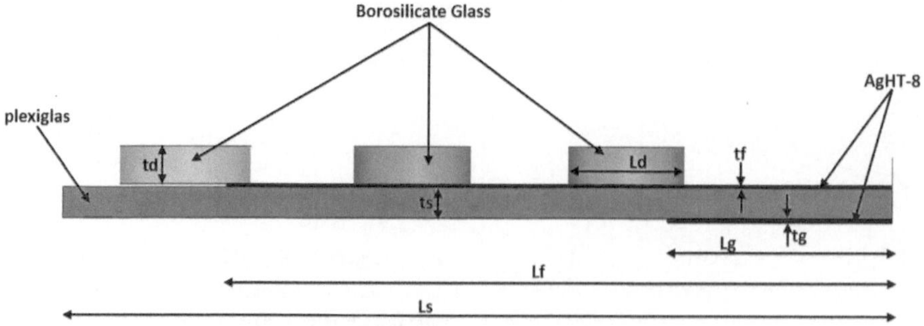

FIGURE 4.2 Proposed DRA based transparent antenna (side view).

Ld = 8 mm; td = 1.84 mm; ts = 1.48 mm; tf = 0.177 mm; tg = 0.177 mm; Lg = 13.5 mm; Lf = 4 mm; Ls = 50 mm

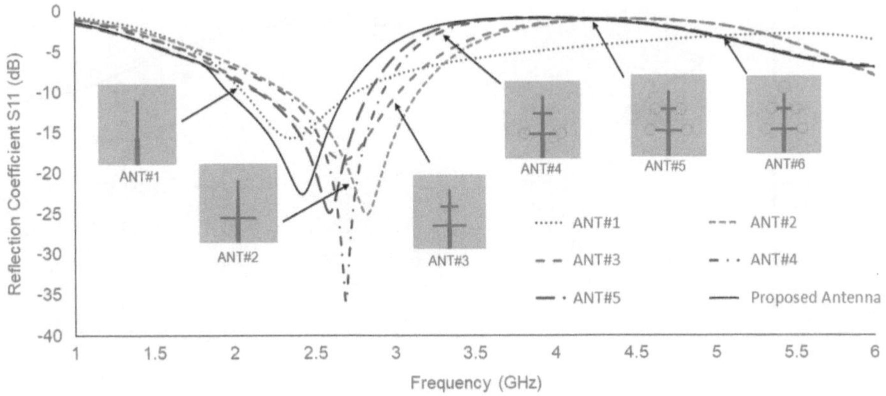

FIGURE 4.3 Antenna evolution process.

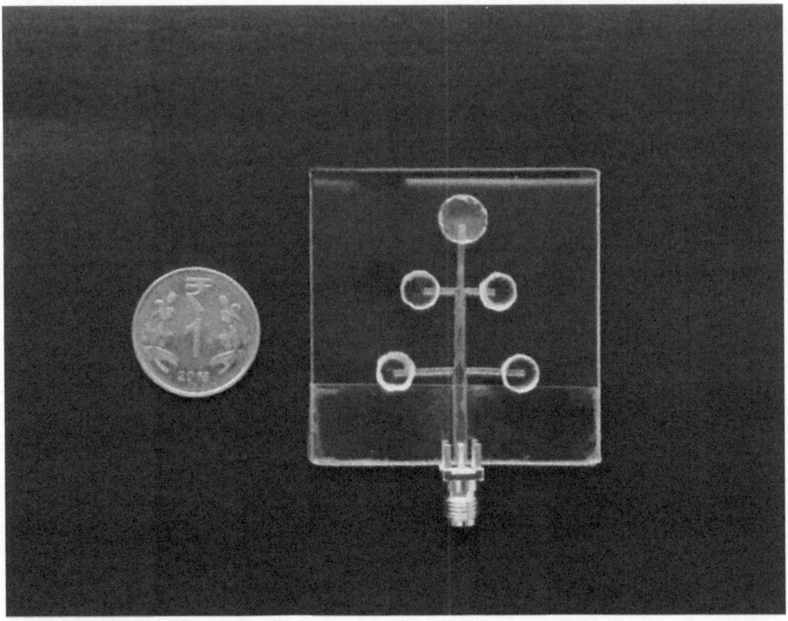

FIGURE 4.4 Fabricated DR based transparent antenna (top view).

line is optimized and then the dielectric resonators are arranged on the top of the feed line to achieve the proposed bands.

The antenna evolution is observed for different values of the reflection coefficient in Figure 4.3. The selection of feedline and DRA elements leads to wide impedance bandwidth along with WLAN frequency band operation.

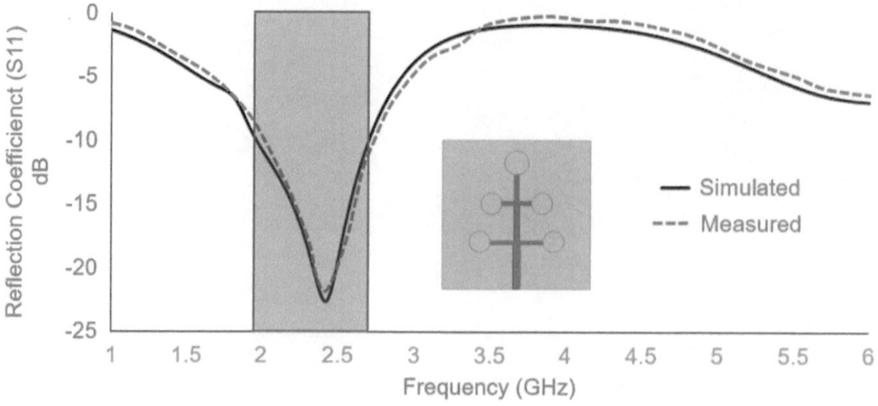

FIGURE 4.5 Simulated (solid) and measured (dashed) return loss of the proposed antenna.

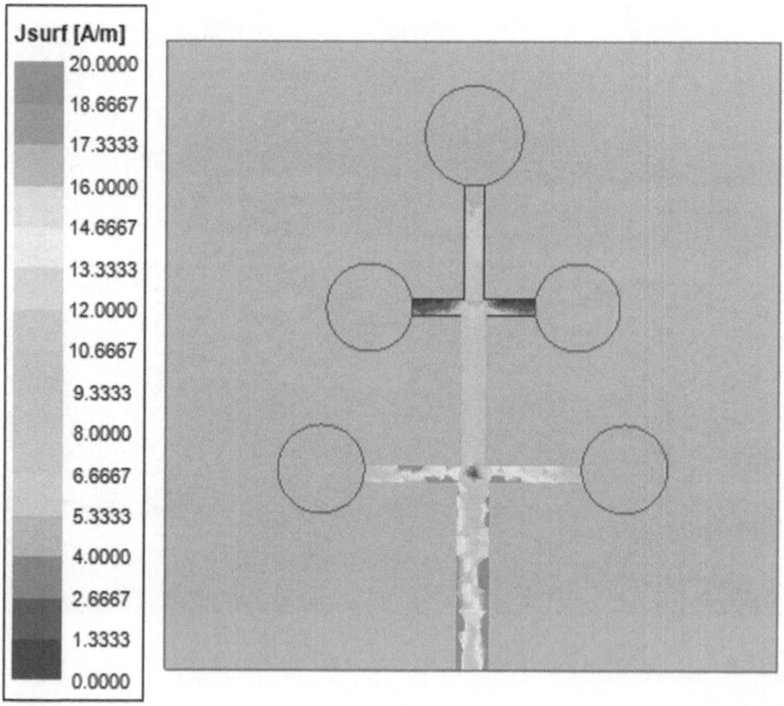

FIGURE 4.6 Current distribution of the proposed antenna at 2.42 GHz.

Figure 4.4 shows the fabricated prototype of the transparent DRA where the transparency of the antenna is depicted by the clear visibility of the background underneath the antenna. The feed is provided using an SMA connector connected to the feedline optimized at 50 Ω.

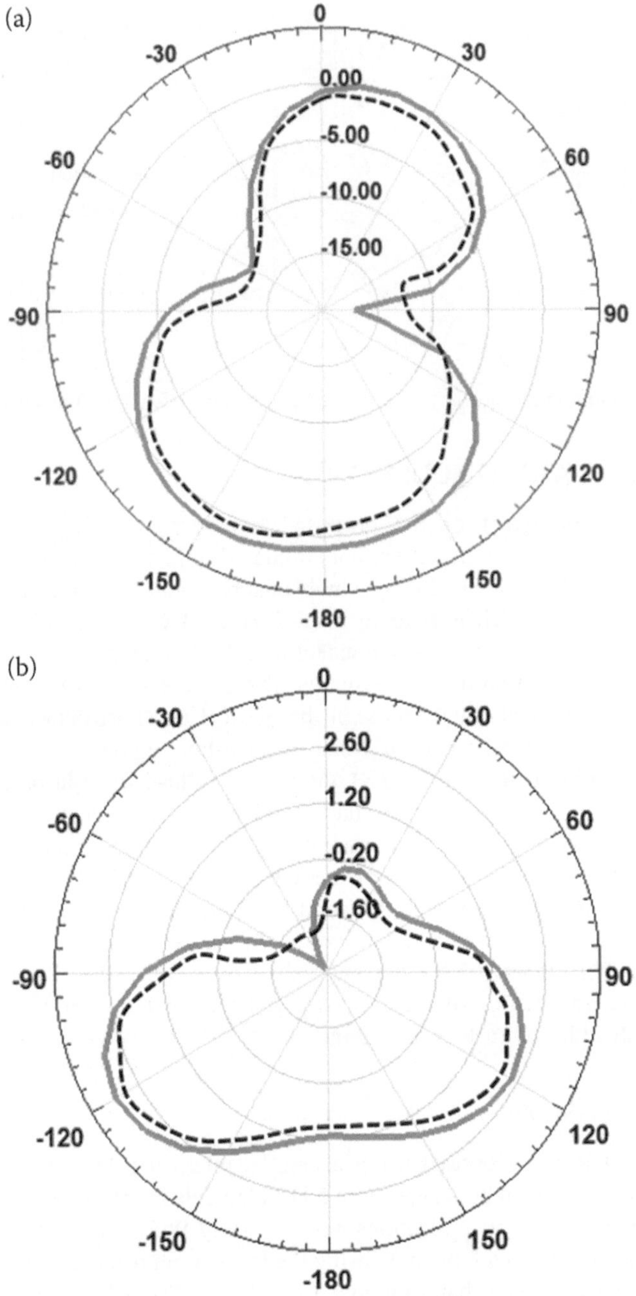

FIGURE 4.7 Simulated (solid) and measured (dashed) 2D radiation pattern at 2.42 GHz (a) E plane and (b) H plane.

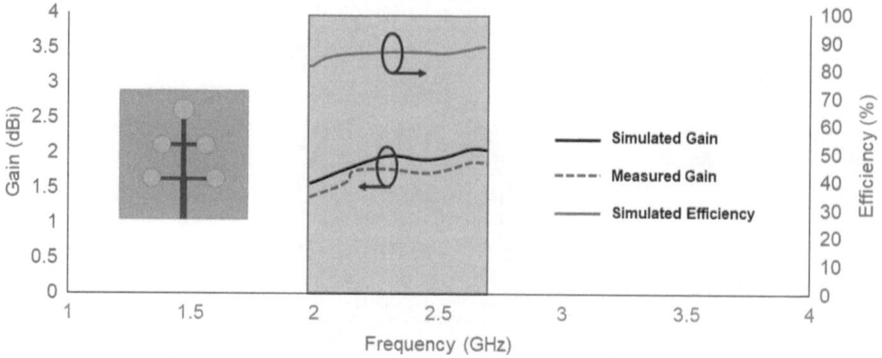

FIGURE 4.8 Gain and efficiency of proposed transparent DRA based antenna.

4.3 RESULTS AND DISCUSSION

The reflection coefficient of the proposed transparent DRA is illustrated in Figure 4.5 where it can be observed that simulated impedance bandwidth is in the range of 1.97–2.70 GHz (31.26%), which matches well with the measured impedance bandwidth, which is ranging from 2.01–2.73 GHz (30.37%). The slight variation is due to the tolerances in connector and fabrication.

The current distribution of the transparent DRA is depicted in Figure 4.6 where the current is distributed evenly through the vertical and horizontal sections with more concentration near the bottom T shaped feed-line region.

The 2D radiation pattern in E (YZ) and H (XY) Plane for transparent DRA is illustrated in Figure 4.7 where dipole shaped and the directional pattern is observed. The antenna placement plays an important role as the main lobe is more towards the bottom side than on the top portion of the antenna.

The gain and efficiency of the transparent DRA are shown in Figure 4.8 where it is observed that the simulated and measured value of gain is more than 1.7 dBi for the entire band of interest with efficiency values exceeding 80%. Table 4.2 shows the antenna characteristics of the proposed antenna in terms of frequency, impedance bandwidth, return loss, and gain.

4.4 CONCLUSION

A transparent DRA is proposed using a tapered microstrip line having two horizontal arms extending on both sides with 5 DR elements interfaced at the end of the feeding structure for smart applications working using WiFi, Bluetooth, and Zigbee protocols. The DR antenna fabricated using the transparent patch, substrate, and DR resonates in the frequency band ranging from 1.97–2.70 GHz achieving 31.26% impedance bandwidth with directional radiation pattern. Gain and efficiency of the antenna is more than 1.7 dBi and 80%, respectively. The inclusion of DR structure leads to bandwidth and gain improvement of the transparent antenna making the antenna suitable for its use in smart applications.

TABLE 4.2
Proposed Antenna Characteristics

Frequency (GHz)		Impedance bandwidth (%)		Return loss (dB)		Gain (dBi)	
Simulated	Measured	Simulated	Measured	Simulated	Measured	Simulated	Measured
2.42	2.41	31.26%	30.37%	–22.58	–22.19	2.04	1.62

REFERENCES

[1]. Long, S., Mark McAllister, and Liang Shen. "The resonant cylindrical dielectric cavity antenna." *IEEE Transactions on Antennas and Propagation* 31, no. 3 (1983): 406–412.

[2]. K. M. Luk and K. W. Leung (Eds.). *Dielectric Resonator Antennas*. London: Research Studies Press, 2003.

[3]. Petosa, Aldo. *Dielectric Resonator Antenna Handbook*. Aldo: Artech, 2007.

[4]. Desai, Arpan, Trushit K. Upadhyaya, Rikikumar Hasmukhbhai Patel, Sagar Bhatt, and Parthesh Mankodi. "Wideband high gain fractal antenna for wireless applications." *Progress in Electromagnetics Research* 74 (2018): 125–130.

[5]. Skrivervik, A. Ks, J-F. Zurcher, O. Staub, and J. R. Mosig. "PCS antenna design: The challenge of miniaturization." *IEEE Antennas and Propagation Magazine* 43, no. 4 (2001): 12–27.

[6]. Patel, RikikumarHasmukhbhai, Arpan Desai, and Trushit K. Upadhyaya. "An electrically small antenna using defected ground structure for RFID, GPS and IEEE 802.11 a/b/g/s applications." *Progress in Electromagnetics Research* 75 (2018): 75–81.

[7]. Desai, Arpan, Riki Patel, Trushit Upadhyaya, Hemani Kaushal, and Vigneswaran Dhasarathan. "Multiband inverted e and u shaped compact antenna for digital broadcasting, wireless, and sub 6 GHz 5G applications." *AEU-International Journal of Electronics and Co mmunications* 123 (2020): 153296.

[8]. Guan, Ning, Hirotaka Furuya, Kuniharu Himeno, Kenji Goto, and Koichi Ito. "Basic study on an antenna made of a transparent conductive film." *IEICE Transactions on Communications* 90, no. 9 (2007): 2219–2224.

[9]. Mias, C., C. Tsakonas, N. Prountzos, D. C. Koutsogeorgis, S. C. Liew, C. Oswald, R. Ranson, W. M. Cranton, and C. B. Thomas. "Optically transparent microstrip antennas." IEE Colloquium on Antennas for Automotives (Ref. No. 2000/002) (2000): 8/1-8/6. 10.1049/ic:20000008.

[10]. Lim, Eng Hock, Kwok Wa Leung, Xiaosheng Fang, and Yongmei Pan. "Transparent antennas." in *Wiley Encyclopedia of Electrical and Electronics Engineering*. New York: Wiley, pp. 1–23, 1999.

[11]. Desai, Arpan, and Trushit Upadhyaya. "Transparent dual band antenna with μ-negative material loading for smart devices." *Microwave and Optical Technology Letters* 60, no. 11 (2018): 2805–2811.

[12]. Desai, Arpan, Trushit Upadhyaya, Merih Palandoken, and Cem Gocen. "Dual band transparent antenna for wireless MIMO system applications." *Microwave and Optical Technology Letters* 61, no. 7 (2019): 1845–1856.

[13]. Peter, T., T. I. Yuk, R. Nilavalan, and S. W. Cheung. "A novel technique to improve gain in transparent UWB antennas." in *2011 Loughborough Antennas & Propagation Conference*. New York: IEEE, pp. 1–4, 2011.

[14]. Fang, Xiao Sheng, and Kwok Wa Leung. "Design of wideband omnidirectional two-layer transparent hemispherical dielectric resonator antenna." *IEEE Transactions on Antennas and Propagation* 62, no. 10 (2014): 5353–5357.

[15]. Lim, Eng Hock, and Kwok Wa Leung. "Transparent dielectric resonator antennas for optical applications." *IEEE Transactions on Antennas and Propagation* 58, no. 4 (2010): 1054–1059.

[16]. Yang, Nan, Kwok Wa Leung, Kai Lu, and Nan Wu. "Omnidirectional circularly polarized dielectric resonator antenna with logarithmic spiral slots in the ground." *IEEE Transactions on Antennas and Propagation* 65, no. 2 (2016): 839–844.

[17]. Yang, Nan, Kwok Wa Leung, and Nan Wu. "Pattern-diversity cylindrical dielectric resonator antenna using fundamental modes of different mode families." *IEEE Transactions on Antennas and Propagation* 67, no. 11 (2019): 6778–6788.
[18]. Esselle, Karu P., and Trevor S. Bird. "A hybrid-resonator antenna: Experimental results." *IEEE Transactions on Antennas and Propagation* 53, no. 2 (2005): 870–871.
[19]. Leung, Kwok Wa, Eng Hock Lim, and Xiao Sheng Fang. "Dielectric resonator antennas: From the basic to the aesthetic." *Proceedings of the IEEE* 100, no. 7 (2012): 2181–2193.
[20]. Iqbal, Javed, Usman Illahi, Mohamad Ismail Sulaiman, Muha mmad Mansoor Alam, Mazliham MohdSu'ud, Mohd Najib, Mohd Yasin, and Mohd HaizalJamaluddin. "Bandwidth enhancement and generation of CP by using parasitic patch on rectangular DRA for wireless applications." *IEEE Access* 7 (2019): 94365–94372.
[21]. Chen, Zhijiao, Changan Shen, Haiwen Liu, Xiuzhu Ye, Limei Qi, Yuan Yao, Junsheng Yu, and Xiaodong Chen. "Millimeter-wave rectangular dielectric resonator antenna array with enlarged DRA dimensions, wideband capability, and high-gain performance." *IEEE Transactions on Antennas and Propagation* 68, no. 4 (2019): 3271–3276.
[22]. Bahreini, Batul, Homayoon Oraizi, Narges Noori, and Saeed Fakhte. "Design of a circularly polarized parasitic array with slot-coupled dra with improved gain for the 5G mobile system." *IEEE Antennas and Wireless Propagation Letters* 17, no. 10 (2018): 1802–1806.
[23]. Rana, Biswarup, and Susanta Kumar Parui. "Microstrip line fed wideband circularly-polarized dielectric resonator antenna array for microwave image sensing." *IEEE Sensors Letters* 1, no. 3 (2017): 1–4.

Part III

Multiple Input Multiple Output (MIMO) Antenna Design and Uses

5 Design and Analysis of 2 × 2/4 × 4 MIMO Antenna Configurations for High Data Rate Transmission

Manish Sharma
Chitkara University Institute of Engineering and Technology, Chitkara University, Punjab, India

5.1 INTRODUCTION

The Federal Communication Commission (FCC) released the usage of ultrawideband (UWB) technology in 2002 that utilizes very low power spectral density for different fields of imaging applications. Hence, planar antenna design technology kick-started and several planar antennas have been designed that are reported in the literature. However, when single antenna radiation is concerned, it finds several disadvantages such as multiple path fading, which in turn reduces operating bandwidth and hence the efficiency of the antenna. In today's scenario, enhancement characteristics of wireless communication are needed such as high reliability, speeding data rate transmission with reduces co-channel interference, and improvement in the capacity of the channel. The aforementioned problems in single element antenna are overcome by using multiple input multiple output (MIMO) technology where there are more than one radiating element forming 2 × 2 or 4 × 4 MIMO configuration. This configuration ensures a higher data rate of transmission without compromising the operating bandwidth. In the MIMO system, when planar antennas are considered, isolation is the key challenging factor that one has to encounter for better diversity performance. In the latest literature, several 2 × 2 or 4 × 4 MIMO configuration planar antennas [1–71] are reported utilizing different methodologies to obtain isolation and diversity performances. MIMO antenna for 5G application is reported [1] with a size of 40 × 42 mm^2 and consists of two planar antenna elements placed orthogonally for better isolation. Two identical P-type radiators oriented by 180° to each other provide UWB bandwidth of 2.50 GHz–12.0 GHz [2]. Radiating elements placed orthogonal to each other [3] result in better isolation of more than 15 dB in operating bandwidth of interest. Also, when two radiating elements are placed adjacent to each other [4–11] with isolation elements between them, UWB antenna with high isolation MIMO configuration is obtained. Isolation is better achieved when the T-type stub is placed on the ground [4–6] or by placing an F-type stub [8]. Using different patches

DOI: 10.1201/9781003187325-5

with isolation elements, also multiband MIMO configuration and dual-polarization MIMO antenna result [12–20]. Electromagnetic Band Gap (EBG) structure, Dielectric Resonator Antennas (DRA) is another mentioned technology used to obtain MIMO configuration [21–23]. These MIMO antennas are further converted to single, dual, triple, and four notched band MIMO antennas obtained by using either stub/slot or other notching techniques that act as band stop filter [24–48]. Super-wideband MIMO antenna is reported [49] where the circular patch is fed by tapered microstrip feed. A series of 4 × 4 MIMO antenna is reported [50–71] where radiating patches are placed orthogonally, placing pair of radiating elements opposite to each other, and by placing the patches on the opposite face of the substrate. Further, different isolation technique is reported for better isolation and they are further converted to notched band characteristics.

5.2 UWB MIMO ANTENNA

Figure 5.1 shows the development of a 2 × 2 MIMO antenna. Initially, a single element unit cell is developed that is applicable for UWB band applications which are shown in

Figure 5.1(a). Further, an identical radiating element is placed adjacent so that there are two radiating patches as observed in Figure 5.1(b), which are shared by common ground. This MIMO antenna is not capable of maintaining the required isolation between radiating elements and modification is necessary. Figure 5.1(c) is the modified version where a T-shaped stub is added to the ground and this stub provides additional current path flow between the radiating elements and hence isolation is achieved.

Further, the MIMO antenna discussed in Figure 5.1 can be converted to dual notched band characteristics MIMO antenna by etching a pair of elliptical slits on the radiating patch. Figure 5.2 provides information that the designed antenna is capable of mitigating WiMAX and WLAN interfering bands without compromising the operational impedance and also maintaining the required isolation between the two input ports.

The above discussed MIMO antenna has to be characterized not only in near and far-field but also needs to be characterized in terms of diversity performance. Envelope Correlation Coefficient (ECC), Directive Gain (DG in dB), Total Active Reflection Coefficient (TARC in dB), Channel Capacity Loss (CCL in bits/s/MHz), and Mean Effective Gain (MEG) are the few MIMO characteristics of the antenna that have to be understood.

Envelope Correlation Coefficient can be calculated by using either radiation characteristics or by extracting S-Parameters. The critical system index for any antenna array is given by

$$ECC = \frac{\left| \iint_{4\pi} \left[\overrightarrow{F_a}(\theta, \varphi) \times \overrightarrow{F_b^*}(\theta, \varphi) \right] d\Omega \right|^2}{\iint_{4\pi} \left| \overrightarrow{F_a}(\theta, \varphi) \right|^2 d\Omega \left| \overrightarrow{F_b}(\theta, \varphi) \right|^2 d\Omega}, \quad (5.1)$$

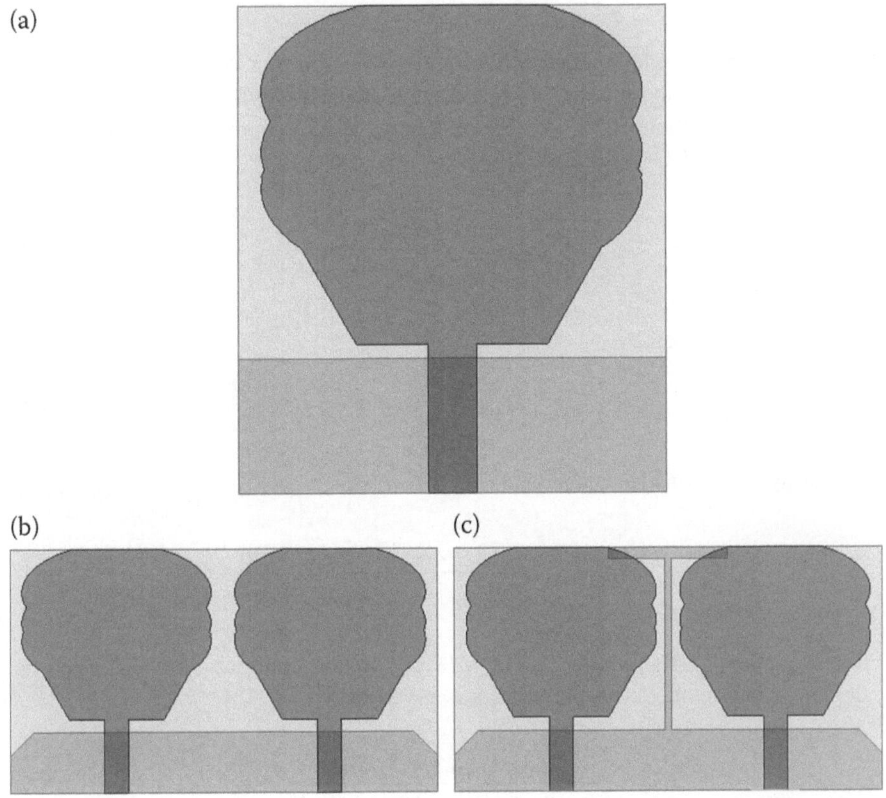

FIGURE 5.1 MIMO Antenna Configuration (a) Single unit cell (b) 2 × 2 MIMO antenna without isolation element (c) 2 × 2 MIMO antenna with T-shaped isolation element.

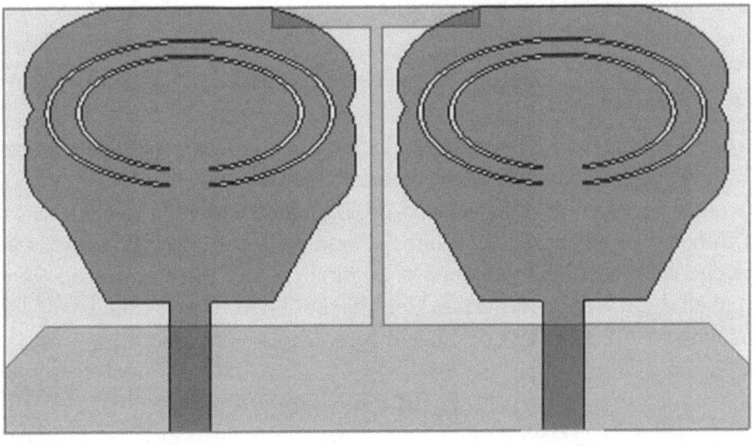

FIGURE 5.2 2 × 2 MIMO antenna with dual notched band characteristics.

where $\overrightarrow{F_j}(\theta, \varphi)$ signifies radiation field pattern for the $N \times N$ MIMO system and indicates that when the jth port is excited while all other ports are matched to 50 Ω impedance. When there is a K number of antenna array system, considering p and q MIMO antenna, ECC is calculated by

$$\rho_e(p, q, K) = \frac{|C_{p,q}(K)|^2}{\prod_{r=p,q}[1 - C_{r,r}(K)]}, \tag{5.2}$$

where $C_{p,q}(r)$ is given by

$$C_{p,q}(r) = \sum_{r=1}^{r} S_{p,r}^{*} S_{r,q}. \tag{5.3}$$

Comparing Equations (5.2) and (5.3)

$$\rho_e(p.\ q.\ r) = \frac{|\sum_{r=1}^{r} S_{p,r}^{*} S_{r,q}|^2}{\prod_{r=i,j}[1 - \sum_{r=1}^{r} S_{n,p}^{*} S_{q,r}]}. \tag{5.4}$$

From Equation (5.4), for any 2 × 2 MIMO antenna configuration, ECC can be calculated from extracted S-parameters is given by

$$\text{ECC} = \frac{|S_{11}^{*} S_{12} + S_{21}^{*} S_{22}|^2}{((1 - |S_{11}|^2 - |S_{21}|^2)((1 - |S_{22}|^2 - |S_{12}|^2)}. \tag{5.5}$$

For any uncorrelated MIMO system, ideally, ECC will be zero but for correlated MIMO system, the values should be less than 0.2. Diversity Gain, which is another parameter for the MIMO system, depends on ECC and is calculated by

$$\text{Directive Gain} = 10\sqrt{1 - ECC^2}. \tag{5.6}$$

It is expected for any MIMO configuration that directive gain should be greater than 9.95 dB. It is expected that when two or more radiating elements are placed adjacent to each other or when they are placed orthogonally, radiating signals tend to interfere with one another. Hence, alone S-Parameters (S_{11}, S_{12}, S_{21}, S_{22}) is not enough to characterize MIMO antenna configuration and there is a need for a new metric called Total Active Reflection Coefficient (TARC) to be calculated, which in turn depends on S-Parameters calculated by

$$\text{TARC} = \sqrt{\frac{(S_{11} + S_{12})^2 + (S_{21} + S_{22})^2}{2}}. \tag{5.7}$$

Ideally, TARC should be less than 0 dB.

Channel capacity indicates the maximum rate of transmission over the channel without any distortion. The merit of the channel is calculated in terms of Channel Capacity Loss (CCL), which is given by

$$C_{Loss} = -\log_2(R^M), \tag{5.8}$$

$$R^M = \begin{bmatrix} R_{11} & R_{12} \\ R_{21} & R_{22} \end{bmatrix}, \tag{5.9}$$

where

$$R_{11} = 1 - [|S_{11}|^2 + |S_{12}|^2], \tag{5.10}$$

$$R_{22} = 1 - [|S_{22}|^2 + |S_{21}|^2], \tag{5.11}$$

$$R_{12} = -[S_{11} * S_{12} + S_{21} * S_{12}], \tag{5.12}$$

$$R_{21} = -[S_{22} * S_{21} + S_{12} * S_{21}]. \tag{5.13}$$

For the MIMO configuration system, CCL should be limited to 0.4 bits/s/Hz.

Mean effective gain is the ratio of power absorption of the designed-patch antenna to the mean-power, which is incident on MIMO configuration, and the reference antenna is considered as an isotropic antenna for calculation of MEG. For consideration of the medium environment and indoor medium, XPR = 0 db is considered, while for outdoor environment XPR = 6 dB is assumed. Mathematically, MEG is given by

$$MEG = \frac{P_{recieved}}{P_{incident}} = \int_0^{2\pi}\int_0^{\pi}\left[\frac{X_{PR}}{1+X_{PR}}F_\theta(\theta,\varphi)P_\theta(\theta,\varphi) + \frac{1}{1+X_{PR}}F_\varphi(\theta,\varphi)P_\varphi(\theta,\varphi)\right]\sin\theta d\theta d\varphi \tag{5.14}$$

In the above equation, F_θ, F_φ represents power gain patterns of the designed MIMO system. F_θ is calculated by changing θ and keeping φ a constant value. Similarly, F_φ is calculated by changing φ and keeping θ a constant value. Power gain measurement can be carried out by using Dual Ridge Horn Antenna as a radiator. It should be noted that MEG should always lie <–3 dB for all band of operation [12].

Table 5.1 shows the comparison of the UWB MIMO antenna configuration with comparison characteristics such as size, ECC, DG, isolation, and operational bandwidth. As per the observation, Reference [8] provides isolation better than 22 dB while References [4] and [8] have better ECC less than 0.005.

TABLE 5.1
Comparison of UWB MIMO Antenna

Reference	Size of Antenna in mm^2	ECC	DG (dB)	Isolation (dB)	Bandwidth (GHz)
[2]	$0.25\lambda_o \times 0.241\lambda_o$	<0.02	>9.50	<−20	2.50–12.0
[3]	$0.309\lambda_o \times 0.309\lambda_o$	<0.04	–	<−15	2.90–12.0
[4]	$0.91\lambda_o \times 1.17\lambda_o$	<0.005	>9.99	<−20	3.10–10.6
[5]	$0.25\lambda_o \times 0.30\lambda_o$	<0.30	–	<−15	3.43–10.1
[6]	$0.23\lambda_o \times 0.29\lambda_o$	<0.15	–	<−15	3.00–12.0
[7]	$0.19\lambda_o \times 0.34\lambda_o$	<0.15	–	<−15	2.70–12.0
[8]	$0.36\lambda_o \times 0.41\lambda_o$	<0.005	–	<−22	3.10–10.6
[9]	$0.21\lambda_o \times 0.25\lambda_o$	–	–	<−12	2.20–10.6
[11]	$0.248\lambda_o \times 0.434\lambda_o$	<0.10	>9.80	–	3.10–10.9

5.3 MULTIBAND/DUAL POLARIZATION/UWB 2 × 2 MIMO ANTENNA WITH NOTCHED BAND CHARACTERISTICS

Multiband MIMO antenna configuration is reported [12–14] where dome-type monopole resonates at two bands of frequencies covering LTE, Wi-Fi/WLAN/WiMAX, and Bluetooth bands. Isolation in these types of antennas is maintained by using T-stub or also by using a rectangular stub in the ground. Dual polarization antennas are discussed [15–20] where a single radiating element is fed by two feeding lines that are placed orthogonally and connected to patch. As discussed in Section 5.2, MIMO antenna configurations without mitigation of interfering bands were discussed. However, impedance bandwidths of these antennas do suffer interference caused by WiMAX, WLAN, and DSS bands. Different techniques have been reported to encounter this interference were quarter-wavelength stub/slot/slits are used as bandstop filters, extruding T-shaped stub in-ground [25], two modified split-ring resonators with two L-type branches [26], adding a single folded stub with the patch, two strips in-ground [31] and adding metal strips to the ground which is connected via from patch [32]. Also, isolation between the radiating elements is maintained by placing rectangular stub in the ground by 45° [24], using T-type stub in-ground [25,26,32], etching an H-shaped slot in the ground [29], placing the radiating elements orthogonally [30], inverted L-type stub [34] are different methods employed for isolation.

Tables 5.2 and 5.3 compare MIMO antenna configuration in terms of different physical, near field, and diversity performance techniques. Table 5.2 shows a comparison of MIMO-Multiband, antenna which are useful for different wireless communication such as LTE, Bluetooth, WiMAX, WLAN, and X Band applications. From the comparison, the entire MIMO antenna maintains good diversity performance such as ECC < 0.2, DG > 9.95 dB, MEG < −3 dB, and TARC < −15 db in the multiband bandwidth. Table 5.3 shows a comparison of MIMO antenna configuration in terms of polarization diversity and shows dual-polarization either in multiband bandwidth or in wideband applications. Designed antennas are fabricated on different

TABLE 5.2
Comparison of Multiband MIMO Antenna.

Ref.	Isolation Technique Used	Isolation (dB)	Size of Antenna (mm^2)	No. of Bands	Bandwidth (GHz)	ECC	DG (dB)	MEG (MEG1 & MEG2)	CCL (bits/s/Hz)	TARC (dB)
[12]	T-stub with combed shape slots	>21	20 × 34	2	2.11–4.19 4.98–6.81	<0.004	>9.97	<–3 dB	<0.32	<–15 dB
[13]	Flag shaped stub in ground	>18 >20	22 × 36	2	2.50–2.85 4.82–6.10	<0.05	>9.992 >9.996	<–3 dB	–	–
[14]	Rectangular stub in ground	>15	36 × 40	3	2.40–2.48 43.40–3.69 5.15–5.825 3.10–10.60	<0.02	>9.456 >9.637 >9.793	<–3 dB	–	–

TABLE 5.3
Comparison of Dual Polarization MIMO Antenna

Ref.	Size (mm^2)	Isolation Technique Used	Bandwidth (GHz)	Isolation (dB)	Substrate	ECC	Radiator Sharing
[15]	25 × 27	T-type slot in patch	3.00–11.0	>15	Rogers TMM4	<0.01	One Radiator
[16]	22 × 24.3	Etched slot and connected stub	3.20–10.60	>15	Rogers TMM4	–	One Radiator
[17]	40 × 40	T-type slot on patch and stub in ground	3.00–11.0	>30	FR4	–	Two Radiator
[18]	30 × 30	Y-slot in ground	5G WLAN 8G X Band	>15	FR4	<0.03	Two Radiator
[19]	35 × 35	NIL	3.00–12.0	>20	FR4	<0.5	Four Radiator
[20]	64 × 64	Etched circular cut in ground	2.00–6.00	>15	Rogers RTDuroid5880	–	One Radiator

substrates including Rogers TMM4, FR4, and Rogers RTDuroid5880. Dual polarization is obtained by using either one or two radiators. Each radiator is fed by two feeding lines which are orthogonal and hence providing dual polarization. The designed antenna also provides good diversity performance with maximum ECC < –0.5.

Table 5.4 shows the comparison of 2 × 2 MIMO configuration with the introduction of notched bands. As per the observation, radiating elements are placed either adjacent to each other. or they are placed orthogonally to each other. In Table 5.4, a single notched band antennas comparison is carried where [25] finds larger impedance bandwidth of 2.90 GHz–20.0 GHz. To obtain isolation between radiating elements, different techniques have been reported which include using of T-type stub in the ground, U-shaped branches in-ground, Funnel type stub in-ground, stepped structure in-ground, or sometimes no requirement of any isolation element. For two notched band MIMO antennas, both WiMAX/WLAN/X band interference is mitigated with better isolation of 20 dB and ECC < 0.50. Similarly, all the above interfering bands are removed by using bandstop filters for 2 × 2 MIMO configuration with good isolation <–18 dB and ECC < 0.02 with radiating elements adjacent to each other.

5.4 4 × 4 MIMO ANTENNA CONFIGURATION

Massive demand for higher data rate transmission over the communication channel has become important. 2 × 2 MIMO configuration satisfies this need to some extent

TABLE 5.4
Comparison of 2 × 2 MIMO Configuration with Single/Dual/Triple Notched Bands

Ref.	Size of Antenna (mm^2)	Operational Bandwidth (GHz)	Isolation Technique Used	Isolation (dB)	No. of Notched bands	Bandwidth of notched Bands (GHz)	ECC	Orientation of Radiators
[24]	48 × 48	2.50–12.0	Rectangle Stub at 45° in ground	<–15	1	5.10–6.00	–	Orthogonal
[25]	18 × 36	2.90–20.0	T-type stub in ground	<–20	1	3.62–4.77	<0.5	Adjacent
[26]	26 × 36	2.80–12.0	Stepped T-type stub in ground	<–15	1	5.00–5.95	<-	Adjacent
[27]	28 × 50	2.80–11.5	NIL	<–15	1	3.30–3.90	<0.02	Orthogonal
[28]	38.5 × 38.5	3.08–11.8	Slot and Slit	<–15	1	5.03–5.97	–	Orthogonal
[29]	22 × 28	2.90–11.8	H-type slot in ground	<–20	1	4.90–6.00	<0.04	Adjacent
[30]	23 × 39.8	2.00–12.0	NIL	<–20	1	4.80–6.20	–	Orthogonal
[31]	22 × 36	3.10–11.0	T-type stub in ground with vertical slot	<–15	1	5.15–5.85	<0.1	Adjacent
[32]	22 × 29	3.10–10.6	T-type stub in ground	<–17	1	5.15–5.85	<0.0015	Adjacent
[35]	45 × 58	3.10–10.6	Slotted ground plane	<–15	2	3.30–3.60 5.00–6.00	<0.02	Adjacent
[36]	25 × 39	2.82–11.2	Two U-shaped branches in ground	<–20	2	5.12–5.81 7.02–7.98	<0.01	Orthogonal
[38]	18 × 34	2.90–20.0		<–22	2		<0.01	Adjacent

(*Continued*)

TABLE 5.4 (continued)
Comparison of 2 × 2 MIMO Configuration with Single/Dual/Triple Notched Bands

Ref.	Size of Antenna (mm^2)	Operational Bandwidth (GHz)	Isolation Technique Used	Isolation (dB)	No. of Notched bands	Bandwidth of notched Bands (GHz)	ECC	Orientation of Radiators
			Slotted T-type stub in ground			5.15–5.80 6.70–7.10		
[39]	15 × 26	3.10–35.0	Funnel type stub in ground	<–24	2	5.10–5.80 6.70–7.10	<0.20	Adjacent
[41]	22 × 26	3.10–11.8	T-type slot in ground	<–20	2	5.40–5.86 7.60–8.40	<0.50	Adjacent
[44]	45 × 64	2.38–10.68	Stepped structure in ground	<–15	3	3.30–3.60 5.00–6.00 7.10–7.90	<0.02	Adjacent
[47]	22 × 28	3.10–10.6	T-type stub in ground	<–18	3	3.40–3.90 5.05–5.85 7.90–8.40	<0.02	Adjacent

but an alternative solution is needed. Hence, 4 × 4 MIMO antenna configuration has taken a breakthrough in fulfilling the above need in the wireless communication environment. To understand the terminology, a 4 × 4 MIMO antenna is considered [66]. Figure 5.3 shows a 4 × 4 MIMO antenna configuration where radiating elements are placed orthogonally to each other. Each antenna is fed by a 50 Ω SMA connector for characterization. Figure 5.3(a) shows a MIMO configuration with four radiating elements placed orthogonally. It is to be noted that no isolation element or technique is used to maintain the required isolation between them. Also, Figure 5.3(b) signifies 4 × 4 MIMO with notched bands which are obtained by using stub and slots on the radiating patch. This MIMO antenna when excited by all the four ports, provides the required fractional bandwidth with mitigation of WiMAX and WLAN bands as shown in Figure 5.3(c). Figure 5.3(d) shows the ECC of the MIMO antenna which is less than 0.02 and DG > 9.95 dB. Figure 5.3(e) shows the VNA measurement arrangement of the fabricated prototype.

Table 5.5 compared 4 × 4 MIMO antenna configuration with different parameters such as size, bandwidth, Isolation, ECC, and orientation of radiating

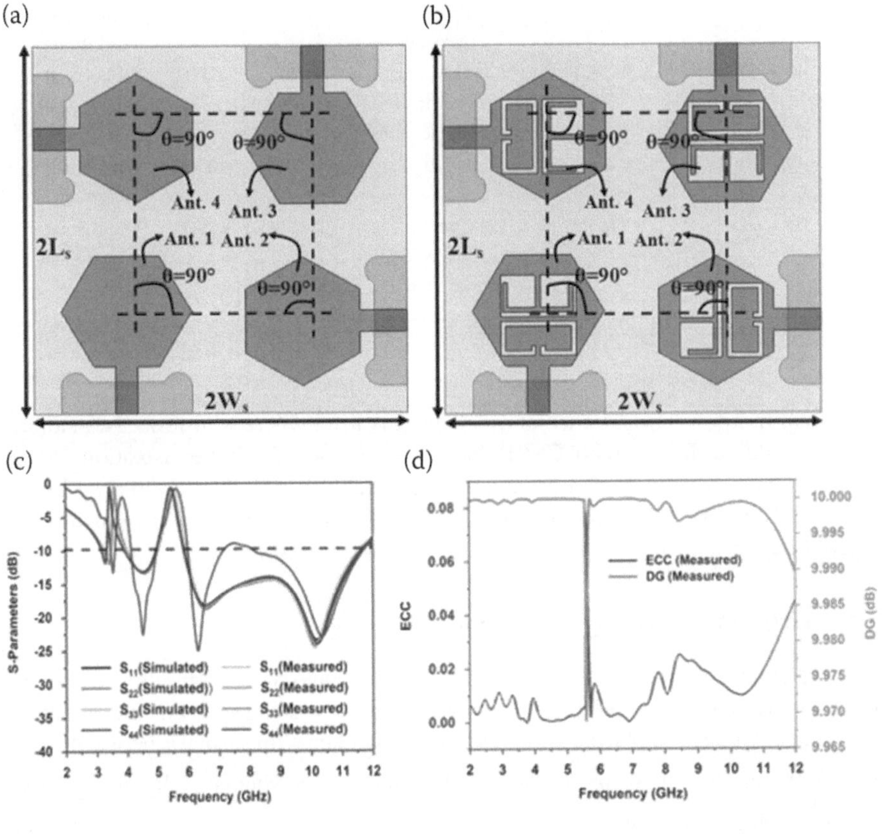

FIGURE 5.3 4 × 4 MIMO antenna configuration (a) Without notched bands (b) With interfering bands (c) S-Parameters (d) ECC & DG (e) VNA measurement of S_{12}.

TABLE 5.5
Comparison of 4 × 4 MIMO Configuration.

Ref.	Size of Antenna (mm^2)	Operating Bandwidth (GHz)	Isolation Technique Used	Isolation (dB)	ECC	Orientation of Radiating Elements
[51]	80 × 80	3.18–11.5	NIL	<–20	<0.005	Orthogonal to each other
[53]	40 × 40	3.10–11.0	NIL	<–20	<0.02	Facing 180° to each other
[55]	25 × 45	3.00–12.0	Slotted circle	<–15	–	Facing 180° to each other on either face of substrate
[57]	24 × 24	3.00–10.9	Stepped slot	<–20	<0.002	Orthogonal to each other
[60]	32 × 38	2.30–2.62 3.46–10.3	Modified T-type stub and CSRR in ground	<–20	<0.003	Facing 180° to each other on same face of substrate
[62]	40 × 40	3.0–18.0	45° placement of pair of rectangle stubs	<–20	<0.003	Orthogonal to each other
[67]	40 × 40	2.6–10.6	NIL	<–20	<0.07	Orthogonal to each other
[71]	63 × 63	0.97–35	Common connected ground	<–17	<0.01	Orthogonal to each other

elements. In a few cases, no isolation element is used whereas in few cases slotted circle, modified T-stub with CSRR in-ground is used for better isolation. Also, radiating elements are placed orthogonally to each other or they are placed by an angle of 180°. By using modified elliptical self-complementary and by using the tapered coplanar feed, a large bandwidth ratio of 36:1 is obtained.

5.5 FAR-FIELD DISCUSSION OF RESULTS

2-D radiation pattern and gain of 4 × 4 MIMO antenna-configuration observed from Figure 5.4(a,b) is plotted in far-field region by using two antenna method. Both the antennas are placed inside anechoic chamber with distance between them a = 250 mm $(=\frac{2a^2}{\lambda})$. Gain of the MIMO antenna is calculated by using Friss transmission given below

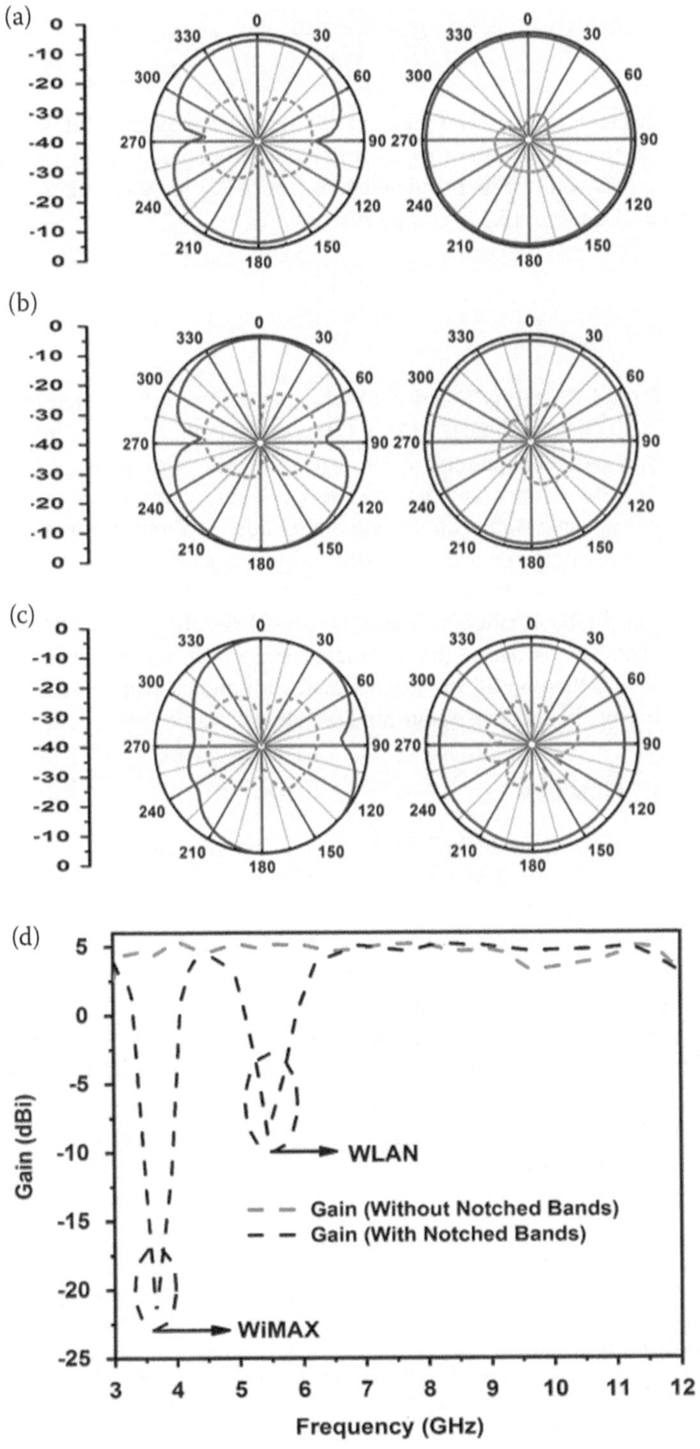

FIGURE 5.4 Far field radiation pattern of proposed antenna at (a) 4.50 GHz (b) 6.85 GHz (c) 11.0 GHz (d) Gain of the antenna (Without and With Notched bands).

$$P_r = \frac{P_t G_t G_r \lambda^2}{(4\pi a)^2}, \quad (5.15)$$

where Pr is transmitted power (watts), Pr is received power (watts), Gt gain of transmitter antenna, Gr is gain of receiver antenna, λ is operating wavelength (m), a is the distance between two identical antennas (mm). For identical antennas, $G_t = G_r = G$ and hence Equation (5.15) can be rewritten as

$$G = \frac{4\pi a\sqrt{P_r}}{\lambda\sqrt{P_t}}. \quad (5.16)$$

It can be observed from Figure 5.4(a–c) which is 2-D plot for different operating frequencies (4.50 GHz, 6.85 GHz & 11.0 GHz) that antenna offers omni-directional and dipole type radiation pattern in Electric- and Magnetic-plane. Figure 5.4(d) shows plot between realized gain and frequency. These gain curves are plotted for antenna mitigating with and without notched bands. Without any notched bands, antenna offers maximum gain of 5.11 dBi, which is calculated at 5.44 GHz. An important observation can be recorded for MIMO antenna including notched bands that there is sharp fall of gain corresponding to −21.69 dBi at 3.66 GHz and −8.66 dBi at 5.44 GHz. This is due to the fact that mismatch of impedance is recorded at notched bands and thereby all the input signals at notched bands are not radiated. For other radiating frequencies, gain remains almost stable.

5.6 CONCLUSIONS

This chapter focuses on the MIMO configuration antenna with either 2 × 2 or 4 × 4 MIMO antenna configuration. Several comparisons were discussed with 2 × 2 MIMO antenna configuration and it was found that this configuration exhibited good diversity performance. Also, 4 × 4 MIMO antenna configurations were compared in terms of size, bandwidth, and diversity performance. All the designed antennas in terms of MIMO performance were well within permissible limits such as Isolation < −15 dB, ECC < 0.50, DG > 9.95 dB, and CCL < 0.2 *b/s*/Hz. The above said characteristics of the MIMO antenna are highly useful in applications including imaging systems that include applications such as ground-penetrating RADAR system, wall imaging system, through-wall imaging system, surveillance system, and medical systems.

REFERENCES

[1]. T. Meng Guan and S. K. A. Rahim, "Compact monopole MIMO antenna for 5G application," *Microwave and Optical Technology Letters*, vol. 59, no. 5, pp. 1074–1077, 2017.

[2]. P. C. Nirmal, A. Nandgaonkar, S. Nalbalwar, and R. K. Gupta, "Compact wideband MIMO antenna for 4G WI-MAX, WLAN and UWB applications," *AEU* -

International Journal of Electronics and Communications, vol. 99, pp. 284–292, 2019.
[3]. R. Jian, H. Wei, Y. Yingzeng, and F. Rong, "Compact printed MIMO antenna for UWB applications," *IEEE Antennas and Wireless Propagation Letters*, vol. 13, pp. 1517–1520, 2014.
[4]. M. A. Ul Haq and S. Koziel, "Ground plane alterations for design of high-isolation compact wideband MIMO antenna," *IEEE Access*, vol. 6, pp. 48978–48983, 2018.
[5]. W. Li, Y. Hei, P.M. Grubb, X. Shi, and R.T. Chen, "Compact inkjet-printed flexible MIMO antenna for UWB applications," *IEEE Access*, vol. 6, pp. 50290–50298, 2018.
[6]. M. S. Khan, A.-D. Capobianco, S. M. Asif, D. E. Anagnostou, R. M. Shubair, and B. D. Braaten, "A compact CSRR-enabled UWB diversity antenna," *IEEE Antennas and Wireless Propagation Letters*, vol. 16, pp. 808–812, 2017.
[7]. L. Liu, S. W. Cheung, and T. I. Yuk, "Compact multiple-input–multiple-output antenna using quasi-self-complementary antenna structures for ultrawideband applications," *IET Microwaves, Antennas & Propagation*, vol. 8, no. 13, pp. 1021–1029, 2014.
[8]. Z. Wani and D. K. Vishwakarma, "An ultrawideband antenna for portable MIMO terminals," *Microwave and Optical Technology Letters*, vol. 58, no. 1, pp. 51–57, 2016.
[9]. Y. Wu and Y. Long, "Compact MIMO antenna for LTE 2500 and UWB applications," *Microwave and Optical Technology Letters*, vol. 57, no. 9, pp. 2046–2049, 2015.
[10]. J. Tao and Q. Feng, "Compact ultrawideband MIMO antenna with half-slot structure," *IEEE Antennas and Wireless Propagation Letters*, vol. 16, pp. 792–795, 2017.
[11]. M. G. N. Alsath and M. Kanagasabai, "Compact UWB monopole antenna for automotive communications," *IEEE Transactions on Antennas and Propagation*, vol. 63, no. 9, pp. 4204–4208, 2015.
[12]. R. N. Tiwari, P. Singh, B. K. Kanaujia, S. Kumar, and S. K. Gupta, "A low profile dual band MIMO antenna for LTE/Bluetooth/Wi-Fi/WLAN applications," *Journal of Electromagnetic Waves and Applications*, vol. 34, no. 9, pp. 1239–1253, 2020.
[13]. M. O. Katie, M. F. Jamlos, A. S. Mohsen Alqadami, and M. A. Jamlos, "Isolation enhancement of compact dual-wideband MIMO antenna using flag-shaped stub," *Microwave and Optical Technology Letters*, vol. 59, no. 5, pp. 1028–1032, 2017.
[14]. H.-F. Huang and S.-G. Xiao, "Compact MIMO antenna for Bluetooth, WiMAX, WLAN, and UWB applications," *Microwave and Optical Technology Letters*, vol. 58, no. 4, pp. 783–787, 2016.
[15]. M. S. Khan, A.-D. Capobianco, A. Iftikhar, S. Asif, and B. D. Braaten, "A compact dual polarized ultrawideband multiple-input- multiple-output antenna," *Microwave and Optical Technology Letters*, vol. 58, no. 1, pp. 163–166, 2016.
[16]. M. S. Khan, A.-D. Capobianco, A. Iftikhar, R. M. Shubair, D. E. Anagnostou, and B. D. Braaten, "Ultra-compact dual-polarised UWB MIMO antenna with meandered feeding lines," *IET Microwaves, Antennas & Propagation*, vol. 11, no. 7, pp. 997–1002, 2017.
[17]. C.-X. Mao and Q.-X. Chu, "Compact coradiator UWB-MIMO antenna with dual polarization," *IEEE Transactions on Antennas and Propagation*, vol. 62, no. 9, pp. 4474–4480, 2014.
[18]. J. Zhu, B. Feng, B. Peng, L. Deng, and S. Li, "A dual notched band MIMO slot antenna system with Y-shaped defected ground structure for UWB applications," *Microwave and Optical Technology Letters*, vol. 58, no. 3, pp. 626–630, 2016.

[19]. J. Zhu, S. Li, B. Feng, L. Deng, and S. Yin, "Compact dual-polarized UWB quasi-self-complementary MIMO/diversity antenna with band-rejection capability," *IEEE Antennas and Wireless Propagation Letters*, vol. 15, pp. 905–908, 2016.

[20]. H. Nawaz and I. Tekin, "Dual port disc monopole antenna for wide-band MIMO-based wireless applications," *Microwave and Optical Technology Letters*, vol. 59, no. 11, pp. 2942–2949, 2017.

[21]. Q. Li, A. P. Feresidis, M. Mavridou, and P. S. Hall, "Miniaturized double-layer EBG structures for broadband mutual coupling reduction between UWB monopoles," *IEEE Transactions on Antennas and Propagation*, vol. 63, no. 3, pp. 1168–1171, 2015.

[22]. N. K. Sahu, G. Das, and R. K. Gangwar, "Dielectric resonator-based wide band circularly polarized MIMO antenna with pattern diversity for WLAN applications," *Microwave and Optical Technology Letters*, vol. 60, no. 12, pp. 2855–2862, 2018.

[23]. M. Abedian, S. K. A. Rahim, C. Fumeaux, S. Danesh, Y. C. Lo, and M. H. Jamaluddin, "Compact ultrawideband MIMO dielectric resonator antennas with WLAN band rejection," *IET Microwaves, Antennas & Propagation*, vol. 11, no. 11, pp. 1524–1529, 2017.

[24]. P. Gao, S. He, X. Wei, Z. Xu, N. Wang, and Y. Zheng, "Compact printed UWB diversity slot antenna with 5.5-GHz band-notched characteristics," *IEEE Antennas and Wireless Propagation Letters*, vol. 13, pp. 376–379, 2014.

[25]. R. Chandel and A. K. Gautam, "Compact MIMO/diversity slot antenna for UWB applications with band-notched characteristic," *Electronics Letters*, vol. 52, no. 5, pp. 336–338, 2016.

[26]. H.-F. Huang and S.-G. Xiao, "Mimo antenna with high frequency selectivity and controllable bandwidth for band-notched UWB applications," *Microwave and Optical Technology Letters*, vol. 58, no. 8, pp. 1886–1891, 2016.

[27]. A. A. Ibrahim, J. Machac, and R. M. Shubair, "Compact UWB MIMO antenna with pattern diversity and band rejection characteristics," *Microwave and Optical Technology Letters*, vol. 59, no. 6, pp. 1460–1464, 2017.

[28]. L. Kang, H. Li, X. Wang, and X. Shi, "Compact offset microstrip-fed MIMO antenna for band-notched UWB applications," *IEEE Antennas and Wireless Propagation Letters*, vol. 14, pp. 1754–1757, 2015.

[29]. L. Kang, H. Li, X.-H. Wang, and X.-W. Shi, "Miniaturized band-notched UWB MIMO antenna with high isolation," *Microwave and Optical Technology Letters*, vol. 58, no. 4, pp. 878–881, 2016.

[30]. M. S. Khan, A. Naqvi, B. Ijaz, M. F. Shafique, B. D. Braaten, and A. D. Capobianco, "Compact planar UWB MIMO antenna with on-demand WLAN rejection," *Electronics Letters*, vol. 51, no. 13, pp. 963–964, 2015.

[31]. L. Liu, S. W. Cheung, and T. I. Yuk, "Compact MIMO antenna for portable UWB applications with band-notched characteristic," *IEEE Transactions on Antennas and Propagation*, vol. 63, no. 5, pp. 1917–1924, 2015.

[32]. N. Malekpour, M. Amin Honarvar, A. Dadgarpur, B. S. Virdee, and T. A. Denidni, "Compact UWB mimo antenna with band-notched characteristic," *Microwave and Optical Technology Letters*, vol. 59, no. 5, pp. 1037–1041, 2017.

[33]. A. Toktas, "G-shaped band-notched ultra-wideband MIMO antenna system for mobile terminals," *IET Microwaves, Antennas & Propagation*, vol. 11, no. 5, pp. 718–725, 2017.

[34]. S. Tripathi, A. Mohan, and S. Yadav, "A compact octagonal fractal UWBMIMO antenna with WLAN band-rejection," *Microwave and Optical Technology Letters*, vol. 57, no. 8, pp. 1919–1925, 2015.

[35]. N. Jaglan, B. K. Kanaujia, S. D. Gupta, and S. Srivastava, "Dual band notched EBG structure based UWB MIMO/diversity antenna with reduced wide band electromagnetic coupling," *Frequenz*, vol. 71, no. 11-12, pp. 555–565, 2017.
[36]. Z. Tang, J. Zhan, X. Wu, Z. Xi, L. Chen, and S. Hu, "Design of a compact UWB-MIMO antenna with high isolation and dual band-notched characteristics," *Journal of Electromagnetic Waves and Applications*, vol. 34, no. 4, pp. 500–513, 2020.
[37]. T. Tzu-Chun and L. Ken-Huang, "An ultrawideband MIMO antenna with dual band-notched function," *IEEE Antennas and Wireless Propagation Letters*, vol. 13, pp. 1076–1079, 2014.
[38]. R. Chandel, A. K. Gautam, and K. Rambabu, "Tapered fed compact UWB MIMO-diversity antenna with dual band-notched characteristics," *IEEE Transactions on Antennas and Propagation*, vol. 66, no. 4, pp. 1677–1684, 2018.
[39]. A. K. Gautam, S. Yadav, and K. Rambabu, "Design of ultra-compact UWB antenna with band-notched characteristics for MIMO applications," *IET Microwaves, Antennas & Propagation*, vol. 12, no. 12, pp. 1895–1900, 2018.
[40]. H. Huang, Y. Liu, S.-S. Zhang, and S.-X. Gong, "Compact polarization diversity ultrawideband mimo antenna with triple band-notched characteristics," *Microwave and Optical Technology Letters*, vol. 57, no. 4, pp. 946–953, 2015.
[41]. C. R. Jetti and V. R. Nandanavanam, "Trident-shape strip loaded dual band-notched UWB MIMO antenna for portable device applications," *AEU - International Journal of Electronics and Communications*, vol. 83, pp. 11–21, 2018.
[42]. W. T. Li, Y. Q. Hei, H. Subbaraman, X. W. Shi, and R. T. Chen, "Novel printed filtenna with dual notches and good out-of-band characteristics for UWB-MIMO applications," *IEEE Microwave and Wireless Components Letters*, vol. 26, no. 10, pp. 765–767, 2016.
[43]. M. Sharma, "Design and analysis of MIMO antenna with high isolation and dual notched band characteristics for wireless applications," *Wireless Personal Communications*, vol. 112, no. 3, pp. 1587–1599, 2020.
[44]. N. Jaglan, S. D. Gupta, E. Thakur, D. Kumar, B. K. Kanaujia, and S. Srivastava, "Triple band notched mushroom and uniplanar EBG structures based UWB MIMO/Diversity antenna with enhanced wide band isolation," *AEU - International Journal of Electronics and Communications*, vol. 90, pp. 36–44, 2018.
[45]. J. Banerjee, A. Karmakar, R. Ghatak, and D. R. Poddar, "Compact CPW-fed UWB MIMO antenna with a novel modified Minkowski fractal defected ground structure (DGS) for high isolation and triple band-notch characteristic," *Journal of Electromagnetic Waves and Applications*, vol. 31, no. 15, pp. 1550–1565, 2017.
[46]. Y. Kong, Y. Li, and K. Yu, "A minimized MIMO-UWB antenna with high isolation and triple band-notched functions," *Frequenz*, vol. 70, no. 11-12, 2016.
[47]. Z.-X. Yang, H.-C. Yang, J.-S. Hong, and Y. Li, "A miniaturized triple band-notched MIMO antenna for UWB application," *Microwave and Optical Technology Letters*, vol. 58, no. 3, pp. 642–647, 2016.
[48]. L. Wu, H. Lyu, H. Yu, and J. Xu, "Design of a miniaturized UWB-MIMO antenna with four notched-band characteristics," *Frequenz*, vol. 73, no. 7-8, pp. 245–252, 2019.
[49]. B. T. Ahmed, P. S. Olivares, J. L. M. Campos, and F. M. Vázquez, "(3.1–20) GHz MIMO antennas," *AEU - International Journal o f Electronics and Communications*, vol. 94, pp. 348–358, 2018.
[50]. M. Shehata, M. S. Said, and H. Mostafa, "Dual notched band quad-element MIMO antenna with multitone interference suppression for IR-UWB wireless applications," *IEEE Transactions on Antennas and Propagation*, vol. 66, no. 11, pp. 5737–5746, 2018.

[51]. M. N. Hasan, S. Chu, and S. Bashir, "A DGS monopole antenna loaded with U-shape stub for UWB MIMO applications," *Microwave and Optical Technology Letters*, vol. 61, no. 9, pp. 2141–2149, 2019.

[52]. R. N. Tiwari, P. Singh, B. K. Kanaujia, and K. Srivastava, "Neutralization technique based two and four port high isolation MIMO antennas for UWB communication," *AEU - International Journal of Electronics and Communications*, vol. 110, 2019. https://doi.org/10.1016/j.aeue.2019.152828

[53]. W. A. E. Ali and A. A. Ibrahim, "A compact double-sided MIMO antenna with an improved isolation for UWB applications," *AEU - International Journal of Electronics and Communications*, vol. 82, pp. 7–13, 2017.

[54]. T. Dabas, D. Gangwar, B. K. Kanaujia, and A. K. Gautam, "Mutual coupling reduction between elements of UWB MIMO antenna using small size uniplanar EBG exhibiting multiple stop bands," *AEU - International Journal of Electronics and Communications*, vol. 93, pp. 32–38, 2018.

[55]. R. Mathur and S. Dwari, "Compact CPW-Fed ultrawideband MIMO antenna using hexagonal ring monopole antenna elements," *AEU - International Journal of Electronics and Communications*, vol. 93, pp. 1–6, 2018.

[56]. D. Sipal, M. P. Abegaonkar, and S. K. Koul, "Compact planar 2 × 2 and 4 × 4 UWB mimo antenna arrays for portable wireless devices," *Microwave and Optical Technology Letters*, vol. 60, no. 1, pp. 86–92, 2018.

[57]. G. Srivastava, A. Mohan, and A. Chakraborty, "A compact multidirectional UWB MIMO slot antenna with high isolation," *Microwave and Optical Technology Letters*, vol. 59, no. 2, pp. 243–248, 2017.

[58]. Z. Wani and D. Kumar, "A compact 4 × 4 MIMO antenna for UWB applications," *Microwave and Optical Technology Letters*, vol. 58, no. 6, pp. 1433–1436, 2016.

[59]. B. Yang, M. Chen, and L. Li, "Design of a four-element WLAN/LTE/UWB MIMO antenna using half-slot structure," *AEU - International Journal of Electronics and Communications*, vol. 93, pp. 354–359, 2018.

[60]. S. Kumar, R. Kumar, R. Kumar Vishwakarma, and K. Srivastava, "An improved compact MIMO antenna for wireless applications with band-notched characteristics," *AEU - International Journal of Electronics and Communications*, vol. 90, pp. 20–29, 2018.

[61]. S. M. Khan, A. Iftikhar, S. M. Asif, A.-D. Capobianco, and B. D. Braaten, "A compact four elements UWB MIMO antenna with on-demand WLAN rejection," *Microwave and Optical Technology Letters*, vol. 58, no. 2, pp. 270–276, 2016.

[62]. Z. Tang, J. Zhan, X. Wu, Z. Xi, and S. Wu, "Simple ultra-wider-bandwidth MIMO antenna integrated by double decoupling branches and square-ring ground structure," *Microwave and Optical Technology Letters*, vol. 62, no. 3, pp. 1259–1266, 2019.

[63]. M. M. Hassan *et al.*, "A novel UWB MIMO antenna array with band notch characteristics using parasitic decoupler," *Journal of Electromagnetic Waves and Applications*, vol. 34, no. 9, pp. 1225–1238, 2019.

[64]. S. Tripathi, A. Mohan, and S. Yadav, "A compact koch fractal UWB MIMO antenna with WLAN band-rejection," *IEEE Antennas and Wireless Propagation Letters*, vol. 14, pp. 1565–1568, 2015.

[65]. R. Gomez-Villanueva and H. Jardon-Aguilar, "Compact UWB uniplanar four-port MIMO antenna array with rejecting band," *IEEE Antennas and Wireless Propagation Letters*, vol. 18, no. 12, pp. 2543–2547, 2019.

[66]. V. Dhasarathan, T. K. Nguyen, M. Sharma, S. K. Patel, S. K. Mittal, and M. T. Pandian, "Design, analysis and characterization of four port multiple-input-multiple-output UWB-X band antenna with band rejection ability for wireless network applications," *Wireless Networks*, vol. 26, no. 6, pp. 4287–4302, 2020.

[67]. G. Kan, W. Lin, C. Liu, and D. Zou, "An array antenna based on coplanar parasitic patch structure," *Microwave and Optical Technology Letters*, vol. 60, no. 4, pp. 1016–1023, 2018.
[68]. H. Liu, G. Kang, and S. Jiang, "Compact dual band-notched UWB multiple-input multiple-output antenna for portable applications," *Microwave and Optical Technology Letters*, vol. 62, no. 3, pp. 1215–1221, 2020.
[69]. A. Gorai, A. Dasgupta, and R. Ghatak, "A compact quasi-self-complementary dual band notched UWB MIMO antenna with enhanced isolation using Hilbert fractal slot," *AEU - International Journal of Electronics and Communications*, vol. 94, pp. 36–41, 2018.
[70]. S. Kumar, G. H. Lee, D. H. Kim, W. Mohyuddin, H. C. Choi, and K. W. Kim, "Multiple-input-multiple-output/diversity antenna with dual band-notched characteristics for ultra-wideband applications," *Microwave and Optical Technology Letters*, vol. 62, no. 1, pp. 336–345, 2019.
[71]. D. K. Raheja, B. K. Kanaujia, and S. Kumar, "Low profile four-port super-wideband multiple-input-multiple-output antenna with triple band rejection characteristics," *International Journal of RF and Microwave Computer-Aided Engineering*, vol. 29, no. 10, p. e21831, 2019.

6 Four Port MIMO Antenna with Swastika Slot for 5G Environment

E. L. Dhivya Priya[1], *T. Poornima (Ph.D)*[2], *and K. R. Gokul Anand*[3]

[1]Assistant professor, Dept. of ECE, Sri Krishna College of Technology, Coimbatore, Tamilnadu, India
[2]Research Associate, Department of ECE, Amrita Vishwa Vidyapeetham, Coimbatore, Tamilnadu, India
[3]Assistant professor, Dept. of ECE, Dr Mahalingam College of Engineering & Technology, Coimbatore, Tamilnadu, India

6.1 INTRODUCTION

Antenna is the most important device that made global communication possible. It act as both transmitter and receiver. There are various types of antennas supporting different sets of applications. The parameters associated with the antenna justifies its application and the variations in them provides better results. Initially, single input single output (SISO) was supporting the data transmission with a single input signal, and the system will be evaluated only for that signal. This SISO works with no diversity and no additional signal processing is required. This system is well known for its simplicity as there is no additional processing. Later advancements in the technology made this SISO to get smarter as MIMO [1,2].

To overcome throughput limitations, MIMO helped the engineers a lot. The MIMO is multiple input multiple output, which accepts multiple input signal and as a result multiple output signals are generated. MIMO offers various advantages because of processing multiple data streams at a time with an improved efficiency and so it is also called smart antennas [3]. The main advantage of MIMO is the increased data throughput, which makes it suitable for real time multimedia applications [3]. The total active reflection coefficient (TARC) for multi-port antenna systems, correlation coefficient, and diversity gain and channel capacity are the characteristics used for defining the MIMO antenna. Using multi-input different yield (MIMO) frameworks [4,5] gives an expansion in channel limit without the requirement for additional radio recurrence range at the transmitter. MIMO and additionally cluster radio wires can be a superior arrangement particularly when managing minimal battery powered gadgets. Considering the power and transfer

DOI: 10.1201/9781003187325-6

speed in the current age of cell frameworks, MIMO procedures offer a few preferences including upgraded bit rate, diminished multipath impacts, and expanded limit [6]. With these advantages a 4 port MIMO antenna was designed. Instead of conventional patch dimensions, H shaped patch are introduced to have a low profile multiband, high gain, and compact antenna element [7]. The 4 ports are excited with a coaxial feed [8]. In order to achieve better results, a swastika shaped slot antenna was introduced [9].

Patch is the radiating element in antenna that holds various shape and size. The geometry of patch of the antenna can be varied for achievement of better results. The patches are usually rectangular [10] in conventional antenna types and later advancements proved that patch can have shapes as square, circular, F shaped [11], L shaped [12], H shaped [7], E shaped [13,14], inverted shapes [11], and sometimes irregular shapes with the introduction of cut slots. The advanced shapes help in achieving various advantages like better isolation and mutual coupling [9]. The designed antenna justifies this statement.

The impact of mutual coupling on limit of MIMO remote channels are contemplated in [15] and the partition between numerous receiving wire components ought to be 0.5λ. The mutual coupling must be on the demand basis if unwanted couplings between the antenna elements are addressed; then, radiation efficiency can be greatly reduced and sometimes the signal correlation can be severely increased. In spite of the fact that the port-to-port coupling can be decreased viably by means of this methodology, the utilization of various component types will bring about various addition and radiation designs, which will debase the consistency of the presentation over the whole angle range [15]. Symmetrically finding antenna elements components to frame symmetrical eigen modes in a MIMO framework is another methodology ordinarily used to lessen the shared coupling. This methodology will be more viable in a MIMO framework radio wire having a mandate radiation design. The implementation of a large number of antennas will make the antenna system more desirable for the generation of most directive beams. The implementation of multiple antennas on a common ground plane helps in achieving better applications in wireless communication systems. There are different types of diversity such as spatial diversity and polarization diversity. The introduction of four antenna makes the system more powerful for improved communication link than the diversity switching. This makes the system more robust and further can be improved by adding more antennas. The four antenna patch are excited individually with a coaxial feed. The coaxial feed is chosen as they can be placed at any desired position and also are supported for maintaining the impedance matching. The space between the 4 ports antenna is utilized for placing the slot. Though many slots are available, the designed antenna uses swastika slot as it provides mutual coupling and better isolation between the elements. The designed antenna are simulated for results with CST microwave studio and analyzed with the S parameters and return loss in dB. The implemented antenna has wider bandwidth and supports for faster data rate, which can be utilized for wireless multimedia applications.

6.2 MIMO

Multiple input multiple output is the smartest antenna today that helps in reducing the demand of power. It also supports transmitting multiple data streams along a same channel over a longer distance. The multiple transmitted data streams will be received at multiple receiver ends. This methodology helps in building a smarter wireless communications system. The advanced MIMO systems helps in offering various advantages such as correlation coefficient, diversity gain, efficiency, and improved throughout. This MIMO antenna, shown in the Figure 6.1, will be well suited in the 5G environment when the data rate and throughput are improved. Thus the proposed 4 port antenna with the swastika slot can be further optimized in the way that suits 5G applications.

6.2.1 Swastika Slot

The introduction of a slot with regular geometrical shapes provides conventional output. To make the slots provide better results, a new irregular shaped antenna called a swastika slot are introduced. This swastika shaped slot proves to be different among the exciting slots by providing better isolation and also supports mutual coupling. Thus the proposed antenna was introduced with a swastika slot.

6.2.2 Four Port Antenna

The implementation of different antennas on a common ground helps to meet better results. The array antennas, though they look complex in design, are proved to offer better efficiency and improved gain. The proposed four-port antenna are evaluated to give better applications in the 5G environment such as TV broadcast and two way radios. The four ports are implemented as H shaped patch or a dumbbell patch on the common ground plane so as to make it unique among the existing ones. The center space between the ports is utilized for the implementation of swastika shaped slot. They are excited with the help of a coaxial feed.

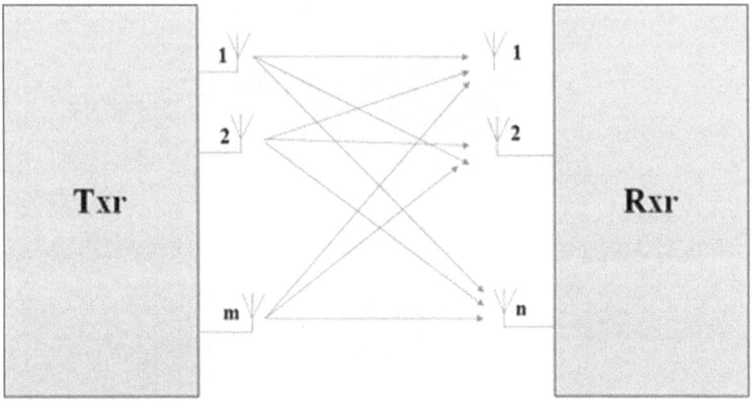

FIGURE 6.1 General MIMO system.

6.2.3 Coaxial Feed

The conventional antenna design prefers micro-strip feed line for easy implementation and fabrication. But because of the better advantages in the coaxial feed, the proposed antenna are implemented with it, as shown in Figure 6.2. They have better impedance matching by offering the designer to select the desirable position for providing the feed line.

Thus the proposed four port are excited with the coaxial feed. The input excitation signal is the Gaussian pulse shown in Figure 6.3.

6.2.4 H Shaped Patch

The patch is the radiating element in an antenna. There are many conventional patch antennas available with unique characteristics. The dimension change in the patches provides different characteristics. The increase in throughput, efficiency, and gain can be achieved through this dimension changes of patch. The return loss for the designed operating frequency can also be varied with the patch dimensions. The different types of

FIGURE 6.2 Pictorial representation of a swastika slot.

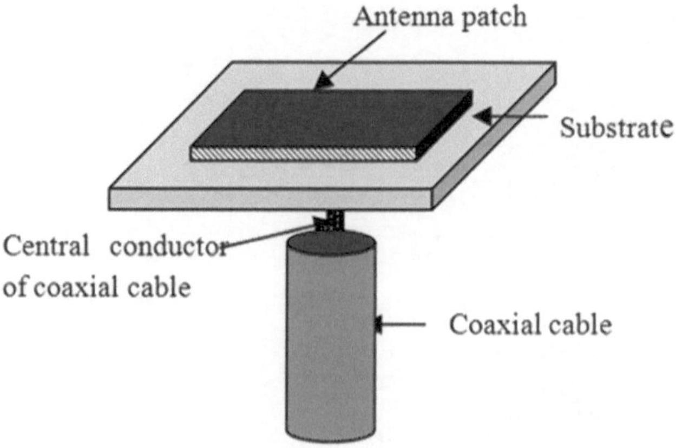

FIGURE 6.3 Coaxial feed representation in conventional antenna.

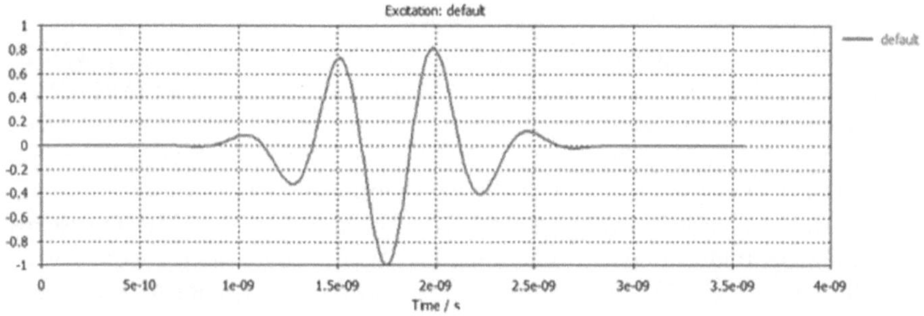

FIGURE 6.4 Input excitation signal—The Gaussian pulse.

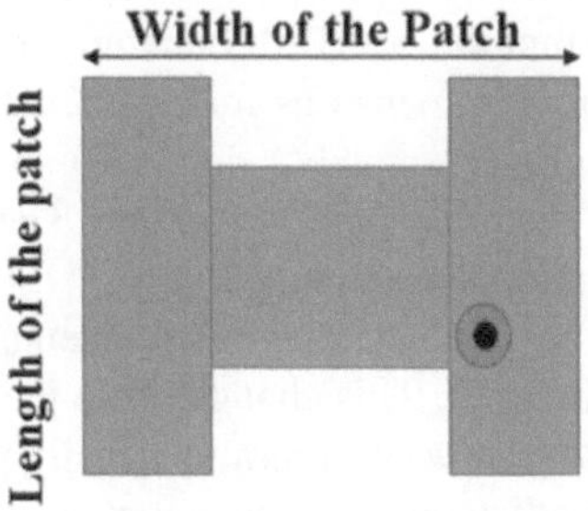

FIGURE 6.5 H shaped patch.

patches are H shaped, L shaped, U shaped, and irregular shapes. The proposed idea uses H shaped or dumbbell shaped patch with coaxial feed. The advantages of H shaped patch are low profile, light weight, low cost, ease of fabrication, and integration with RF devices. The H shaped patch is represented in Figure 6.5.

6.3 LITERATURE SURVEY

MIMO involves multiple antennas placed on a common ground plane. The advantages of MIMO makes it suitable for 5G applications. As many antenna elements are arranged in an array, isolation between the elements is difficult. The size of the antenna cannot be increased much, due to size constraints of the antenna. Thus a new technique was introduced to implement with the four and six element antenna array [15]. This technique proved that the isolation between the elements was improved and also yields better pattern diversity [16] in the MIMO antenna. The antenna designed in the paper "MIMO Antennas for Smart 5G Devices" gives better return loss and radiation performance [5], making it reasonable for future 5G gadgets, for example, shrewd watches and dongles and so on. The receiving wires showed a reduced calculation and a wide transfer speed of 4 GHz going from 23.1 to 27.2 GHz. There has been a good understanding between the simulation and the estimation results [5].

MIMO can be used in video intensive applications [1] with high speed. Four port antenna [8,17] was proposed with efficient bandwidth broadening. The proposed antenna is found to accomplish great pattern diversity, low correlation coefficient, high gain, excellent directivity, and sensible transfer speed in the previously mentioned territory, exceptionally reasonable for LTE group's application with 10 dB return loss [1]. Double band numerous information various yield (MIMO) reception apparatus with designed structure, focusing at 2.4 GHz and 1.8 GHz. Diverse radiation patterns are achieved by utilizing the arrays of printed dipoles. These printed dipoles are fed individually with a different input phase values for the achievement of diverse radiation patterns [18]. The applications of MIMO systems in wireless environment and its role in 4G and 5G environments are explained in [19]. In addition to 4G and 5G applications, wireless environment also needs MIMO for better results with improved pattern diversity and isolation of the antenna elements. 5G frequency allocations are used by the wearable device applications by means of designed 2 × 2 MIMO antenna [20–24].

The design and implementation of MIMO antenna are studied in [7,10–14,25]. The patch analysis for various dimensions makes the antenna suitable for various applications. The patch can be H shaped or dumbbell shaped, L shaped antenna, E shaped patch antenna, and inverted patch antennas. These antennas offer various advantages over the conventional antenna.

The conventional micro strip rectangular patch antenna are modified to E shaped patch along with two parallel slots. This changes help the antenna to work in a dual band [13]. Low volume, low profile configuration, easily mounted, light weight, and low fabrication cost are the advantages of this antenna. The antenna for wimax/wifi applications are designed with this E shaped patch antenna [14]. The other type of antenna called H shaped patch antenna offers better isolation and pattern diversity. The design and implementation of H shaped in [7,25] analysis of H shaped antenna with the conventional antenna types are performed [7]. The analyzed antenna well suits the following applications such as SAR, wireless local area network (WLAN), mobile communication system, and global positioning system (GPS) [7].

H shaped compact MIMO antenna designed with FR4 epoxy substrate material are found to operate with different sets of frequencies. This designed antenna well suits the frequency range of Wi-fi. To provide high isolation with high gain, Swastika shaped antenna was introduced between the ports [9]. The L shaped patch antenna helps in reducing the antenna size, when compared with the conventional micro strip patch antenna [12]. The disadvantages of conventional patch antennas such as narrow bandwidth and non-uniform coverage range can be overcome with the U shaped inverted F shaped antenna. The inverted F–U shaped antenna holds the following advantages such as increased bandwidth, better coverage, and less interaction with reduced loss [11].

The 5G applications are analyzed to perform well with MIMO antenna systems. The antenna array elements are to be placed with better isolation and improved pattern diversity. The MIMO in 5G applications also offers an important advantage, which is the increased throughput. The video intensive applications highly demands increased throughput to maintain better synchronization. Thus MIMO antennas are found to have space in 5G environment. The increased antenna elements with better isolation will definitely increase the throughput. The design of 5G antenna with the basic circuit structure are proposed in [26]. The needs of other supportive elements are proposed and verified for results. MIMO is one of the key empowering strategies for 5G remote innovation, giving increments in throughput and signal to noise ratio. MIMO adds to expanded limit first by empowering 5G NR organization in the higher recurrence go in Sub-6 GHz [26]. The massive MIMO antennas offer various advantages in wireless communication such as increased network capacity, improved coverage, and better user experience. Not only in wireless communications, MIMO antenna can be used in virtual gaming, video conferencing, and TV broadcasting.

5G MIMO Rectangular Microstrip Patch Antenna is expanding because of the high information rate required. Because of their smaller size, it is broadly utilized in the plan of portable wireless equipment and also in wearable devices. With the fast development of the remote correspondence framework, future advances need an extremely little and multiband radio wire. Radio wire takes a fundamental job in the field of remote application. The reflector based receiving wires are ordinarily utilized in light of the fact that they fulfil all the prerequisites; however, they are not useful because of their moderately enormous size and their 3D math. So we are moving to MIMO receiving wire. MIMO receiving wire is one of the promising innovations for 5G [26,27]. MIMO innovation can improve information transmission speed and be impervious to different way blurring, which has been broadly researched. The MIMO remote framework demonstrates the increased efficiency in the wireless environment, which includes multipath efforts. By utilizing MIMO innovation [12], designed rectangular microstrip antenna is utilized in remote correspondence due to its position of safety, size, and light weight. A microstrip patch antenna comprised of a transmitting patch on one side of a dielectric substrate, which has a ground plane on the opposite side [20,21,28,28,29]. This radio wire relies on the line feeding technique and its performance attributes.

The limits of wireless digital communications are important in designing a better MIMO based wireless applications. The limits of bandwidth efficient delivery of higher data rates are explained in [4]. Exploitation of multi-element array (MEA) technology is processing the spatial dimension to improve wireless capacities in certain applications. The basic information theory in wireless communications is explained [4]. At the point when the channel characteristic isn't accessible at the transmitter, the recipient knows (tracks) the trademark, which is dependent on Rayleigh fading. Fixing the transmitted power, communicating the limit offered by MEA innovation and the limit scales with expanding SNR for a huge however n, of receiving wire components at both transmitter and collector.

There are two central points driving the advancement of 5G: initially a need to help expanding interest for broadband administrations of numerous sorts conveyed over versatile systems, and furthermore a craving to help or make administrations for the Internet of Things (IoT) including for machine-to-machine (M2M) applications [30].

6.4 DESIGNED PARAMETERS

The four port MIMO antenna are designed with the dimensions of 104 X 104 mm^2. Antenna design consists of four patches placed above the substrate of the antenna design. The substrate is assigned with a FR4 epoxy substrate with dielectric constant of 4.4 and loss tangent 0.02. The geometrical top view of the proposed antenna and the top view of the antenna in the simulation environment are represented in Figure 6.6 and Figure 6.7.

The design consists of four H shaped radiators fed by coaxial probe feeding technique. To provide a better impedance matching, the location of the coaxial feed point is considered. The main advantage by using coaxial probe feed is that the

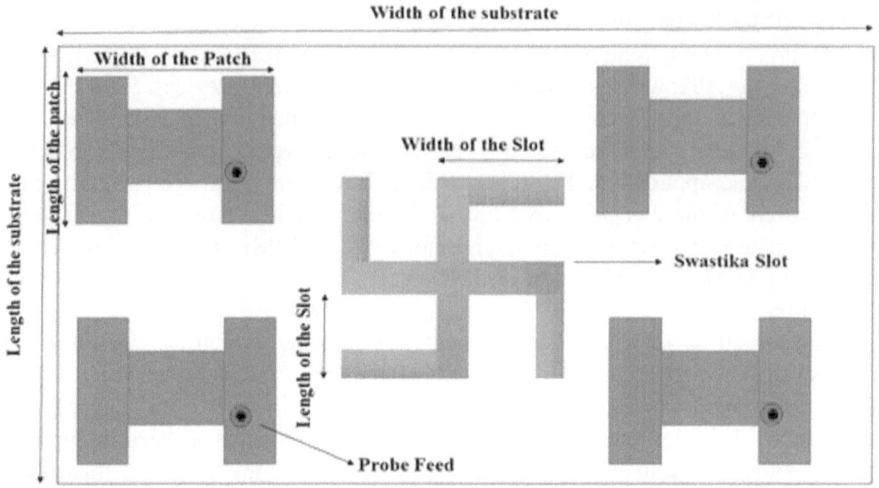

FIGURE 6.6 Top view of the geometrical structure of proposed MIMO antenna.

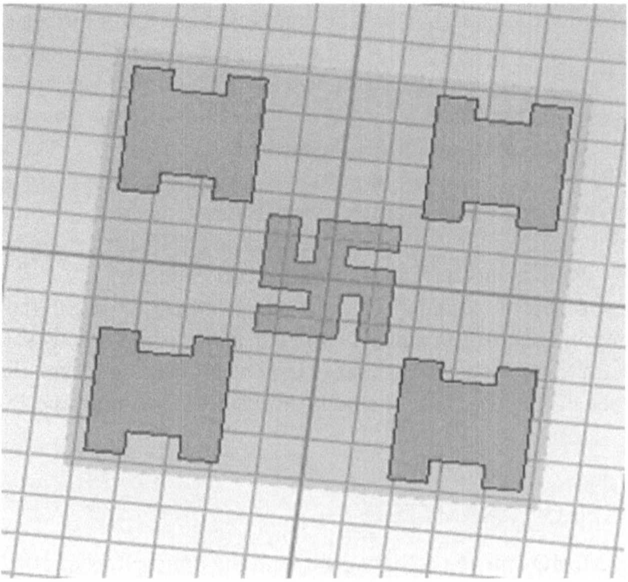

FIGURE 6.7 Top view of the proposed MIMO antenna in CST microwave studio.

inner conductor can be positioned at any preferred location so as to have a better impedance matching.

All the major parameters are studied to find their influence on the impedance matching of the proposed antenna structure. With this arrangement, the antenna covers frequency range with less isolation. To provide high isolation between the elements and to reduce the mutual coupling between the radiating elements, a swastika slot is introduced. Figure 6.8 represents the coaxial feed for the four ports of the proposed antenna. Figure 6.9 shows the single coaxial feed in the CST microwave simulation environment.

6.5 ANTENNA DESIGN EQUATIONS

Design parameters [9] are calculated by using the following equations.

FIGURE 6.8 Coaxial feed representation in the proposed antenna.

FIGURE 6.9 A single coaxial feed.

$$W_p = \frac{c}{2f_0\sqrt{\frac{(\varepsilon_r+1)}{2}}}. \tag{6.1}$$

- **Patch width**

The effective dielectric constant (ε_{reff}) of the antenna

$$\varepsilon_{reff} = \left(\frac{\varepsilon_r + 1}{2}\right)\left(\frac{\varepsilon_r - 1}{2}\right)\left(1 + 12\frac{h}{w}\right)^{-1/2}, \tag{6.2}$$

where ε_{reff} = effective dielectric constant
ε_r = dielectric constant of the substrate
h = height of the dielectric substrate
W = the patch width

- **Extension of the length (ΔL)**

$$\Delta L = h * 0.412\frac{(\varepsilon_{reff} + 0.3)\left(\frac{w}{h} + 0.264\right)}{(\varepsilon_{reff} - 0.258)\left(\frac{w}{h} + 0.8\right)}. \tag{6.3}$$

The fringing effect is used to enhance the effective electrical length of the patch longer than its physical length. Thus, the resonance condition depends on Leff.

- **Effective length of the patch**

$$L_{\text{eff}} = L_p + 2\Delta L. \tag{6.4}$$

- **The actual length of the patch (Lp)**

$$L_p = \frac{c}{2f_0\sqrt{\varepsilon_{reff}}} - 2\Delta L. \tag{6.5}$$

- **Substrate length (Lsb)**

$$L_{\text{sb}} = 12h + L_p. \tag{6.6}$$

- **Width of the substrate**

$$W_{\text{sb}} = 12h + W. \tag{6.7}$$

- **Length of Slot (Lsl)**

$$L_{\rm sl} = \frac{L_p}{\varepsilon_{r\rm eff}}. \tag{6.8}$$

- **Width of Slot (Wsl)**

$$W_{\rm sl} = \frac{w}{2}. \tag{6.9}$$

The proposed antenna design parameter values are calculated from equations 1 to 9. The calculated dimensions of the proposed antenna are represented in Table 6.1.

The MIMO behavior and the mutual coupling between the adjacent elements of the proposed antenna is studied and evaluated in terms of ECC. The ECC curve can be evaluated using S-parameters.

6.6 SIMULATION ENVIRONMENT

CST is the high performing 3D environment that helps in designing, analyzing, and optimizing electromagnetic (EM) components and systems. This includes the antenna design, filter design, and its parameter calculation such as reflection coefficient, VSWR, radiation pattern, efficiency, E field and H Field. Regular subjects of EM examination incorporate the exhibition and productivity of radio wires and channels, electromagnetic similarity and obstruction (EMC/EMI), presentation of the human body to EM fields, electro-mechanical impacts in engines and generators, and warm impacts in high-power gadgets.

TABLE 6.1
Geometrical Parameters of the Proposed MIMO Antenna

S.No	Parameter	Value (mm)
1	Length of the ground	104
2	Width of the ground	104
3	Length of the substrate	104
4	Width of the substrate	104
5	Patch length	29
6	Patch width	30.5
7	Slot length	12
8	Slot width	5
9	Length of the swastika slot	14.5
10	Width of the swastika slot	6.5

6.7 RESULTS AND DISCUSSIONS

The proposed antenna is simulated by placing FR4 epoxy as substrate material over a ground plane. Four H shaped patches are presented over the substrate with a swastika slot between the radiating patches as shown in Figure 6.10. Coaxial feed lines are used to excite all the four patches designed. The proposed antenna is simulated in CST microwave studio.

The simulated antenna operating at a frequency of 1.961 GHz with a return loss of about −20 dB is shown in Figure 6.11.

The values of return loss at the operating frequency of 1.96 GHZ is represented in Table 6.2. These values meet the design requirement, which is less than −10 dB, indicating that more than 90% of the fed power is absorbed.

The reflection coefficient, which is the frequency dependent return loss of the antenna, is in Figure 6.12, which is simulated by the transient solver in CST Microwave Studio.

The simulation results for MIMO isolation parameters at the operating frequency are listed in Table 6.3.

Figure 6.13 Depicts the frequency dependent correlation coefficients for all values of isolation parameters between antennas, with their return loss in dB. Each

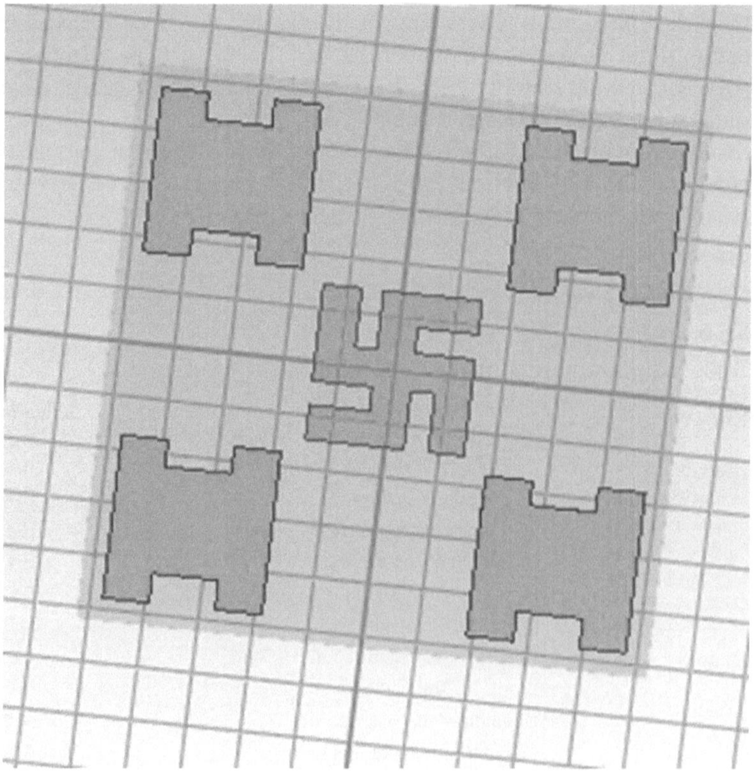

FIGURE 6.10 Proposed four port antenna with a swastika slot.

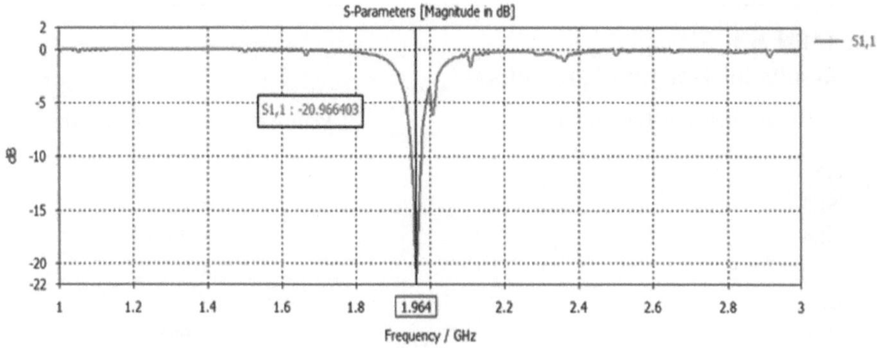

FIGURE 6.11 Reflection coefficient of the proposed antenna.

TABLE 6.2
Value of Return Loss at Operating Frequency

Operating Frequency	Return Loss (dB)			
	S11	S22	S33	S44
1.96 GHz	−20.496	−20.5082	−20.180	−19.093

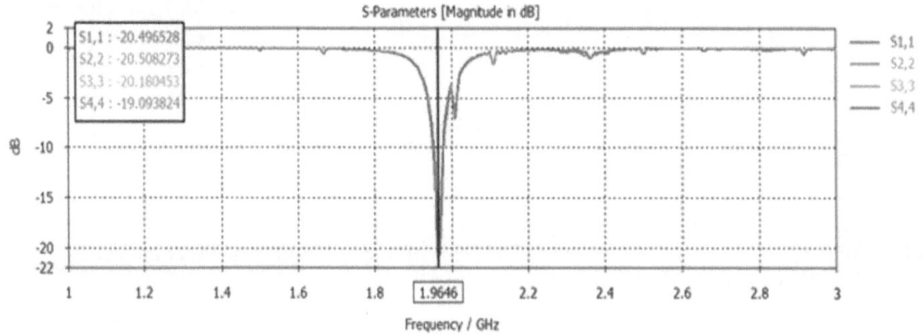

FIGURE 6.12 Simulated reflection coefficients for each port of the proposed MIMO antenna.

return loss is represented in different colors during the simulation in CST microwave studio.

The gains of the antenna at its operating frequency are simulated in CST microwave studio and are represented in Figure 6.14.

TABLE 6.3
Results for Isolation Parameters

Isolation Parameter Description	Return Loss in dB at 1.96 GHz
S1,2	–40.947
S1,3	–21.667
S1,4	–31.262
S2,1	–40.947
S2,3	–31.230
S2,4	–21.695
S3,1	–21.667
S3,2	–31.230
S3,4	–40.910
S4,1	–31.262
S4,2	–21.695
S4,3	–40.910

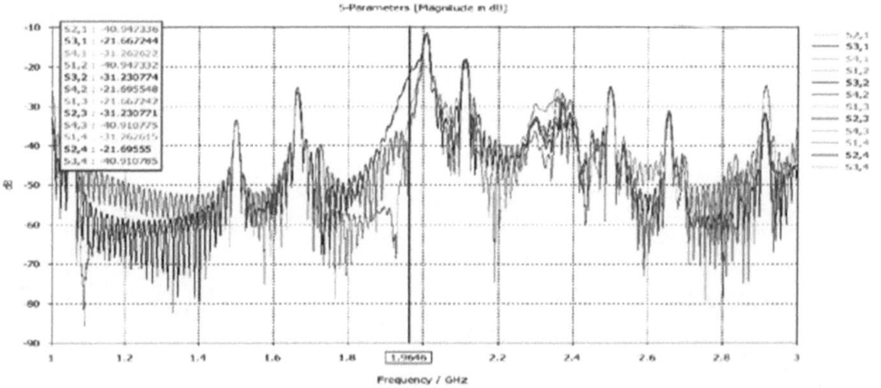

FIGURE 6.13 Frequency dependent reflection coefficients for all isolation parameters of 4-port MIMO antenna.

6.8 CONCLUSION

Thus the proposed antenna are simulated and verified for results in CST microwave studio. The reflection coefficient of the proposed antenna at all four ports are simulated and are found to meet the design considerations. To overcome the problem of high mutual coupling with low gain in MIMO antennas, a swastika slot is introduced among the four ports of the proposed antenna. The coaxial feed is used to avoid the surface wave radiation and also supports better impedance matching. The

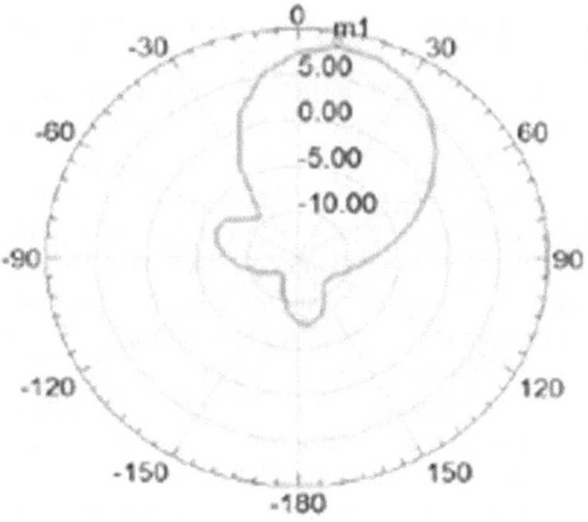

FIGURE 6.14 Gain of optimized compact MIMO antenna.

simulated antenna works with the resonant frequency of about 1.96 GHz, which can be utilized for two way radio, TV broadcasts, and microwave ovens and also in global positioning systems. Further optimization can make the antenna suitable for other applications. The better gain, mutual coupling, and bandwidth broadening can be achieved with further optimization of the proposed antenna.

REFERENCES

[1]. K.-S. Min, D.-J. Kim, and M.-S. Kim, *Multi-Channel MIMO Antenna Design for WiBro/PCS Band.* New York: IEEE, 2007. 1-4244-0878-4/07

[2]. Garg, R., Bhartia, P., Bahl, I., Ittipiboon, A., *Microstrip Antenna Design Handbook.* Norwood: Artech House, Inc. 2001

[3]. K. Mowafak et al., "Novel design and implementation of MIMO antenna for LTE Application," *Journal of Telecommunication, Electronic and Computer Engineering*, vol. 10, no. 2–8, pp. 43–49.

[4]. G. J. Foschini, and M. J. Gans, "On limits of wireless communications in a fading environment when using multiple antennas," *Wireless Personal Communications*, vol. 6, pp. 311–335, 1998.

[5]. N. Shoaib, S. Shoaib, et al., "MIMO antennas for smart 5G devices," doi:10.1109/ACCESS.2018.2876763

[6]. J. Malik, A. Patnaik, and M. Kartikeyan, "Novel printed MIMO antenna with pattern and polarization diversity," *IEEE Antennas and Wireless Propagation Letters*, vol. 14, pp. 739–742, 2015.

[7]. I. Pal, and P. Aggarwal, "Study of H shaped antenna for the improvement of resonance frequency," *International Journal of Advanced Technology in Engineering and Science*, vol. 2, no. 5, pp. 31–35, 2014

[8]. Y.-C. Lu, Y.-C. Chan, H.-J. Li, Y.-C. Lin, S.-C. Lo, and G. C.-H. Chuan, *Design and System Performances of A Dual-Band 4-Port MIMO Antenna for LTE.* New York: IEEE, 2011. 978-1-4244.

[9]. T. Vidhyavathi, G. S. N. Raju, and M. Satyanarayana, "Design and implementation of MIMO antenna using Swastika Slot for wireless applications," *International*

Journal of Recent Technology and Engineering (IJRTE) ISSN: 2277-3878, vol. 8, no. 3, pp. 3972–3976, 2019.

[10]. K.-L. Wong, and W. H. Hsu, "A broad-band rectangular patch antenna with a pair of wide slits," *IEEE Transactions on Antennas Propagation*, vol. 49, no. 9, pp. 1345–1347, 2001

[11]. M. Manteghi, S. S. Naeini, and A.Abbaspur, "A U-shaped inverted f microstrip patch antenna for portable terminals," in *IEEE Conference on Antennas and Propagation for Wireless Communications (APS)*, pp. 31–34, 2000.

[12]. P. Moghe, and P. K. Singhal, "Design of a single layer L-shaped microstrip patch antenna," in *International Conference on Emerging Trends in Electronic and Photonic Devices & Systems, Electro' 09*, pp. 307–309, 2009.

[13]. A. Jayakumar, and A. Saranyakumari, "Design of E-shaped nano patch dual band antenna for 5G applications," *International Journal of Engineering and Advanced Technology (IJEAT)*, vol. 8, no. 5, pp. 2802–2807, 2019.

[14]. H.-T. Hsu, F.-Y. Kuo, and P.-H. Lu, "Design of WiFi/WiMAX dual-band E-shaped patch antennas through cavity model approach," *Microwave and Optical Technology Letters*, vol. 52, no. 2, pp. 471–474, 2009.

[15]. Y. Kabiri, A. L. Borja, et.al, "A technique for MIMO antenna design with flexible element number and pattern diversity," doi:10.1109/ACCESS.2019.2910822.

[16]. J. Malik, D. Nagpal, and M. V. Kartikeyan, "Mimo antenna with omnidirectional pattern diversity," *Electronics Letters*, vol. 52, no. 2, pp. 102–104, 2016.

[17]. J. Babu, S. R. Krishna, and L. Pratap Reddy, "A review on the design of MIMO antennas for upcoming 4G," *International journal of Advanced Engineering Research,* vol. 1, no. 4, pp. 85–93.

[18]. D. Sarkar, K. Saurav, and K. V. Srivastava, "Dual band complementary split-ring resonator-loaded printed dipole antenna arrays for pattern diversity multiple-input_multiple-output applications," *IET Microwaves, Antennas & Propagation*, vol. 10, no. 10, pp. 1113–1123, 2016.

[19]. J. Sharony. (Nov. 15, 2006), "Introduction to Wireless MIMO_Theory and Applications," https://ieee.li/pdf/viewgr-aphs/introduction_to_wireless_mimo.pdf

[20]. L. Zhu, H. S. Hwang, E. Ren, and G. Yang, "High performance MIMO antenna for 5G wearable devices," in *Proceedings of the IEEE International Symposium on Antennas and Propagation and USNC-URSI Radio Science Meeting*, San Diego, CA, USA, Jul. 2017, pp. 1869–1870

[21]. E. Biglieri, R. Calderbank, A. Constant inides, A. Goldsmith, Arogy as wami Paulraj, H. Vincent Poor, *MIMO Wireless Communications*. Cambridge University Press: New York, pp. 1–3, 2007

[22]. H. Wang, L. Liu, Z. Zhang, Y. Li, and Z. Feng, "Ultra-compact three port MIMO antenna with high isolation and directional radiation patterns," *IEEE Antennas and Wireless Propagation Letters*, vol. 13, pp. 1545–1548, 2014.

[23]. X. Zhao, S. P. Yeo, and L. C. Ong, "Planar UWB MIMO antenna with pattern diversity and isolation improvement for mobile platform based on the theory of characteristic modes," *IEEE Transactions on Antennas and Propagation*, vol. 66, no. 1, pp. 420–425, 2018.

[24]. Q. Rao and D. Wang, "A compact dual-port diversity antenna for long term evolution handheld devices," *IEEE Transactions on Vehicular Technology*, vol. 59, no. 3, pp. 1319–1329, 2010

[25]. D. Singh, C. Kalialakis, P. Gardner, and P.S. Hall, "Small H-shaped antennas for MMIC applications," *IEEE Transactions on Antennas and Propagation*, vol. 48, pp. 1134–1141, 2000.

[26]. L. Ge, J. Wang, M. Li, T.-Y. Shih, and S. Chen, "Antenna and circuits for 5G Mobile Communications," *Wireless Communications and Computing*, doi:10.1155/2018/3249352
[27]. C. A. Balanis, *Antenna Theory*.2nd edn. New York: John Wiley, 1982.
[28]. G. Oliveri et al., "Design of compact printed antennas for 5G base stations," in *Proceedings of the 11th European Conference on Antennas and Propagation (EUCAP)*, Paris, France, Mar. 2017, pp. 3090–3093.
[29]. W. Hong, K. Baek, and Y. Lee, "Quantitative analysis of the effects of polarization and pattern reconfiguration for mmWave 5G mobile antenna prototypes," in *Proceedings of the IEEE Radio and Wireless Symposium (RWS)*, Phoenix, AZ, USA, Jan. 2017, pp. 68–71.
[30]. A GSA Executive Report from Ericsson, Huawei and Qualcomm, *The Road to 5G: Drivers, Applications, Requirements and Technical Development*. Stockholm, Sweden: Global Mobile Suppliers Association, 2015.

Part IV

Fractal and Defected Ground Structure Microstrip Antenna

7 Multiband Circular Disc Monopole Metamaterial Antenna with Improved Gain for Wireless Application

S. Prasad Jones Christydass[1], *Dr R. Saravanakumar*[2], *and M. Saravanan*[3]

[1]Assistant Professor, Electronics and Communication Engineering, K. Ramakrishnan College of Technology, Trichy, Tamilnadu, India

[2]Assistant Professor, Electronics and Communication Engineering, Hindusthan Institute of Technology, Coimbatore, Tamilnadu, India

[3]Assistant Professor, Electronics and Communication Engineering, Annamalai University, Cuddalore, Tamilnadu, India

7.1 INTRODUCTION

In the past few decades, researchers have been highly motivated in designing a multiband antenna with low profile characteristics. The reason for the above is a single antenna with multiband capabilities replaces multiple antennas, which results in an increase in space efficiency. Many methods of obtaining multiband characteristics are proposed, but the major drawback with the proposed methods is the degradation of other antenna performance. Therefore a significant trust area is available for the design of a single antenna with multiband characteristics and optimum antenna parameter performance. Such an antenna can be designed by incorporating the metamaterials in the antenna structure. In the last few years, the multiband antenna with metamaterials and defected ground structures are widely used to obtain the desired performance.

DOI: 10.1201/9781003187325-7

7.2 LITERATURE REVIEW

In the recent past, modern communication devices require compact antennas with reasonable gain and bandwidth. Some other requirements are the cost-effectiveness, ease of fabrication, and analysis. The microstrip patch antenna is the antenna that can overcome the above shortcomings, but the major disadvantage is its narrow bandwidth and gain. This disadvantage is due to the surface waves. To reduce surface waves, EBG, multi substrate, partial, and defected ground techniques are widely used [1], which demands high fabrication costs. In a multilayer substrate, the bandwidth can be improved, but the efficiency and gain will be reduced [2]. To enhance gain, array, surface mounted horn, corrugation, and lens are widely used [3–5], but the shortcomings of the above methods are larger space and high fabrication cost. In [6], an antenna with a microstrip feed for WLAN is proposed to operate at 2.4 GHz and 5.2 GHz. A dual-band microstrip patch antenna is proposed in [7]. Some of the other techniques to achieve multiband are defected ground structure, the introduction of a slot in-ground, and patch, fractal structures [8,9]. But the major issue concerning all the proposed methods is complicated in fabrication, higher cost, and reduction in the performance.

All the above shortcomings are resolved by utilizing the metamaterial structure in the antenna. Metamaterials [10,11] are the artificial structure that has unique properties [12–14], such as negative permittivity, negative permeability, negative refractive index, and negative impedance. Metamaterials are the periodic structure with maximum cell size less than the quarter of the guided wavelength. Many types of antenna structures are proposed for wireless applications [15,16]. The metamaterial structure is used patch in order to reduce the surface waves, which increases the gain and bandwidth [15–17]. Metamaterial superstrates [12] are widely used to rise the antenna gain. Metamaterial loading [18–20] techniques are used for achieving multiband characteristics. Various types of metamaterial, such as SRR, CSRR, OCSRR, S, and Eighth shaped resonators [21–24], are widely used in the metamaterial loading to achieve multiband characteristics.

In this article, a circular monopole with elliptical SRR (ESRR) metamaterial is proposed for both multiband characteristics and gain improvement. The proposed structure has three stages of evolution. Its design stages, along with its impact on the parameters, are explained in Section 7.2. With the parametric analysis help, the optimum value of the critical parameters is decided and presented in Section 7.3; in Section 7.4, the results are discussed. In Section 7.5, the research findings of the articles are concluded.

7.3 DESIGN OF ESSR INSPIRED CIRCULAR MONOPOLE ANTENNA

The proposed antenna is having a size of 40 mm × 68.5 mm × 1.6 mm, which is made-up on an FR4 substrate. The antenna evolution is depicted in Figure 7.1. The parameter values and proposed structure are presented in Table 7.1 and Figure 7.2, respectively. The proposed structure has three stages of evolution, namely Antenna i, ii, and iii. Antenna A is a simple circular monopole with partial ground. Antenna B is

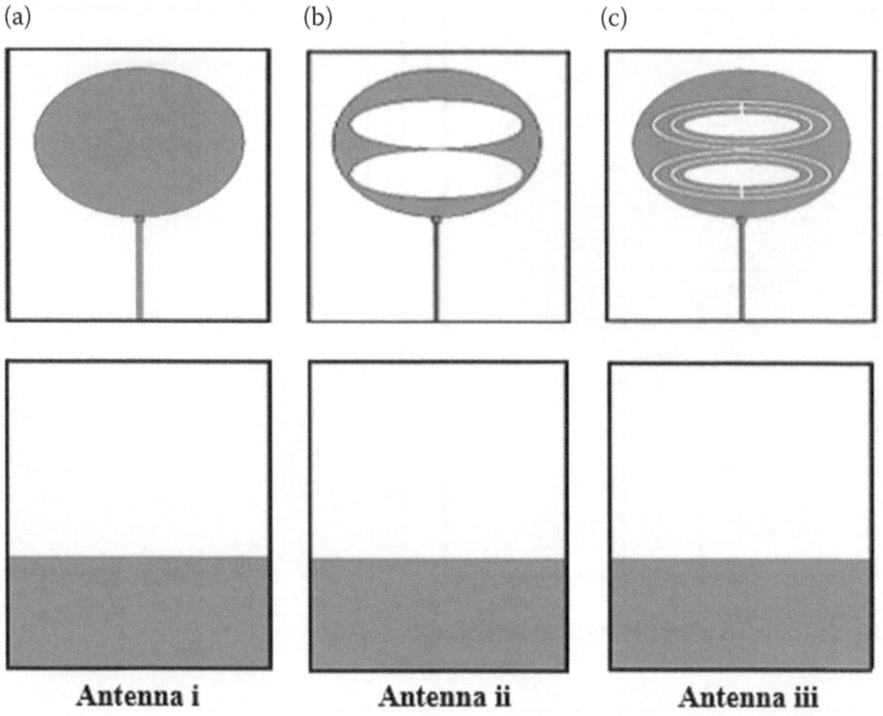

FIGURE 7.1 Front and back view of Antenna i, ii, and iii, (a) Antenna i, (b) Antenna ii, (c) Antenna iii.

TABLE 7.1
Parameter Values in mm

w	l	lf	lf1	wf	wf1	r
40	68.5	32.5	2	1	2	24
r1	**r2**	**r3**	**s**	**t**	**lg**	**h**
20	19	15	0.5	0.035	32.5	1.6

designed by including two elliptical slots in the circular patch. The elliptical slot dimension is *x* radius is 20 mm, and *y* radius is 8 mm, and two elliptical slots are etched such that the entire etched region looks like eight. Further, the ESRR is introduced in each of the elliptical slots to design the antenna C. The ESSR outer ring has 19 mm and 7 mm as major and minor radius, while the 15 mm and 5 mm are the major and minor radius of the ESRR inner ring.

Antenna A is a simple circular patch antenna. The following design equations are used to calculate the radius of the circular patch antenna.

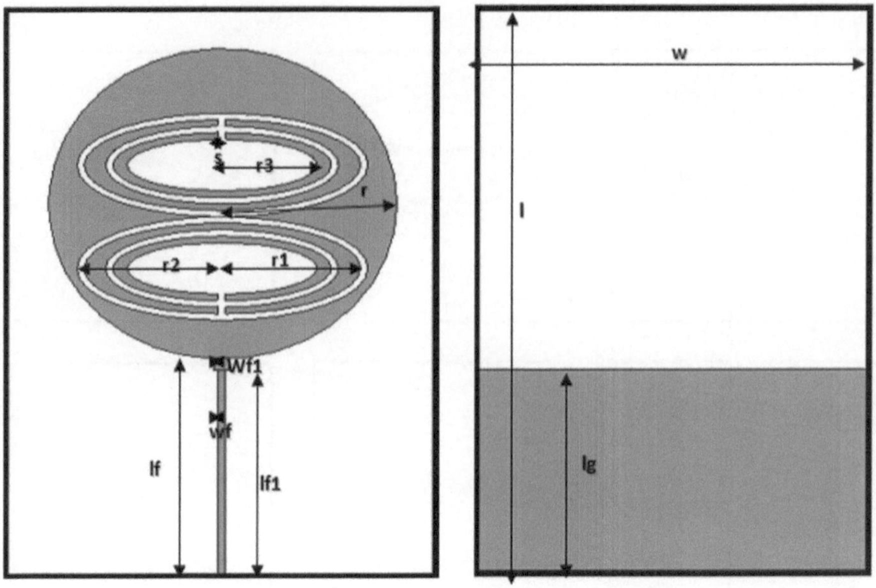

FIGURE 7.2 Proposed ESRR circular monopole with its parameters.

TABLE 7.2
Simulated Result Comparison Antenna A, B, and C

Antenna	Techniques	No. of Resonating Bands	Frequency of Resonance GHz	Operating Range GHz	Bandwidth MHz	Return Loss dB
i	Circular Patch	1	2.56	2.42–2.95	532	–26.35
ii	Antenna I with elliptical slot	2	1.49	1.38–1.61	232	–15.96
			5.32	5.02–5.51	496	–18.25
iii	Antenna II with elliptical SRR	7	1.46	1.30–1.62	320	–21.56
			2.28	2.24–2.33	83	–22.07
			2.61	2.48–2.76	277	–18.45
			2.85	2.76–2.89	133	–13.91
			3.68	3.46–3.89	430	–15.73
			4.31	4.20–4.57	371	–16.12
			5.31	4.87–5.64	769	–21.42

$$r = \frac{g}{\left\{1 + \frac{2h}{\pi\varepsilon_r g}\left[\ln\left(\frac{\pi g}{2h} + 1.7726\right)\right]\right\}^{1/2}}, \tag{7.1}$$

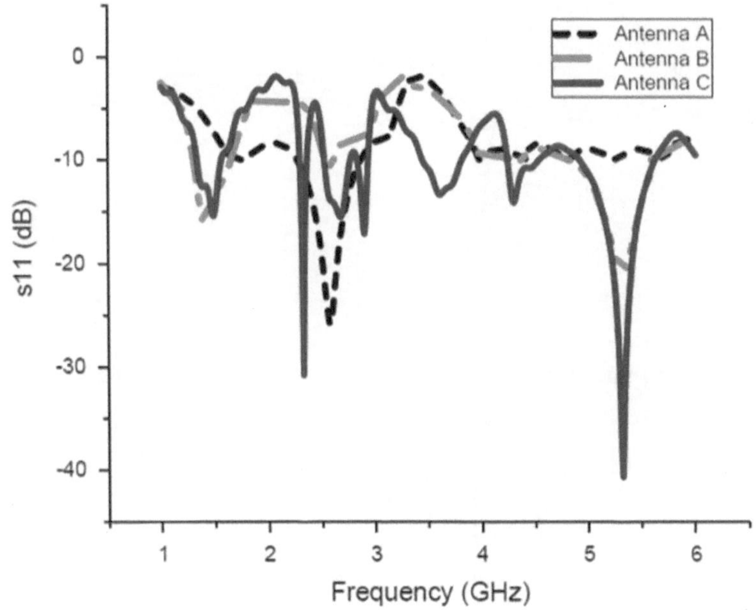

FIGURE 7.3 s11 Comparision for evolved antenna.

where

$$g = \frac{8.791 * 10^9}{f_r \sqrt{\varepsilon_r}}. \quad (7.2)$$

The above equation is modified by considering the fringing effects

$$F_r = \frac{1.8412 * c}{2\pi a_e \sqrt{\varepsilon_r}} \quad (7.3)$$

and the effective radius is calculated as

$$r_e = r * \left\{1 + \frac{2h}{\pi\varepsilon_r r}\left[\ln\left(\frac{\pi r}{2h} + 1.7726\right)\right]\right\}^{1/2}, \quad (7.4)$$

where h is the thickness of the FR4, ε_r is the dielectric constant, C is equal to $3 * 10^8$m/sec, r and r_e are the circular patch radius and effective radius, respectively. The calculated radius based on the equation is about 24 mm, and the structure is optimized to operate in 2.5 GHz.

Then the elliptical slot is introduced, which alters the current direction in the circular patch. The elliptical slot has an X radius 20 mm, and the Y radius is 8 mm. The modified structure Antenna ii resonates in two bands due to the increase in the

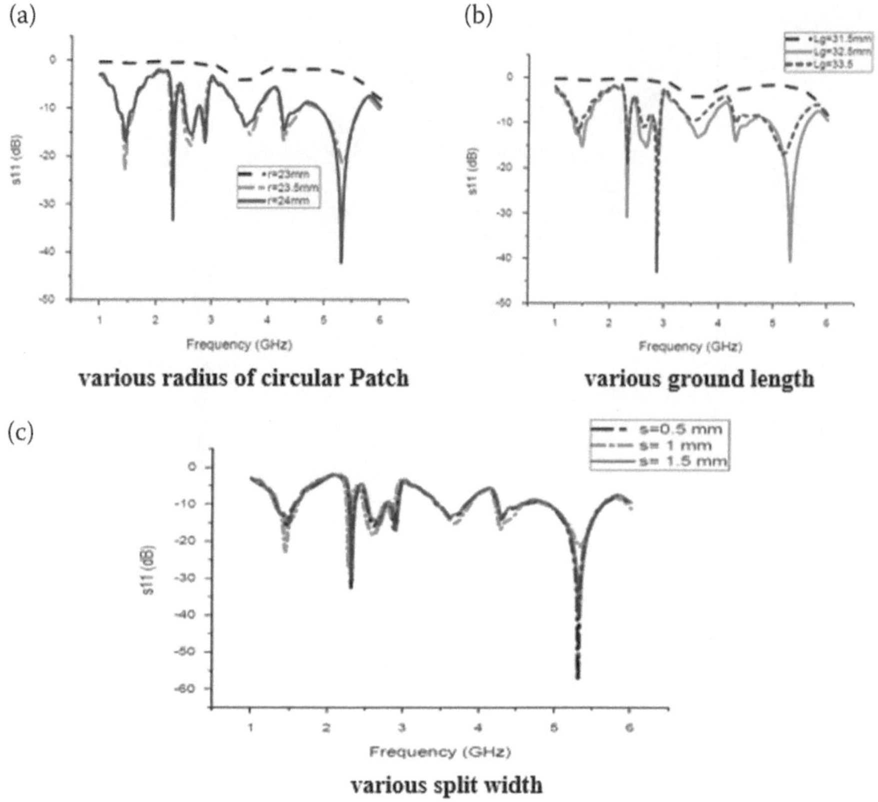

FIGURE 7.4 Parametric analysis.

capacitance; the upper 2.5 GHz band is moved downward, and another resonating band is created at 5.3 GHz. The antenna is then included with elliptical SRR, which makes the antenna resonate at seven different bands. Antenna i is an antenna with a circular radiating element, as depicted in Figure 7.1a. A single band antenna resonates at 2.56 GHz with the operating band from 2.42 GHz to 2.95 GHz. Antenna ii resonates at dual-band at 1.49 GHz and 5.32 GHz. The operating band of Antenna ii is from 1.38 GHz to 1.61 GHz and from 5.02 GHz to 5.51 GHz. Finally, Antenna iii with the inclusion of the metamaterial ESRR is capable of operating at seven different bands at 1.46 GHz, 2.28 GHz, 2.60 GHz, 2.85 GHz, 3.68 GHz, 4.31 GHz, and 5.31 GHz.

7.4 PARAMETRIC ANALYSIS

The critical parameters of the proposed ESRR based circular antenna is subjected to the parametric analysis to choose an optimum value. The critical parameters chosen are radius of the circular patch (r), Ground height (Lg), and then the split width (s) of the ESRR. The paramteric analysis is done with the help of CST software. First the radius of the patch is increased in steps of 0.5 mm from 23 mm to 24 mm. The s11 charactristics of the proposed structure with respect to various circular patch

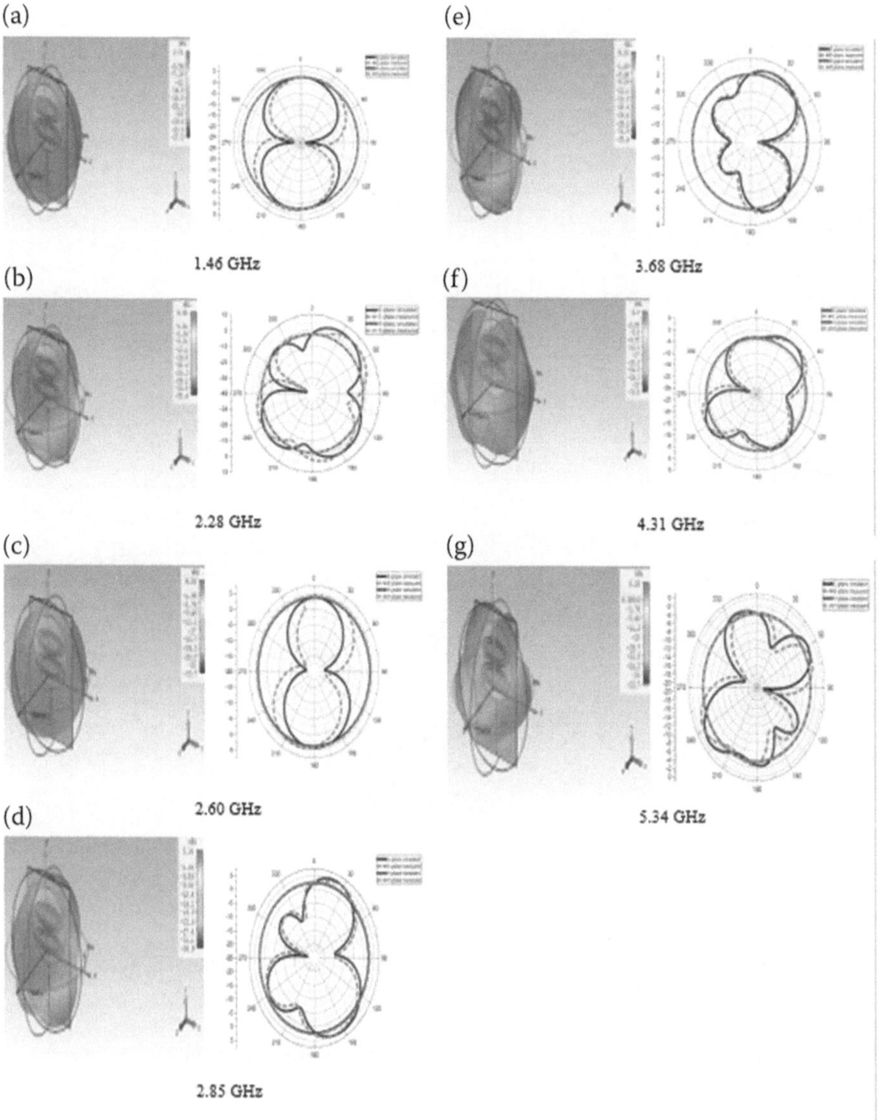

FIGURE 7.5 3D radiation pattern, *E*-plane (measured & simulated) and *H*-plane (measured & simulated) at various resonating frequency.

radius is plotted in Figure 7.4a. It is pragmatic from Figure 7.4a that radius of 24 mm is having good impedance matching with reasonable bandwidth. So, 24 mm is chosen as the final value for the fabrication. Next, the length of the ground is increased in steps of 1 mm from 31.5 mm to 33.5 mm. Its performance is depicted in Figure 7.4b. We can observe that the ground length has a more significant effect on the impedance matching from Figure 7.4b. And the ground length of 32.5 mm 9s chooses for the final fabrication based on the parametric analysis.

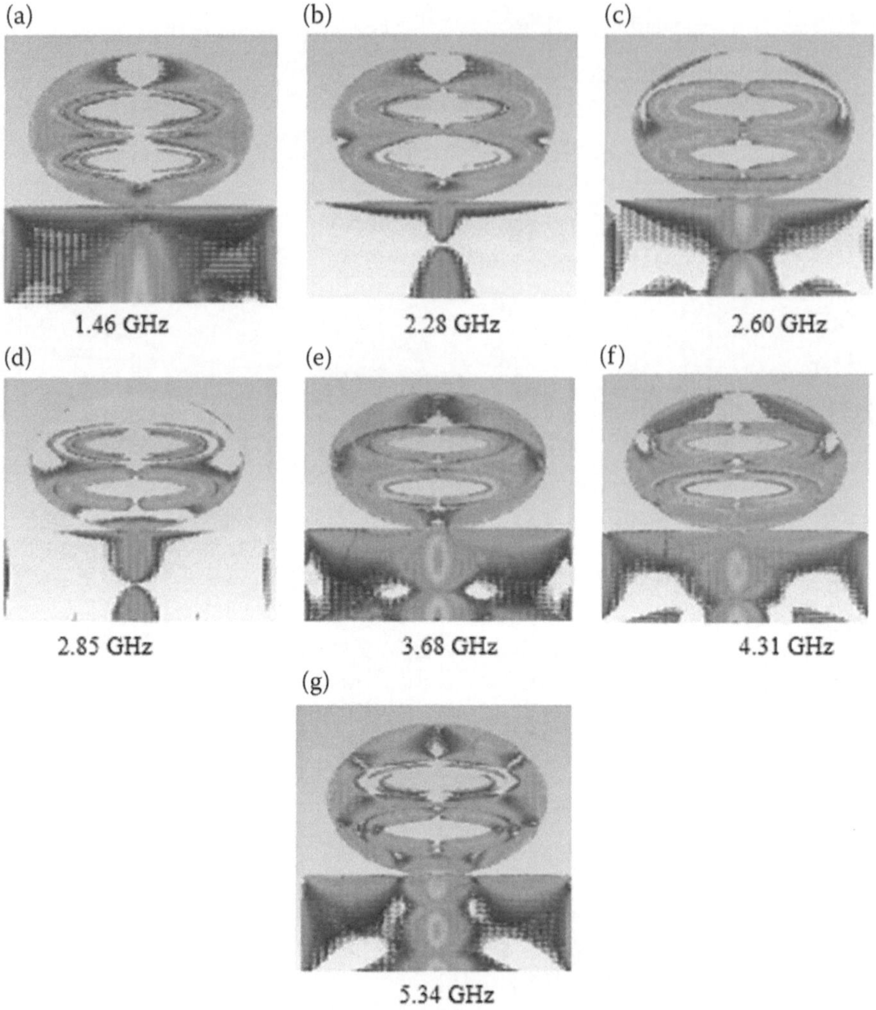

FIGURE 7.6 Surface current density at the various resonating frequency.

The split width is increased in steps of 0.5 mm from 0.5 mm to 1.5 mm. The impact of the step width on the s11 characteristics is shown in Figure 7.4c, and as the split width increases, there is a slight deviation in the resonant frequency created by the ESRR. The decrease is due to an increase in capacitance. And $S = 0.5$ mm is chosen for the final fabrication since it is able to have good impedance matching in all the resonating bands.

7.5 DISCUSSION OF RESULTS

In Figure 7.3, the return loss plot for the three evolution stages of the proposed structure is presented. It is clearly observed that in the introduction of the elliptical slot, the resonant frequency is shifted to 1.49 GHz, and another resonance is created

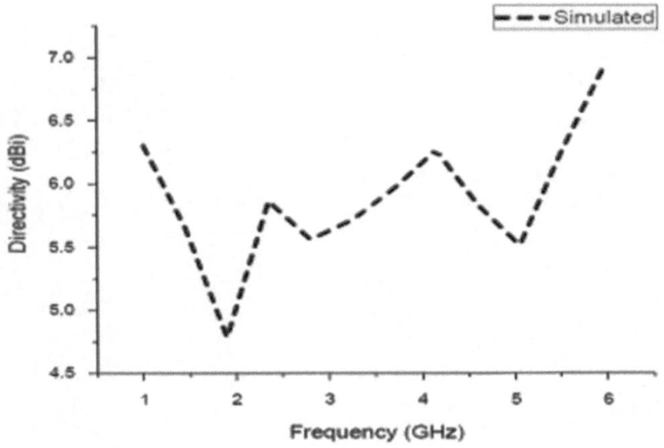

FIGURE 7.7 Directivity vs frequency.

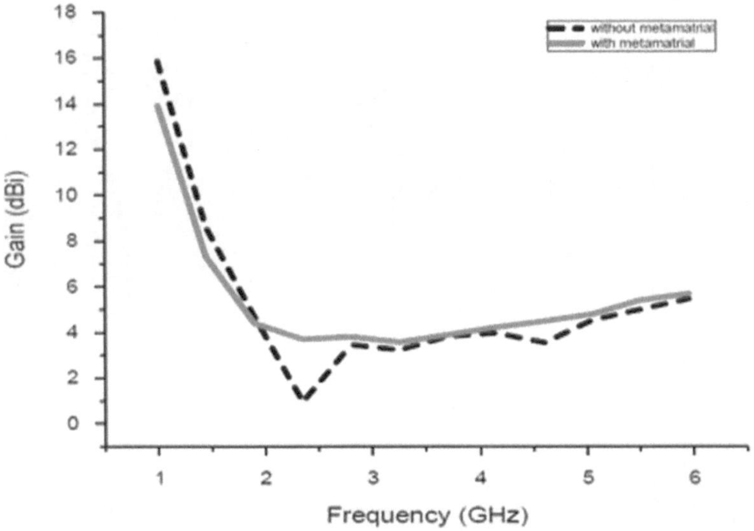

FIGURE 7.8 Gain vs frequency.

at 5.31 GHz. Then the proposed structure is designed by introducing a set of dual ring elliptical SRR at both the elliptical slots. The proposed structure is capable of resonating at seven bands. The s11 is measured with the help of VNA Anritsu S820E.

In Figure 7.5, the 3D radiation pattern is presented. The comparison patterns of the simulated E and H plane at various resonating frequencies are also compared with measured and presented in Figure 7.5. The shape of the pattern is eight-speed, which is similar to a dipole pattern. The H plane has an omnidirectional pattern. In Figure 7.6, the surface current density at various resonating frequency is depicted.

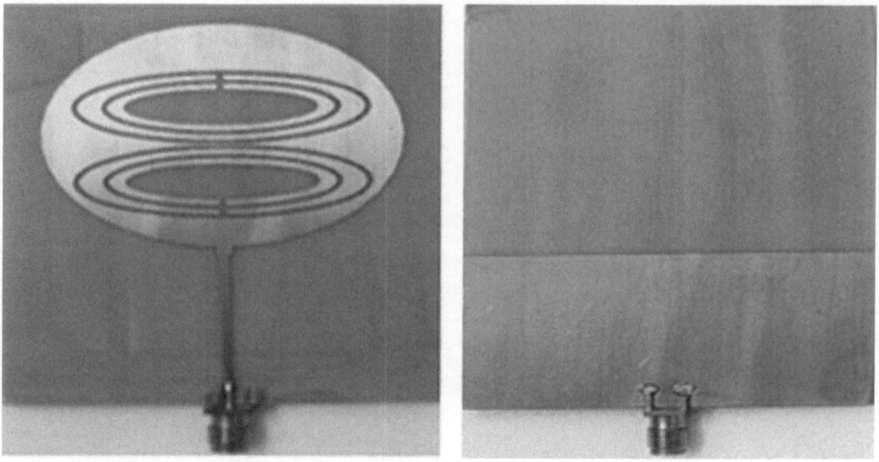

FIGURE 7.9 Fabricated antenna.

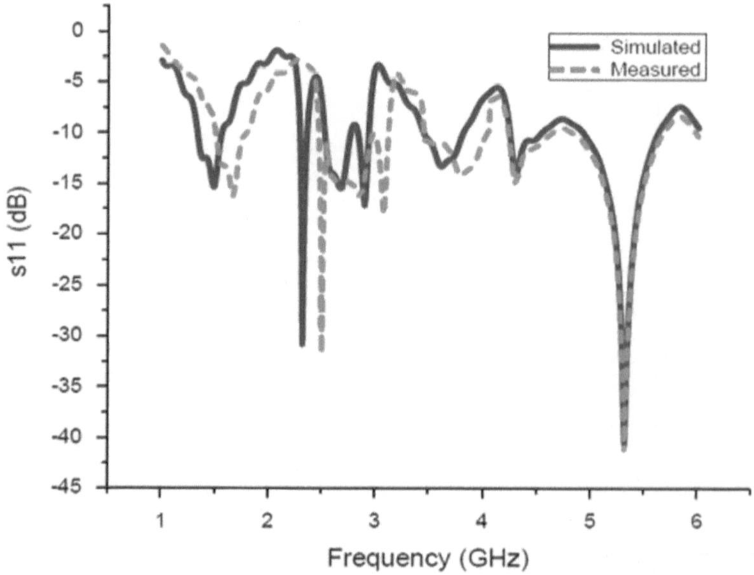

FIGURE 7.10 Measured vs simulated s11.

The figure is observed at 1.46 GHz and 5.34 GHz, the surface current is maximum over the entire circular patch, and it is responsible for those dual bands.

The band 2.28 GHz is due to the upper elliptical slot, and band 2.85 GHz is due to the lower elliptical slot since the surface current is maximum at the upper edges and lower elliptical slot, respectively. It is observed from the figure that bands 2.60, 3.68, and 4.31 GHz are created due to the inclusion of ESRR, which is evident from the accumulation of surface current in the rings of ESRR.

TABLE 7.3
Simulated vs Measured Results

Simulated			Measured		
Frequency of Resonance GHz	Operating Range GHz	Bandwidth MHz	Frequency of Resonance GHz	Operating Range GHz	Bandwidth MHz
1.46	1.30–1.62	320	1.52	1.42–1.78	362
2.28	2.24–2.33	83	2.5	2.48–2.53	51
2.61	2.48–2.76	277	2.81	2.54–2.92	384
2.85	2.76–2.89	133	3.2	2.92–3.29	372
3.68	3.46–3.89	430	3.81	3.46–4.2	748
4.31	4.20–4.57	371	4.32	4.21–4.61	408
5.31	4.87–5.64	769	5.32	4.87–5.65	782

In Figure 7.7, the directivity of the proposed structure is plotted with respect to frequency. The maximum directivity of the proposed ESRR based circular Patch is 6.85 dBi. The gain is plotted with respect to the frequency in Figure 7.8. In the figure, the gain of antenna i and antenna iii is compared. The figure shows that the gain of antenna iii is above 3 dBi in all the resonating antenna. This increment is gained due to the metamaterial inclusion, which has negative permeability in the ESRR resonating bands. In Figure 7.9, the prototype antenna is depicted. The simulated results are compared with measured s11 and it is depicted in Figure 7.10. In Table 7.3, the measured and simulated results are related.

7.6 CONCLUSION

A circular disc monopole with an elliptical split-ring resonator is proposed for wireless communication. The proposed structure has three stages of evolution antenna i, a simple circular disc operating at a single band, antenna ii with elliptical slot it operates in dual-band, and antenna iii with elliptical SRR operating at seven bands. The parameters are optimized using the CST parametric analysis. The proposed structure is fabricated and measured. The simulated and measured results are compared, and the slight deviation in the return loss characteristics is due to fabrication error. The simulated and measured radiation pattern is presented, which are nearly equal to each other. The proposed structure is operating at seven different bands. This band covers the GPS, LTE33, Mobile-WiMax, LTE42/43, 5G sub 6 GHz band, WiMAX, WAIC, and WLAN applications. The compact size, good radiation pattern, multiband characteristics, and improved gain make this proposed ESRR based circular patch antenna the right choice for wireless applications.

REFERENCES

1. Thai TT, DeJean GR, Tentzeris MM. Design and development of a novel compact soft-surface structure for the front-to-back ratio improvement and size reduction of a microstrip Yagi array antenna. *IEEE Antennas Wirele Propag Lett* 2008;7:369–373.
2. Lier E, Jakobsen K. Rectangular microstrip patch antennas with infinite and finite ground plane dimensions. *IEEE Trans Antennas Propag* 1983;31(6):978–984.
3. Zhu H, Cheung SW, Yuk TI. Enhancing antenna boresight gain using a small metasurface lens: reduction in half-power beamwidth. *IEEE Antennas Propag Mag* 2016;58(1):35–44.
4. Latif SI, Shafai L, Shafai C. Gain and efficiency enhancement of compact and miniaturised microstrip antennas using multi-layered laminated conductors. *IET Microwa Antennas Propag* 2011;5(4):402–411.
5. Munk BA. *Metamaterial: Critique and Alternatives*. New York: Wiley, 2009.
6. Sun XL, Cheung SW, Yuk TI. Design of compact antenna for 2.4/4.9/5.2/5.8-GHz WLAN. *Microw Opt Technol Lett* 2014;56:1360–1366.
7. Kaur J, Khanna R. Development of dual band microstrip patch antenna for WLAN/MIMO/WiMAX/AMSAT/WAVE applications. *Microw Opt Technol Lett* 2014;56: 988–993.
8. Rajabloo H, Kooshki VA, Oraizi H. Compact microstrip fractal koch slot antenna with ELC coupling load for triple band application. *AEU Int J Electron C* 2017;73:144–149.
9. Saber Dakhli, Hatem Rmili, Jean-Marie Floc'h, Muntasir Sheikh, Abdallah Dobaie, Kourosh Mahdjoubi, Fethi Choubani, Ziolkowski Richard W. Printed multiband metamaterial inspired antennas. *Microw Opt Technol Lett* 2016;58(6):1281–1289.
10. Caloz C., Itoh T. *Electromagnetic Metamaterials: Transmission Line Theory and Microwave Applications*. New York: Wiley-IEEE Press, 2005.
11. Huang J. Circularly polarized conical patterns from circular microstrip antennas. *IEEE Antennas Propag Mag* 1984;32(9):991–994.
12. Li D, Szabó Z, Qing X, Li EP, Chen ZN. A high gain antenna with an optimize metamaterial inspired superstrate. *IEEE Trans Antennas Propag* 2012;60 (12):6018–6023.
13. Ghatak R, Mishra RK, Poddar DR. Perturbed Sierpinski carpet antenna with CPW feed for IEEE 802.11a/b WLAN application. *IEEE Antennas Wirele Propag Lett* 2008;7:742–745.
14. Xu Y, Jiao YC, Luan YC. Compact CPW-fed printed monopole antenna with triple band characteristics for WLAN/WiMAX applications. *Electron Lett* 2012;48(24): 1519–1520.
15. Singh AK, Abegaonkar MP, Koul SK. High-gain and high-aperture-efficiency cavity resonator antenna using metamaterial superstrate. *IEEE Antennas Wirele Propag Lett* 2017;16:2388–2391.
16. Rajak N, Chattoraj N. A bandwidth enhanced metasurface antenna for wireless applications. *Microwave Opt Technol Lett* 2017;59(10):2575–2580.
17. Zheng YJ, Gao J, Zhou YL, Cao XY, Xu LM, Li SJ. Metamaterial-based patch antenna with wideband RCS reduction and gain enhancement usingimproved loading method. *IET Microw Antennas Propag* 2017;11(9): 1183–1189.
18. Saha C, Siddiqui JY. Versatile CAD formulation for estimation of the resonant frequency and magnetic polarizability of circular split ring resonators. *Int J RF Microw Comput Aided Eng*. 2011;21:432–438.
19. Daniel S, Pandeeswari R, Raghavan S. A compact metamaterial loaded monopole antenna with offset-fed microstrip line for wireless applications. *AEU–Int J Elect Com* 2017;83:88–94.

20. Rao MV, Madhav BTP, Anilkumar T, Nadh BP. Metamaterial inspired quad band circularly polarized antenna for WLAN/ISM/-Bluetooth/WiMAX and satellite communication applications. *Int J Electron and Commun (AEÜ)* 2018;97:229–241.
21. Pirooj, A, Moghadasi, MN, Zarrabi, FB. Design of compact slot antenna based on split ring resonator for 2.45/5 GHz WLAN applications with circular polarization. *Microw Opt Technol Lett* 2016;58(1):12–16.
22. Pandeeswari R, Raghavan S. A CPW-Fed triple band OCSRR embedded monopole antenna with modified ground for WLAN and WiMAX applications. *Microw Opt Technol Lett* 2015;57:2413–2418.
23. Patel Shobhit K, Yogeshwar Kosta. Multiband meandered miniaturized patch antenna loaded with split ring resoantor and thin wire arrays. *Microw Opt Technol Lett* 2014;56(2):306–310.
24. Rajabloo H, Kooshki VA, Oraizi H. Compact Microstrip fractal koch slot antenna with ELC coupling load for triple band application. *AEU Int J Electron C* 2017;73:144–149.

8 Fractal Based Ultra-Wideband Antenna Design

A Review

C. Muthu Ramya[1] *and R. Boopathi Rani*[2]
[1]National Institute of Technology Puducherry, Puducherry, India
[2]Assistant Professor, Department of ECE, National Institute of Technology Puducherry, Puducherry, India

8.1 INTRODUCTION

The Ultra-wideband has become a promising technology which meets the requirements of high speed communication. As the recent year wireless communication is in need of extremely high transmission rate, a high data rate and low power with compact design leads UWB technology preferred in the short range, fast multimedia, and voice communications. The Ultra-wideband (UWB), Cognitive Radio (CR) are the technologies utilized to mitigate the problem of fully occupied and still not fully used spectrum by means of re-use of frequency spectrum. Cognitive Radio (CR) senses the unused spectrum and allocates for data transformation. Hence no interference to the main spectrum users found. On the other hand UWB uses a broadband signal, with very narrow pulses of low average power for communication. Therefore, the system provides extremely high data rate communication with the wideband.

The interference avoidance capability of the transmitter and receiver mainly depends on the power spectral density of the pulses. The narrow line spacing in the spectrum gives a chance to spread the power uniformly throughout the spectrum. Therefore, the power in each spectral line is much reduced [1]. Since UWB uses an extremely wideband and low average transmit power, the disturbance to other narrowband systems is limited. It is possible due to fact that the UWB signal naturally overlaps with the services using the primary spectrum. They can be shown as white noise to the conventional narrow band systems [2].

The chapter is organized as follows: Section 8.1 presents the various bandwidth definitions, FCC approval and UWB definition, and some applications where the technology is currently in use and band notch characteristics. Section 8.2 explains about the fractal geometry used in antennas, an iterative function system. Sections

DOI: 10.1201/9781003187325-8

8.3–8.7 presents the iterative generation process of fractals such as Sierpinski (gasket, carpet, knopp fractal curve) fractal structures, Koch fractal structure, Hilbert curve fractal, tree fractal structure, and Apollonian fractal structures and their impact on UWB antenna design through literatures respectively. Section 8.8 presents the literature of conventional shaped fractal antennas utilized in the UWB antenna designs. Section 8.9 gives the survey on multiple fractal structures in a single antenna, that is, how hybrid fractal structures are applied on the design of UWB antenna. Finally, Section 8.10 presents the conclusion of the chapter. The figures are given in relation to the fractal concept discussed in the respective sub heading. The chapter is mainly focusing on planar antennas. Hence, the figures are shown as it is made on some substrate material. Summary on state of the art UWB antenna using fractal structure is presented in Table 8.1.

8.1.1 BANDWIDTH DEFINITIONS

The bandwidth of an antenna is the range of spectrum over which the desired properties are exhibited by the antenna. The radiation characteristics, gain, and input impedance are the properties to be satisfied in this spectral range. The general bandwidth definitions are absolute bandwidth, fractional bandwidth, and ratio bandwidth. They are described by the following expressions.

The absolute bandwidth is described as

$$BW = f_h - f_l \tag{8.1}$$

The fractional bandwidth is described as

$$BW = \frac{f_h - f_l}{f_c} * 100\% \tag{8.2}$$

$$= \frac{2 * (f_h - f_l)}{f_h + f_l} * 100\% \tag{8.3}$$

The ratio bandwidth is described as

$$BW = \frac{f_h}{f_l} \tag{8.4}$$

where

fh is the upper frequency of UWB spectrum

fl is the lower frequency of UWB spectrum

fc is the center frequency of UWB spectrum, which is calculated by taking arithmetic mean of upper and lower frequency of operating frequency range.

TABLE 8.1
Summary on State of the Art UWB Antenna using Fractal Structure

Ref. no.	Fractal Technique Used	Dimension (mm^2)	Operating Band (GHz)	Bandwidth (GHz)	Peak Gain	Notch Band Frequency (GHz)	Notch Band Application
[25]	Sierpinski gasket-based fractal	36 × 48	3–12	9	2–4.5 dBi	–	–
[26]	Modified Sierpinski square fractal antenna	34 × 34	3.1–10.6	7.5	2–5 dBi	5.5 (5–6)	WLAN
[27]	Sierpinski carpet fractal with circular boundary	40 × 38	2.7–12	9.3	1.85–6 dBi	(5.15–5.825)	IEEE802.11a & HIPERLAN/2
[28]	Hexagonal boundary Sierpinski carpet fractal	33 × 32	3–12	9	1.25–6 dBi	(5.15–5.825)	IEEE802.11a & HIPERLAN/2
[29]	Octagonal Sierpinski fractal	25 × 19	3.73 to more than 20	>16	1.4–6.85 dB	–	–
[30]	Sierpinski square fractal slot on decagonal shaped monopole	28 × 28	3.50–15.1	11.6	1.4–5.95 dBi	–	–
[31]	Modified Sierpinski carpet fractal	24 × 30	3–12.6	9.6	2–4.8 dBi	–	–
[32]	Sierpinskiknopp fractal	40 × 40	2.6–10.6	8		3.5 5.4	WiMAX WLAN
[36]	Koch fractal slot	28 × 24	2.85–12	9.15	2–3 dBi	(4.65–6.40)	WLAN

(Continued)

TABLE 8.1 (continued)
Summary on State of the Art UWB Antenna using Fractal Structure

Ref. no.	Fractal Technique Used	Dimension (mm^2)	Operating Band (GHz)	Bandwidth (GHz)	Peak Gain	Notch Band Frequency (GHz)	Notch Band Application
[37]	Koch fractal	31 × 28	3–12.8	9.8	6–4 dBi	–	–
[38]	Fractal Koch and T-shaped stub	50 × 50	1.78–11	9.22	−3.5–6.5 dBi	2(1.95–2.25) 3.5(3.15–3.65) 5.8(5.4–6)	PCS WiMAX WLAN
[39]	Koch fractal boundary	18.5 × 39	3.2–12	8.8	4 dBi	5.5	WLAN
[42]	Hilbert curve slot	40 × 40	3.1–10.6	7.5	–	5.5 (5.3 –5.8)	WLAN
[6]	Hilbert curve slot	25 × 45.75	2.5–12	9.5	3–5 dBi	(5.15–5.85) (7.9–8.4)	IEEE 802.11a & HYPERLAN/2 X–band uplink satellite communication systems.
[43]	Hilbert curve	39 × 30	3.1–10.6	7.5	2.2–4 dBi	–	–
[44]	Hilbert fractal shaped parasitic resonator element	35 × 35	2.61–13.22	10.61	–	(7.89–8.83)	X–band uplink satellite communications
[45]	Hilbert fractal shaped slot	30 × 41	2.2–11	8.8	4 dBi	5.5 8.1	WLAN X–band uplink satellite communications
[46]		30.75 × 37.8	2.7–11.22	8.52	0.07–3.40 dBi		C–band downlink satellite communications

	Hilbert fractal defected ground structure					(3.7–4.2) (5.15–5.825) (7.9–8.4)	WLAN X–band uplink satellite communications
[48]	Fractal binary tree slot	16 × 22	3.1–10.6	7.5	–	5.65 9.9	WLAN
[49]	Fractal tree shaped radiating patch	14 × 18	2.94–11.17	8.23	–	(3.3–4.2)	WiMAX C–band downlink satellite communications
[54]	Apollonian fractal	100 × 100 (ground plane)	2.65 to more than 20	>17.35	–	–	–
[55]	Apollonian fractal structure	42 × 46	3–18	15	–	–	–
[56]	Apollonian gasket monopole	50 × 60	2.8–15	12.2	–	–	–
[57]	Circular shaped Apollonian fractal	44 × 58	1.8–10.6	8.8	2–6 dBi	(5.125–5.825)	IEEE 802.11a HIPERLAN/2
[58]	Hexagonal patch with small fractals at the corners	25 × 25	Antenna1: 2.8–11.3 Antenna2:3–11.5	8.5 8.5	–	–	–
[59]	Hexagonal fractal integration on the radiating patch	63.2 × 51	3.1–12.7	9	–	–	–
[60]	Dual-reverse-arrow fractal	24 × 22	3.1–10.9	7.8	–1–4 dB	5–6	WLAN
[61]	Circular fractal patch	32 × 36	2.93–9.53	6.57	5.17 dBi	–	–
[62]	Ladder-shaped fractal	16 × 12	4.56–13.1	8.54	2.84 dB (Average)	–	–

(Continued)

TABLE 8.1 (continued)

Summary on State of the Art UWB Antenna using Fractal Structure

Ref. no.	Fractal Technique Used	Dimension (mm^2)	Operating Band (GHz)	Bandwidth (GHz)	Peak Gain	Notch Band Frequency (GHz)	Notch Band Application
[63]	Combination of GiusepePeano and Sierpinski Carpet	25 × 20	1–15	14	–	–	–
[12]	Combination of Koch fractal slot, Sierpinski curve-shaped ring resonator and Minkowski fractal slots.	41 × 45	2.8–12	9.2	1.5–4 dBi	5.5 6.8 8.1	WLAN RFID X-band uplink satellite communication systems
[64]	Combination of fractal binary tree and half Minkowski fractal, hybrid fractal slot, and Sierpinski arrowhead curve fractal slot.	33.3 × 24.4	–	10.6	1–1.5 dBi (Bluetooth) 2.5–6 dBi(UWB)	5.5	WLAN
[65]	Combination of triangular fractals and cantor set fractal	26 × 21	2.8–10.3	4.5	–	(5–6.3)	WLAN
[66]	Hybrid Sierpinski Koch fractal	40 × 20	2.5–11	8.5	9.9(Diversity gain)	5.45 (5–6)	WLAN

8.1.2 UWB DEFINITION

The official release of regulation for UWB system by Federal Communications Commission describes an ultra-wideband device as a specific device that has fractional bandwidth above 20% or absolute bandwidth of 500 MHz [3].

$$BW = \frac{f_h - f_l}{f_c} \geq 0.2, \tag{8.5}$$

or

$$BW = f_h - f_l \geq 500\text{ MHz} \tag{8.6}$$

The FCC allowed unlicensed UWB communication in the spectral band from 3.1 to 10.6 GHz with some limitations. The limitations include that the UWB system is approved for a lower average transmit power in contrast with other narrowband systems. This strongly limits UWB communications to short range, approximately 10 m.

Some antennas are unable to cover the complete UWB spectral range (3.1 –10.6 GHz) licensed by the FCC. Many techniques are utilized for getting better impedance bandwidth. One such bandwidth improvement techniques is the usage of fractal geometry.

8.1.3 UWB APPLICATIONS

UWB applications include radio base stations and portable mobile stations in many applications like [4]

- Ranging and localization

It acts as an indoor GPS that gives high precision in ranging with the very wideband.

- High speed data link

Personal area network becomes popular in communication among home appliances and personal computing applications.

- Body area network

It has number of sensor nodes in the human body or clothes; UWB is used in location finding and short-range communication establishment such as BAN.

- Wireless sensor networks

UWB is used for data communications between a large number of sensor nodes and central server.

- UWB radar

UWB is effectively used in civilian and defence industries and vehicular collision avoidance systems, etc.

- Bio medical imaging

UWB is utilized in medical diagnosis like cancer detection.

8.1.4 BAND NOTCH CREATION

Moreover, it is necessary for the UWB antenna to notch the frequencies used by other narrow band services with which the UWB spectrum overlaps. There are many techniques implemented to avoid the interference with the following conventional communication systems.

- Worldwide interoperability for microwave access WiMAX (3.3–3.7 GHz)
- IEEE 802.11a in the United States (5.15–5.35 GHz, 5.725–5.825 GHz Uplink and Downlink respectively)
- Wireless local area network, WLAN (5.3–5.85 GHz)
- HIPERLAN/2 in Europe (5.15–5.35 GHz, 5.47–5.725 GHz Uplink and Downlink respectively)

The interfering band can be notched out using the traditional filtering mechanism. It is not preferable to use the filter for portable devices as it would increase the complexity, size, and weight. Hence the design of antenna with band notch functionality is more preferred [5]. It is feasible to obtain band rejection characteristics by means of adopting the following techniques:

- Incorporating fractal structure [6]
- Slots of different shapes such as hexagonal [7]
- Parasitic stub [8,9]
- Metamaterials [10]
- Electromagnetic Band Gap (EBG) structures [11]
- Hybrid techniques [12]

This chapter also presents that the fractal structure imprint on an antenna is one of the techniques to create band notch.

8.2 FRACTAL ANTENNAS

8.2.1 Fractal Geometry in Antennas

The original idea of fractal concept emerged from nature like gloomy sky, seacoasts, river networks, leaves, mountains, trees, branches of blood vessels and neurons, snowflakes etc. The nature is irregular and might not be measured with the help of classical geometries like lines, polygons, spheres, etc., until the emergence of fractal concept. Fractal means broken fragments. Fractal geometry development helped to explore the irregularities analytically [13].

Fractal geometry was devised and advanced by Benoit B. Mandelbrot in 1978. These structures are utilized in different mathematics, science, and engineering domains. The name fractal was acquired from *frangere*, a Latin word meaning fractured to make rough fragments [14].

Fractal geometries have the following properties [15].

- Space-filling property defines a method of plotting a 2-dimensional surface area with the use of 1-dimensional curve.
- Self-similarity property defines repeated elements of self-similar objects.
- Lacunarity defines the space in the structure between the similar elements.

The concept of fractal is also applied in electromagnetics especially to design antennas, frequency selective surfaces, for improving the characteristics. As fractal geometry has self-similar structure generated by means of recursive process, it can have long length or perimeter in a very small space. Generally, the operating frequency will be increased along with the antenna miniaturization. The miniaturization is a continuing objective of antenna design. However, the fractal geometry can rise to tiny antennas that produce the radiation characteristics the same as the large-sized antennas.

The most important property of fractal named 'space-filling' gives rise to a design of miniaturized antennas with multiband or broadband operation. The printed monopole antenna bandwidth can be expanded through various ways, like changing the shape of patch and ground or feed, decreasing substrate permittivity 'εr', or by increasing substrate height, etc. [16].

The fractal geometries have the capability to produce multiband and the multiple resonances will be turned into broadband by making the resonances come closer with each other. Fractal structure incorporation in printed monopoles is a good technique for generating ultra-wideband signals since it can operate excellently in different frequencies in parallel [17].

The fractal structure can be used as a radiating patch of UWB antenna. The patch or ground plane with fractal slot can also be used to realize the ultra-wideband or band notch characteristics in UWB applications. The key objective is to show some ultra-wideband (UWB) antenna design that adopted fractal structure and its performance.

There are some limitations in employing fractal structures to produce notch bands. That includes design and manufacturing complexity, low rejection in gain at the notch bands, and irregular radiation pattern at higher frequencies [18,19].

8.2.2 Iterative Function System

Fractal structures are generated by using iterative function system (IFS). IFS provides us with a structure that describes, organizes, and controls the fractal patterns. The idea of IFS is constructed based on related sequence transformations. Basically an iterative fractal generation method starts using a conventional figure called 'initiator'. A 'Euclidean triangle' acts as the initiator for Sierpinski gasket as well as

Koch Snowflake. The Minkowski Island uses a 'Square' as initiator. A 'generator' is a term applied repeatedly on the initiator to bring fractal structure. The generator replaces the straight lines in each iteration operation.

The building process of fractal structure makes use of the initiator (basic structure) along with the generator in a recurring order. The delineation of fractal structure has an important term called the 'self-similarity dimension' (Ds) as reported in [15], given by

$$D_s = \frac{log\ (N)}{log\left(\frac{1}{s}\right)} \tag{8.7}$$

In Equation (8.7), 'N' indicates the total number of self-similar duplicates.

's' denotes scale factor.

It is not necessary for Ds value to be an integer. There are several fractal-structure-based antennas that were reported in the literature. However, some fractal geometries are extensively used to design the ultra-wide bandwidth antennas and notch band design. They are Sierpinski fractal antenna, Koch fractal antenna, Snowflake antenna, Hilbert antenna, fractal tree antennas, Apollonian fractal antenna, some conventional shaped fractal antennas like circular fractal, hexagonal fractal, etc.

8.3 SIERPINSKI FRACTAL STRUCTURE CHARACTERIZED FOR UWB ANTENNA

8.3.1 Sierpinski Gasket Construction

The Sierpinski fractal structure is a widely known mathematical geometry developed by Waclaw Sierpinski in 1915 [14], who was a Polish mathematician. The initial steps of the composition of Sierpinski gasket are shown in Figure 8.1. It takes a Euclidean triangle as an initiator. The construction starts with an initiator. A scaled down initiator whose vertices placed on the centers of edges of the basic structure is then removed from the initiator. The three triangles left will again undergo the above-mentioned procedure. These steps are then repeated to create the Sierpinski gasket. The dark area in the structure shows the metal portion. The bright areas inside the initiator are the slots. This procedure brings a diaphanous veil. The self-similarity dimension for Sierpinski carpet fractal Ds is 1.58496 [20].

FIGURE 8.1 Sierpinski gasket construction.

8.3.2 Sierpinski Carpet Construction

Figure 8.2. shows the constructional steps of Sierpinski carpet fractal structure. A square is used as a basic structure (initiator). In the step 1 iteration, nine squares scaled by a factor of three of the initiator are considered, and the central square is taken out from the initiator. This procedure is repeated on the eight left over small squares in the second step. The subsequent iterations can be done in this way. The self-similarity dimension for Sierpinski carpet fractal Ds is 1.89279 [21–23].

8.3.3 SierpinskiKnopp Fractal Construction

The construction of SierpinskiKnopp fractal Curve structure is shown in Figure 8.3. It takes a diamond as an initiator. The first iteration is done by adding four copies of diamonds of reduced size with the factor of 2 at the corners of the initiator. The subsequent iterations are done in the same way [24].

The Sierpinski fractal is a widely used fractal shape to design UWB antenna as mentioned in the literature [25–32]. A hexagonal fractal configuration created with imprint of 6 samples of Sierpinski gaskets was proposed. Here, the second-iteration triangular Sierpinski gasket was characterized to realize the full UWB spectral range (3.1–10.6 GHz) [25]. A Sierpinski square fractal antenna was presented in [26] with some modification. It realized the entire UWB band by increased number of iterations in the Sierpinski square fractal geometry and rectangular slot in the ground surface. By incorporating ∩ -slot in the feed surface, a band rejection for WLAN communication (5–6 GHz) is produced.

A UWB antenna using Sierpinski carpet fractal done on circular frame was proposed in [27]. A sharp notched band characteristic was attained by using meander shaped slot. The antenna produces the bandwidth of 9 GHz and has band rejection capability from 5.15 to 5.825 GHz band. A hexagonal frame Sierpinski

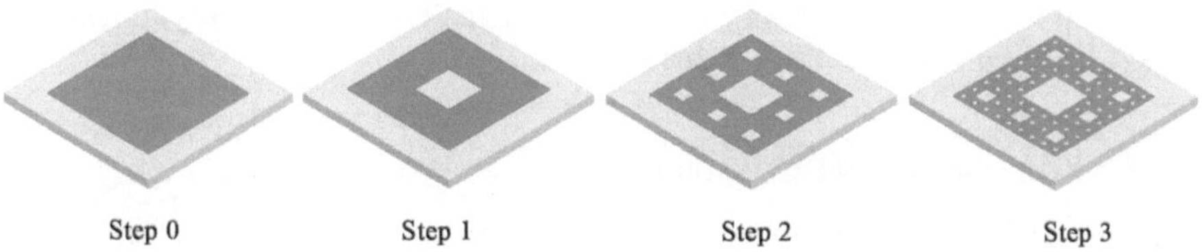

FIGURE 8.2 Sierpinski carpet construction.

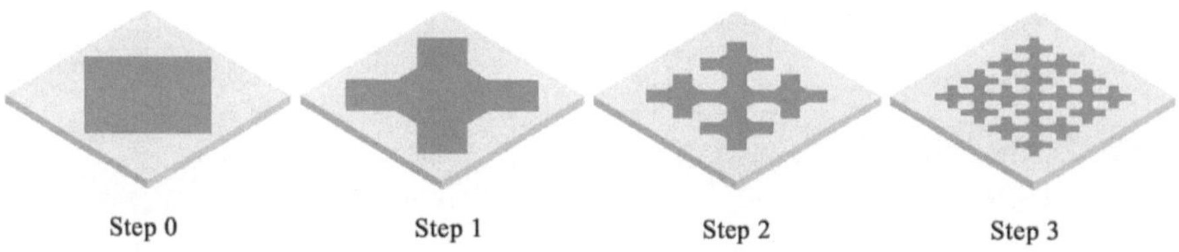

FIGURE 8.3 SierpinskiKnopp fractal construction.

carpet fractal antenna by modifying hexagonal monopole antenna was proposed, which has the capability to reject the bands covering IEEE 802.11a and HIPER-LAN2 by inserting a 'Y' like slot inside the radiating patch surface that extended up to the feed line of CPW feeding. The antenna covers the spectrum from 3 to 12 GHz with notch characteristics from 5.15 to 5.825 GHz range [28].

An octagonal Sierpinski fractal antenna was proposed in [29]. In that, a fractal of 2nd iteration was used as a patch. An asymmetrical CPW feed with truncated ground were used to achieve UWB operation. As a result of etching central slot, it provides ultra-wide band ranges from 4.18 to 18.31 GHz. To bring UWB antenna miniaturization without affecting the antenna performance, a Sierpinski square fractal slot was designed on the decagonal shaped patch. This resulted in the fractional bandwidth of about 124.7% [30].

A two element UWB multiple input multiple output (MIMO) antenna is presented where the miniaturization and broadband characteristics are acquired by the Sierpinski carpet fractal geometry. The author showed that the impedance bandwidth of 9.6 GHz was attained [31]. An UWB MIMO antenna with 4 elements was proposed in which the miniaturization was accomplished with the help of introducing SierpinskiKnopp curve structure on the edges of patches. The antenna achieved bandwidth of 8 GHz. The dual band notch was created by CSRR like element [32].

8.4 KOCH FRACTAL STRUCTURE CHARACTERIZED FOR UWB ANTENNA

8.4.1 Koch Curve Construction

The Koch curve is titled after the Swedish mathematician Helge Von Koch. Von Koch pioneered the Koch curve in 1904. The composition of Koch curve is shown in Figure 8.4. A flat line is subdivided into three portions. The middle third portion is converted into a two edges of a Euclidean triangle. Then, the single line segment is changed into a four-line segment of equal length. The self-similarity dimension for Koch fractal structure Ds is 1.26186 [33,34].

8.4.2 Koch Snowflake Construction

The initiator for both Sierpinski gasket as well as Koch snowflake fractal structures is a Euclidean triangle. However, the generation process differs from each other. In the case of gasket structure, the generation is done by removing scaled down initiators from the initiator in every step. But the Koch fractal generation involves the

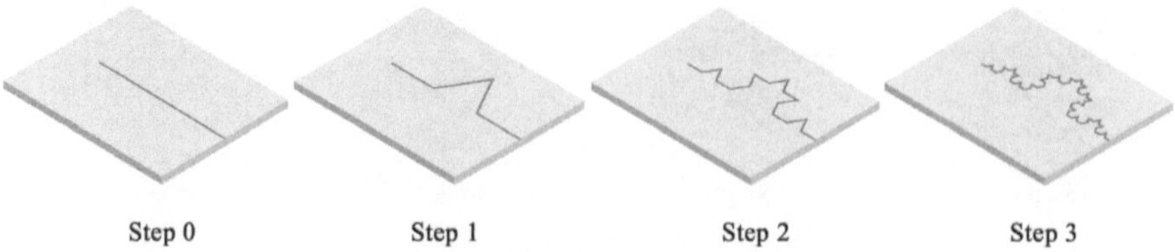

FIGURE 8.4 Koch curve construction.

FIGURE 8.5 Koch snowflake construction.

addition of scaled triangles with the base structure (initiator) in every step [35]. Koch snowflake is a rough boundary with finite area and increasing perimeter as iteration increases. The self-similarity dimension for Koch fractal structure Ds is 1.26186. The construction process of Koch snowflake fractal is shown in Figure 8.5.

The literature reveals that Koch fractal is also popularly used for UWB antenna design [36–39]. The size reduction was done by incorporating a Koch fractal slot in the UWB antenna, which produces bandwidth of 9.15 GHz. It was shown that the frequency notch function was also achieved from 4.65 to 6.40 GHz [36]. The Koch curve introduced on the hexagonal patch as well as ground gave rise to miniaturization and wideband phenomenon in the UWB antenna. The fractional bandwidth achieved by the fractal edges in this work was 122%. The fractal nature of structure produces the stable radiation pattern [37].

By adding a Koch curve and T-shaped stub on the monopole antenna, the bandwidth was obtained from 2 to 11 GHz with three notch band characteristics to reject PCS, WiMAX, and WLAN signals [38]. A planar quasi-self-complementary antenna was designed using a Koch curve boundary, which was embedded on a hexagonal radiating patch. The antenna has achieved frequency bandwidth of 8.8 GHz and band notch characteristics at 5.5 GHz. In addition, this antenna resonates for Bluetooth application at 2.4 GHz [39].

8.5 HILBERT FRACTAL STRUCTURE CHARACTERIZED FOR UWB ANTENNA

8.5.1 Hilbert Curve Construction

The German mathematician David Hilbert introduced an algorithm for Hilbert curve in 1891 [40]. The curve has been utilized as a space-filling structure. In antenna design, the Hilbert curve configuration has been used for reducing the size and producing multiband frequency. The geometry of Hilbert curve is constructed with the procedure shown in Figure 8.6. Every step generated with the four replicas of the previous shape and additional line sections. This shape possesses four basic properties. They are self-avoidance, self-similarity, space-filling, and simplicity [41]. From Figure 8.6, it can be observed that the full length of the Hilbert curve greatly grows with respect to of iteration. However, the overall dimension of the structure is kept unchanged. This is achieved by means of the electrically reducing the size of the structure. The self-similarity dimension for Hilbert curve fractal structure Ds is 2.

The cumulative length of all the line segments is given by (8.8) [34],

FIGURE 8.6 Hilbert curve construction.

$$S = (2^{2n} - 1)d = (2^n + 1)L \tag{8.8}$$

$$d = \frac{L}{(2^n - 1)} \tag{8.9}$$

here,

'L' is the side dimension of the Hilbert-curve.
'd' is the length of each line segment.
'n' represents the order of iteration.

By imprinting Hilbert fractal shape in a UWB antenna, an adequate miniaturization and the band notch characteristic can be achieved. The literature shows that Hilbert fractal structure is extensively used for miniaturization as well as band notch characteristics in UWB antenna design [6,42–46]. By introducing the Hilbert-curve slot, both the miniaturizations as well as the frequency rejection characteristics were attained. The third iteration Hilbert-curve was etched between the center of the patch and feed line to reject a band centered at 5.5 GHz [42].

The etched HCS was positioned near the center of the antenna between a feed line part and a radiating patch. The stop bands pertaining to our frequencies of interest can be tuned by variations of d and w of the HCS.

Multiple number of Hilbert curve slot were incorporated in the radiating patch and in the ground, which exhibits the spectrum from 2.5 to 12 GHz with dual notches. The antenna has an ice-cream cone like monopole [6]. The Hilbert curve is characterized as a radiating patch and a Yagi-shaped portion was etched from it in the proposed work. For optimizing the antenna, a hexagonal slot was taken away from the substrate. The antenna produces a resonance function over the bandwidth of 7.5 GHz [43].

The Hilbert fractal shaped parasitic resonating element incorporated UWB monopole antenna was proposed that exhibits the bandwidth of 10.61 GHz. The resonator was able to produce band notch function for X-band satellite communications [44]. An UWB MIMO antenna was proposed in which quasi-self-complementary monopoles were used to produce an operating band of 8.8 GHz. For getting better isolation in the entire spectral band, a Hilbert curve like slot was removed from the ground surface [45]. Two element UWB, MIMO antenna with triple band rejection function was proposed. The antenna was comprised of three 2nd order Hilbert fractal defected ground structure. On imprinting this Hilbert fractal defects, the antenna gives a peak port to port isolation throughout the UWB spectrum. It showed an increase in port isolation with the increase in the number of fractal defects. The antenna produces bandwidth of 8.52 GHz along with band rejection characteristics to reject C-band, WLAN, and X-band communication signals [46].

8.6 FRACTAL TREE STRUCTURE CHARACTERIZED FOR UWB ANTENNA

8.6.1 Fractal Tree Construction

Fractal tree is like a real tree, in which every branch top will be repeatedly separated into many branches. In binary tree construction, every branch will split into two branches. The self-similarity dimension for fractal tree structure Ds is 1.5849. The Figure 8.7 shows the construction of a fractal tree [47].

The literature shows that fractal binary tree was used in UWB antenna design [48,49]. A fourth iteration fractal binary tree was etched from the antenna structure to obtain two stop band characteristics [48]. A CPW monopole of fractal tree shaped radiating patch was proposed. It has an embedded T-shaped element to produce a band notch characteristic to reject WiMAX and C-band frequency bands [49].

8.7 APOLLONIAN FRACTAL STRUCTURES CHARACTERIZED FOR UWB ANTENNA

8.7.1 Apollonian Fractal Gasket Construction

Apollonian gasket is a fractal created by triple circle, in which every circle is tangent to others. Apollonian gasket is named after Greek mathematician Apollonius of Perga (ca. 262BC–ca.190BC). The construction of Apollonian gasket starts with three mutually tangent unit circles. The scaled-down circle generator, which is tangent to all three circles, etched from the interstice patch formed by the three initial circles. This process is repeated for every newly formed interstice in consecutive iteration steps. These space filling mutually tangent circles are known as Soddy circles. Descartes Circle Theorem (DCT) is used to calculate the radius of Soddy circles. The construction process of 2D Apollonian gasket fractal is shown in Figure 8.8. The self-similarity dimension for Apollonian fractal structure Ds is 1.3058 [50–53].

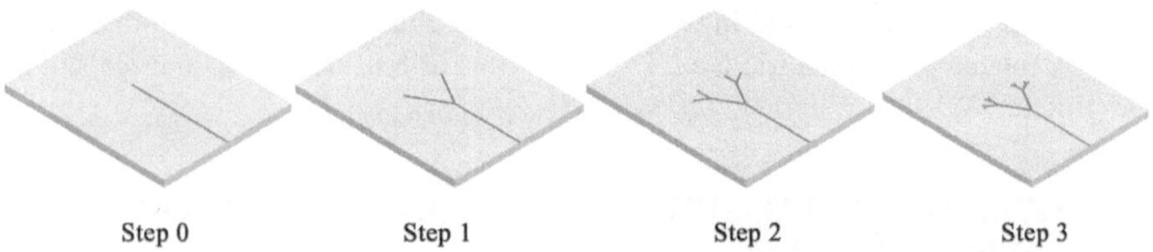

FIGURE 8.7 Fractal tree construction.

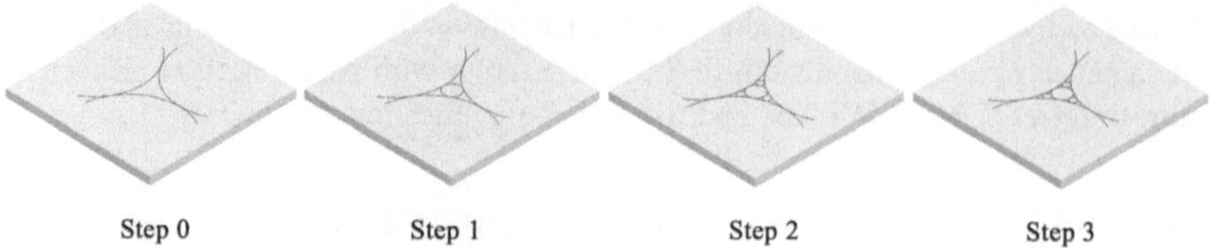

FIGURE 8.8 Apollonian gasket fractal construction.

The literature shows that Apollonian fractal structure was widely used in UWB antenna design [54–57]. A fractal circular antenna by including elliptical iterations named Apollonian was proposed. The antenna design was done by using Descartes Circle Theorem (DCT) [54]. A unique Apollonian UWB fractal antenna with modified ground was presented that exhibited bandwidth of 12.2 GHz [55].

A CPW-fed Apollonian gasket monopole CPW antenna was presented that exhibits the bandwidth ranges from 2.8 to 15 GHz [56]. The Descartes Circle Theorem (DCT) based circular shaped Apollonian fractal UWB antenna was proposed. The antenna achieves band rejection function from 5.125 to 5.825 GHz by means of L-like slots removed from the ground surface. The antenna achieved the bandwidth about 8.8 GHz [57].

8.8 CONVENTIONAL SHAPE FRACTAL MONOPOLE FOR UWB ANTENNA

The literature shows that the conventional shape fractal monopole antenna is also employed in the UWB antenna design [58–62]. A hexagonal monopole antenna is presented that covers UWB band by affixing small fractals at the vertices of the hexagonal patch. A study was carried out for analyzing the different shapes of small fractal elements to achieve the multi resonance characteristics [58]. A CPW-fed hexagonal microstrip fractal antenna was proposed by Kailas. Multi-resonance in the entire UWB frequency range was achieved by the hexagonal fractal integration on the radiating patch. Stepped CPW feed was used to improve the impedance matching [59].

By incorporating a dual-reverse-arrow fractal shape at the edges of a triangular radiating surface, the UWB bandwidth was achieved with notch band from 5 to 6 GHz [60]. A compact circular fractal patch antenna with microstrip feeding was presented in which the operating range is from 2.9–9.5 GHz [61]. A ladder-shaped fractal antenna was proposed in which the antenna achieves the bandwidth of 8.54 GHz by means of asymmetric feed, H-like fractals of four iterations, and the slot at the ground [62].

8.9 HYBRID FRACTAL STRUCTURE CHARACTERIZED FOR UWB ANTENNA

The literature reveals that multiple fractal structures can be utilized to design UWB antennas [12,63–66]. A hybrid fractal of GiusepePeano and Sierpinski Carpet utilized in the monopole for covering UWB frequency band was proposed. The Sierpinski square carpet fractal whose edges are filled with the GiusepePeano fractals of iteration one was characterized as the main radiating patch. The bandwidth achieved by this antenna structure is 14 GHz [63]. A planar elliptical shaped UWB antenna has multi-fractal slots with three notched band (5.5, 6.8, and 8.1 GHz) characteristics was discussed. The bandwidth produced by the antenna is 9.2 GHz. The second order iterated Koch slot on patch surface is contributing the notch band at 5.5 GHz. The rejection at 6.8 GHz was created by adding Sierpinski curve based ring resonator in the rear side. Minkowski slots on ground portion are responsible for the band rejection centered at 8.1 GHz [12].

A modified half annular shaped antenna was presented. Additionally, three fractal shaped slots were introduced individually along the sides of the feed line for interference avoidance characteristics. Three antennas were investigated, which are as follows, one with a half Minkowski fractal, the second and third antennas with a hybrid fractal slot. The third antenna is designed with Sierpinski arrowhead curve fractal slot. The antenna was designed to operate in UWB and Bluetooth frequencies together by inserting a fractal binary tree of order two [64].

The triangular fractals were introduced in cantor set fractal wide slot antenna for UWB applications. The third iteration cantor set fractal structure produces the spectrum from 2.8 to 10.3 GHz. The band notch function exhibited from 5–6.3 GHz [65]. A hybrid Sierpinski Koch fractal structure was characterized as patch of proposed UWB MIMO antenna. Reflecting stub and stepped ground structure were utilized to get better the isolation between antenna elements. The antenna produces the bandwidth of 8.5 GHz in which a notch band characteristics attained by removing a 'U' like slot from patch [66].

8.10 CONCLUSION

Modern wireless communications need high data rate, low power, and compact antenna design for fast multimedia, voice, and data communications. The ultra-wideband (UWB) is one of the prominent technologies that meet the requirements of high-speed communication. The printed UWB antennas are most widely used for easy integration into handheld devices. So miniaturization plays a vital role in UWB antennas design. The fractal type structure makes the design compact by expanding the electrical length of antenna. Self-similar and space-filling are the important properties of fractal geometry, with the fractal structure being able to produce multiband or broadband resonance. Incorporation of fractal structure in printed antennas is one of the most popular techniques utilized in ultra-wideband antennas. Fractal structure can generate broadband resonance for ultra-wideband antenna. In some cases fractal geometry can also be utilized to create band rejection characteristics to mitigate the problem of interference with the conventional wireless applications. They include WLAN, WiMAX, and X-band satellite communication systems. The radiating patch or slots on the radiating patch or slots on the ground surface (defected ground structure) can be characterized with the fractal structure to design the ultra-wideband antenna. This chapter presented the most common fractal geometries and its step-by-step constructional process. This also described the impact of those fractal structures when it is imprinted on the UWB antenna systems. The state-of-the-art fractal based UWB antennas and its performance presented in this chapter will enable the antenna technocrats to speed up their understanding for the design and development of fractal-based UWB antennas and further advancements.

REFERENCES

[1]. Win, M. Z., & Scholtz, R. A. (1998). Impulse radio: How it works. *IEEE Communications Letters*, *2*(2), 36–38.

[2]. Sipal, V., Allen, B., Edwards, D., & Honary, B. (2012). Twenty years of ultra-wideband: Opportunities and challenges. *IET Communications*, *6*(10), 1147–1162.

[3]. FCC. (2002). *First Report and Order on Ultra-Wideband Technology*. Washington, DC.

[4]. Saha, C., Siddiqui, J. Y., &Antar, Y. M. M. (2019). *Multifunctional Ultrawideband Antennas: Trends, Techniques and Applications*. Boca Raton: CRC Press.

[5]. Zarrabi, F. B., Shire, A. M., Rahimi, M., & Gandji, N. P. (2014). Ultra-wideband tapered patch antenna with fractal slots for dual notch application. *Microwave and Optical Technology Letters*, *56*(6), 1344–1348.

[6]. Karmakar, A., Verma, S., Pal, M., & Ghatak, R. (2012). An ultra-wideband monopole antenna with multiple fractal slots with dual band rejection characteristics. *Progress In Electromagnetics Research*, *31*, 185–197.

[7]. Boopathi Rani, R., & Pandey, S. K. (2016). A parasitic hexagonal patch antenna surrounded by same shaped slot for WLAN, UWB applications with notch at vanet frequency band. *Microwave and Optical Technology Letters*, *58*(12), 2996–3000.

[8]. OjaroudiParchin, N., Basherlou, H. J., & Abd-Alhameed, R. A. (2020). UWB Microstrip-fed slot antenna with improved bandwidth and dual notched bands using protruded parasitic strips. *Progress in Electromagnetics Research*, *101*, 261–273.

[9]. Ryu, K. S., & Kishk, A. A. (2009). UWB antenna with single or dual band-notches for lower WLAN band and upper WLAN band. *IEEE Transactions on Antennas and Propagation*, *57*(12), 3942–3950.

[10]. Fertas, K., Ghanem, F., Azrar, A., & Aksas, R. (2020). UWB antenna with sweeping dual notch based on metamaterial SRR fictive rotation. *Microwave and Optical Technology Letters*, *62*(2), 956–963.

[11]. Sanmugasundaram, R., Natarajan, S., & Rajkumar, R. (2020). Ultrawideband notch antenna with EBG structures for WiMAX and satellite application. *Progress in Electromagnetics Research*, *91*, 25–32.

[12]. Gorai, A., Karmakar, A., Pal, M., & Ghatak, R. (2013). Multiple fractal-shaped slots-based UWB antenna with triple-band notch functionality. *Journal of Electromagnetic Waves and Applications*, *27*(18), 2407–2415.

[13]. Mandelbort, B. B. (1977). *Fractals: Form, Chance and Dimension*. M. San Francisco: Freeman.

[14]. Peitgen, H. O., Jürgens, H., & Saupe, D. (2006). *Chaos and Fractals: New Frontiers of Science*. New York: Springer Science & Business Media.

[15]. Ghosh, B., Sinha, S. N., & Kartikeyan, M. V. (2014). *Fractal Apertures in Waveguides, Conducting Screens and Cavities. Springer Series in Optical Sciences*, vol. 187. New York: Springer

[16]. Kumar, G., & Ray, K. P. (2003). *Broadband Microstrip Antennas*. Boston: Artech House.

[17]. Ghatak, R., Biswas, B., Karmakar, A., & Poddar, D. R. (2013). A circular fractal UWB antenna based on descartes circle theorem with band rejection capability. *Progress In Electromagnetics Research*, *37*, 235–248

[18]. Asokan, H., & Gopalakrishnan, S. (2018). A Miniaturized inductive–Loaded narrow strip wide band-notched ultra-wideband monopole antenna with dual-mode resonator. *AEU-International Journal of Electronics and Communications*, *86*, 125–132.

[19]. Ranjan, P., Raj, S., Upadhyay, G., Tripathi, S., & Tripathi, V. S. (2017). Circularly slotted flower shaped UWB filtering antenna with high peak gain performance. *AEU-International Journal of Electronics and Communications*, *81*, 209–217.

[20]. Jaggard, D. L. (1990). On fractal electrodynamics. In *Recent Advances in Electromagnetic Theory* (pp. 183–224). Springer, New York, NY.

[21]. Rahim, M. K. A., Aziz, M. A., & Abdullah, N. (2005, December). Wideband Sierpinski carpet monopole antenna. In *2005 Asia-Pacific Conference on Applied Electromagnetics* (p. 4). New York: IEEE.

[22]. D.H. Werner and R. Mittra, (2000). *Frontiers in Electromagnetics* (pp. 1–787). New York: IEEE Press.

[23]. Muthu Ramya C, Rani, R. B. (2020, February). A compendious review on fractal antenna geometries in wireless communication. In *2020 International Conference on Inventive Computation Technologies (ICICT)* (pp. 888–893). New York: IEEE.

[24]. Reha, A., El Amri, A., & Bouchouirbat, M. (2018). The behavior of CPW-Fed Sierpinski curve fractal antenna. *Journal of Microwaves, Optoelectronics and Electromagnetic Applications*, *17*(3), 366–372.

[25]. Kaka, A. A., & Toycan, M. (2016). Modified hexagonal Sierpinski gasket-based antenna design with multiband and miniaturized characteristics for UWB wireless communication. *Turkish Journal of Electrical Engineering & Computer Sciences*, *24*(2), 464–473.

[26]. Choukiker, Y. K., & Behera, S. K. (2014). Modified Sierpinski square fractal antenna covering ultra-wide band application with band notch characteristics. *IET Microwaves, Antennas & Propagation*, *8*(7), 506–512.

[27]. Ghatak, R., Karmakar, A., & Poddar, D. R. (2011). A circular shaped Sierpinski carpet fractal UWB monopole antenna with band rejection capability. *Progress in Electromagnetics Research*, *24*, 221–234.

[28]. Ghatak, R., Karmakar, A., & Poddar, D. R. (2013). Hexagonal boundary Sierpinski carpet fractal shaped compact ultrawideband antenna with band rejection functionality. *AEU-International Journal of Electronics and Communications*, *67*(3), 250–255.

[29]. Singhal, S., Singh, P., & Kumar Singh, A. (2016). Asymmetrically CPW-fed octagonal sierpinski UWB fractal antenna. *Microwave and Optical Technology Letters*, *58*(7), 1738–1745.

[30]. Ali, T., Bk, S., & Biradar, R. C. (2018). A miniaturized decagonal Sierpinski UWB fractal antenna. *Progress in Electromagnetics Research*, *84*, 161–174.

[31]. Gurjar, R., Upadhyay, D. K., Kanaujia, B. K., & Kumar, A. (2020). A compact modified sierpinski carpet fractal UWB MIMO antenna with square-shaped funnel-like ground stub. *AEU-International Journal of Electronics and Communications*, *117*, 153126.

[32]. Rajkumar, S., Selvan, K. T., & Rao, P. H. (2018). Compact 4 element Sierpinski Knopp fractal UWB MIMO antenna with dual band notch. *Microwave and Optical Technology Letters*, *60*(4), 1023–1030.

[33]. Von Koch, Helge (1904). Sur unecourbe continue sans tangente, obtenue par une construction géométriqueélémentaire". *ArkivförMatematik (in French)*. *1*, 681–704.

[34]. Sagan, H. (2012). *Space-Filling Curves*. New York: Springer Science & Business Media.

[35]. Borja, C., & Romeu, J. (2003). On the behavior of Koch island fractal boundary microstrip patch antenna. *IEEE Transactions on Antennas and Propagation*, *51*(6), 1281–1291.

[36]. Lui, W. J., Cheng, C. H., & Zhu, H. B. (2006). Compact frequency notched ultra-wideband fractal printed slot antenna. *IEEE Microwave and Wireless Components Letters*, *16*(4), 224–226.

[37]. Tripathi, S., Mohan, A., & Yadav, S. (2014). Hexagonal fractal ultra-wideband antenna using Koch geometry with bandwidth enhancement. *IET Microwaves, Antennas & Propagation*, *8*(15), 1445–1450.

[38]. Zarrabi, F. B., Mansouri, Z., Gandji, N. P., & Kuhestani, H. (2016). Triple-notch UWB monopole antenna with fractal Koch and T-shaped stub. *AEU-International Journal of Electronics and Communications*, *70*(1), 64–69.

[39]. Gorai, A., Pal, M., & Ghatak, R. (2017). A Compact fractal-shaped antenna for ultrawideband and bluetooth wireless systems with WLAN rejection functionality. *IEEE Antennas and Wireless Propagation Letters*, *16*, 2163–2166.

[40]. Hilbert, D. (1891). Ueber die stetigeAbbildungeiner Line auf einlächenstück. *MathematischeAnnalen*, *38*(3), 459–460.

[41]. Peitgen, H. O., Henriques, J. M., & Penedo, L. F. (Eds.). (1991). Fractals in the fundamental and applied sciences: Proceedings of the first. In *IFIP Conference on Fractals in the Fundamental and Applied Sciences, Lisbon, Portugal*, 6–8 June, 1990. North Holland.

[42]. Kim, D. O., Kim, C. Y., Park, J. K., & Jo, N. I. (2011). Compact band notched ultra-wideband antenna using the Hilbert-curve slot. *Microwave and Optical Technology Letters*, *53*(11), 2642–2648.

[43]. Kaka, A. O., Toycan, M., Bashiry, V., & Walker, S. D. (2012). Modified Hilbert fractal geometry, multi-service, miniaturized patch antenna for UWB wireless communication. *COMPEL-The International Journal for Computation and Mathematics in Electrical and Electronic Engineering,13*. 10.1108/03321641211267146

[44]. Banerjee, J., Karmakar, A., & Ghatak, R. (2015, December). An ultra wideband monopole antenna using fractal shaped parasitic resonator for band notching characteristics. In *2015 International Conference on Microwave and Photonics (ICMAP)* (pp. 1–2). New York: IEEE.

[45]. Gorai, A., Dasgupta, A., & Ghatak, R. (2018). A compact quasi-self-complementary dual band notched UWB MIMO antenna with enhanced isolation using Hilbert fractal slot. *AEU-International Journal of Electronics and Communications*, *94*, 36–41.

[46]. Banerjee, J., Gorai, A., & Ghatak, R. (2020). Design and analysis of a compact UWB MIMO antenna incorporating fractal inspired isolation improvement and band rejection structures. *AEU-International Journal of Electronics and Communications*, *122*, 153274.

[47]. Gianvittorio, J. P., & Rahmat-Samii, Y. (2002). Fractal antennas: A novel antenna miniaturization technique, and applications. *IEEE Antennas and Propagation magazine*, *44*(1), 20–36.

[48]. Jahromi, M. N., Falahati, A., & Edwards, R. M. (2011). Application of fractal binary tree slot to design and construct a dual band-notch CPW-ground-fed ultra-wide band antenna. *IET Microwaves, Antennas & Propagation*, *5*(12), 1424–1430.

[49]. Naser-Moghadasi, M., Sadeghzadeh, R. A., Sedghi, T., Aribi, T., & Virdee, B. S. (2013). UWB CPW-fed fractal patch antenna with band-notched function

employing folded T-shaped element. *IEEE Antennas and Wireless Propagation Letters*, *12*, 504–507.
[50]. Bourke, P. (2006). An introduction to the Apollonian fractal. *Computers & Graphics*, *30*(1), 134–136.
[51]. Andrade Jr, J. S., Herrmann, H. J., Andrade, R. F., & Da Silva, L. R. (2005). Apollonian networks: Simultaneously scale-free, small world, Euclidean, space filling, and with matching graphs. *Physical Review Letters*, *94*(1), 018702.
[52]. Liu, J. C., Wu, C. Y., Chang, D. C., & Liu, C. Y. (2006). Relationship between Sierpinski gasket and Apollonian packing monopole antennas. *Electronics Letters*, *42*(15), 847–848.
[53]. Doye, J. P., & Massen, C. P. (2008). Self-similar disk packings as model spatial scale-free networks. *Physical Review E*, *71*(1), 016128.
[54]. Khan, S. N., Hu, J., Xiong, J., & He, S. (2008). Circular fractal monopole antenna for low VSWR UWB applications. *Progress in Electromagnetics Research*, *1*, 19–25.
[55]. Kumar, R., & Nikam, P. B. (2012). A modified ground apollonian ultra wideband fractal antenna and its backscattering. *AEU-International Journal of Electronics and Communications*, *66*(8), 647–654.
[56]. Kumar, R., & Srikanth, I. (2012). Design of apollonian gasket ultrawideband antenna with modified ground plane. *Microwave and Optical Technology Letters*, *54*(8), 1793–1796.
[57]. Ghatak, R., Biswas, B., Karmakar, A., & Poddar, D. R. (2013). A circular fractal UWB antenna based on descartes circle theorem with band rejection capability. *Progress in Electromagnetics Research*, *37*, 235–248.
[58]. Fallahi, H., & Atlasbaf, Z. (2013). Study of a class of UWB CPW-fed monopole antenna with fractal elements. *IEEE Antennas and Wireless Propagation Letters*, *12*, 1484–1487.
[59]. Sawant, K. K., & Kumar, C. S. (2015). CPW fed hexagonal micro strip fractal antenna for UWB wireless communications. *AEU-International Journal of Electronics and Communications*, *69*(1), 31–38.
[60]. Orazi, H., & Soleiman, H. (2016). Miniaturisation of UWB triangular slot antenna by the use of DRAF. *IET Microwaves, Antennas & Propagation*, *11*(4), 450–456.
[61]. Gupta, M., & Mathur, V. (2017). Wheel shaped modified fractal antenna realization for wireless communications. *AEU-International Journal of Electronics and Communications*, *79*, 257–266.
[62]. Singhal, S., & Singh, A. K. (2017). Asymmetrically CPW-fed ladder-shaped fractal antenna for UWB applications. *Analog Integrated Circuits and Signal Processing*, *92*(1), 91–101.
[63]. Oraizi, H., & Hedayati, S. (2011). Miniaturized UWB monopole microstrip antenna design by the combination of GiusepePeano and Sierpinski carpet fractals. *IEEE Antennas and Wireless Propagation Letters*, *10*, 67–70.
[64]. Biswas, B., Ghatak, R., & Poddar, D. R. (2015). UWB monopole antenna with multiple fractal slots for band-notch characteristic and integrated Bluetooth functionality. *Journal of Electromagnetic Waves and Applications*, *29*(12), 1593–1609.
[65]. Terlapu, S. K., Chowdary, P. S. R., Jaya, C., Chakravarthy, V. S., & Satpathy, S. C. (2018). On the design of fractal UWB wide-slot antenna with notch band characteristics. In *Microelectronics, Electromagnetics and Telecommunications* (pp. 907–912). Singapore: Springer.
[66]. Sampath, R., & Selvan, K. T. (2020). Compact hybrid Sierpinski Koch fractal UWB MIMO antenna with pattern diversity. *International Journal of RF and Microwave Computer-Aided Engineering*, *30*(1), e22017.

9 Advanced Microstrip Antennas for Vehicular Communication

Dr. B. T. P. Madhav M.Tech., Ph.D[1] *and*
T. Anilkumar M.Tech[2]
[1]Professor & Associate Dean (Research), Antennas & Liquid Crystals Research Center, Department of ECE, K L University, Vaddeswaram, AP, India
[2]Assistant Professor, Department of ECE, Lendi Institute of Engineering and Technology, Vizianagaram, AP, India

9.1 INTRODUCTION TO VEHICULAR COMMUNICATION

From the previous three decades, microstrip patch antennas (MPA) have been developed and researched widely. These microstrip patch antennas are a class of planar antennas. These antennas have been utilized in different applications among wireless communication systems including the commercial sector, military sector, and vehicular communication. The main goal of this chapter is to present the theory, design, and analysis of antennas used for vehicular communication. For the past six years, the author has been working on the research and development of vehicular communication MPAs, and this book also includes the partial work record of the authors' personal journey in this particular field. An outstanding portion of the material is taken from their own work in the last four and a half years.

These days, vehicular communication has increased a lot of consideration because of emerging technologies in mobile networks, which guides to initiate intelligent transportation systems (ITS) for giving active safety along with entertainment related modules. The applications (for example, hazardous location vehicle-to-vehicle [V2V] notifications and general mobile cellular communications along with the emergency braking, etc.) increase the consistency and coherence of transportation systems. Access to the web in vehicle improves amusement and security for travelers [1]. Wireless communication standards such as GSM, WLAN IEEE802.11a/b/g/n, PCS, WiMAX, UMTS, Dedicated Short-Range Communications (DSRC) IEEE802.11p have been used for providing such capabilities. To access these facilities in vehicles, it ought to be equipped with effective electronic and communication hardware and in which the antenna plays a major role for transmitting and receiving the signals. Such antennas with low profile and mounted at lower elevations on the vehicle are constantly attractive with great radiative capability.

DOI: 10.1201/9781003187325-9

The best solution for these cases is a multiband antenna, which can handle the mentioned vehicular applications. Numerous planar antenna design techniques have been developed in the previous literature such as adding the stubs and creating the slots in radiating structures to create resonant modes, etc. The idea of fractals has been playing a major role from past few years in the antenna design field. It is having the capacity of creating multiband performance with their self-comparable and space-filling properties also given as an object generated recursively with fractional dimension [2,3]. In past days, the vehicular applications were found in UHF and VHF ranges and the supposed aerials usually possess big element sizes, which were intended for 10–30 MHz band for military based vehicular applications [4] and a tri-band antenna (TIFA) operating at 27 MHz, 49 MHz, and 53 MHz separately [5]. The notable thumb rule is that the antenna size reduces with increase in the operating frequency. In this way, the vehicular communication applications gradually moved to higher frequencies and structured R-shaped PIFA for 225 MHz and 450 MHz group bands [6].

9.2 APPLICATION BANDS USED IN VEHICULAR COMMUNICATION

In the electromagnetic spectrum, the order from radio to gamma rays occupied at different wavelengths and applications also scattered. It is the range of different types of the electromagnetic radiation, i.e., the energy that goes and spreads out as it goes. The electromagnetic radiation can be expressed in terms of energy, frequency, and wavelength. The radio part of the EM spectrum goes in the range of 1 cm to 1 km, which is 300 Kilo Hz to 30 GHz.

Antenna is the device that will transmit and/or receive the electromagnetic waves. Antennas are being used in the modern vehicles for getting multiple applications. The antennas for vehicular communication platforms will fall under one or many of the following applications (Table 9.1).

To get these attributes in present-day vehicles, it ought to be furnished with productive electronic and communication hardware and in which the antenna plays a major role in transmitting and receiving the signals. Such antennas with low-profile and mounted at lower elevations on the vehicle are constantly attractive with great radiation execution. These antennas should provide good impedance bandwidth and radiation performance characteristics to cater to the services in vehicular communication. Moreover, the antenna should be adaptable to different practical scenarios and flexible to the dynamic system, which can help reduce the frequency filtering and leads to efficient utilization of the electromagnetic spectrum. Frequency reconfigurability will provide a solution to such kind of problems with optimum performance characteristics.

There are tremendous ongoing research attempts happening in flexible antenna technology because these antennas can enable communications in surfaces with curve shape besides the traditional type of rigid antennas, as well as exhibiting low mass density, wide adaptability, small volume, low volume, low cost, and lightweight. A considerable attempt is being carried out on the novel antenna configurations exhibiting tunable bandwidth, agile operating frequencies, reconfigurable radiation pattern, and switchable polarization.

TABLE 9.1
Some Applications Related to Vehicular Communications

Application		Operating Band
GSM-850		824.2–848.8 MHz (Uplink), 869.2–893.8 MHz (Downlink)
GSM-900	E-GSM-900 (Extended GSM-900	880.0–915.0 MHz (Uplink), 925.0–960.0 MHz (Downlink)
	R-GSM-900 or GSM-R (Railways GSM-900)	876.0–915.0 MHz (Uplink), 921.0–960.0 MHz (Downlink)
	T-GSM-900 (Trunking-GSM)	870.4–876.0 MHz (Uplink), 915.4–921.0 MHz (Downlink)
GPS	L5 Band	1164–1189 MHz, centered at 1176.45 MHz
	L2 Band	1215–1239.6 MHz, centered at 1227.6 MHz
	L1 Band	1563–1587 MHz, centered at 1575.42 MHz
GLONASS GNSS	G3 Band	Centered at 1201 MHz
	G2 Band	1242.9375–1248.625 MHz
	G1 Band	1589.0625–1605.375 MHz
GALILEO GNSS	E5 Band	1164–1214 MHz
	E6 Band	1260–1300 MHz
	E1 Band	1559–1591 MHz
PCS band		1.68–2.68 GHz
GSM-1800/DCS 1800		1710–1785 MHz (Uplink), 1805–1880 MHz (Downlink)
GSM-1900/PCS 1900		1850–1910 MHz (Uplink), 1930–1990 MHz (Downlink)
LTE33-37		1.9–2.025 GHz
UMTS		1920–2170 MHz
Mobile satellite service E2S Reverse link		1980–2200 MHz
LTE 2300		2305–2400 MHz
ISM2.4G/WLAN/ISM/Bluetooth, GPS,		2.4–2.4835 GHz
LTE 2500		2500–2690 MHz
WiMAX		3400–3600 MHz
LTE-V2X		3.4–3.8 GHz
WLAN 802.11a/n		5.15–5.875 GHz
DSRC, V2V V2I 802.11p		5.85–5.925 GHz
ITS-G5		5.9 GHz
5G		26–60 GHz

9.3 DESIGNED MODELS FOR VARIOUS VEHICULAR COMMUNICATION APPLICATIONS

Various antenna models for vehicular communication applications have been designed at Antennas and Liquid Crystals Research Center of K L University. The designed models are simulated using Ansys High-Frequency Structure Simulator (HFSS) and CST Microwave Studio. Parametric analysis of the dimensional characteristics and the optimization of the performance characteristics also carried through these tools before the fabrication of the proposed models. All the designed models are tested physically with a combinational analyzer in the anechoic chamber for validation.

9.3.1 Design of a Planar Wheel-Shaped Fractal Antenna for Roof-Top Vehicular Applications

A planar wheel-shaped circular microstrip patch antenna is designed on the FR4 substrate. Initially, the analysis is done with the characteristic parameters of the antenna when the ground plane is partially etched. The patch is etched with the slots in an iterative procedure and has created the overall design looking like a wheel-shaped structure. This was particularly done to impart the additional resonant modes to the antenna and its behavior is analyzed. The impedance bandwidth effects are examined with the semi/quarter-circular defected ground structures. The dimensions of the slot parameters are parametrically tuned to get optimum performance. Further, the antenna is practically characterized by manufacturing the model and tried utilizing with the Anritsu MS2037C combinational analyzer antenna when putting over a vehicular body utilizing the co-simulation tool ANSYS Savant along with HFSS.

The circular patch is taken as the radiating element and initially the dimension is calculated using Eqs. (9.1) and (9.2).

$$R_1 = \frac{F}{\sqrt{1 + \frac{2h}{\pi \varepsilon_r F}\left[\ln \frac{\pi F}{2h} + 1.7726\right]}}, \tag{9.1}$$

$$F = \frac{8.791 \times 10^9}{f_r \sqrt{\varepsilon_r}}, \tag{9.2}$$

where 'f_r' is the antenna resonant frequency, 'ε_r' is the dielectric material relative permittivity, and 'h' is the dielectric layer thickness. The patch of the circular shape is excited by microstrip line feed with a width of 'W_f' = 2.2 mm having 50 ohms impedance. The ground plane is moderately layered on the substrate rear. The scale factors used in making the tree-fractal slots are computed from Eq. (9.3).

$$\mathrm{SF} = \frac{a_n}{a_{n-1}}, \tag{9.3}$$

where 'SF' = scale factor, 'a_n' = radius of nth iteration of the circular slot. Figure 9.1 shows the antenna model in HFSS, prototyped model on FR4, and reflection coefficient S_{11} result in dB.

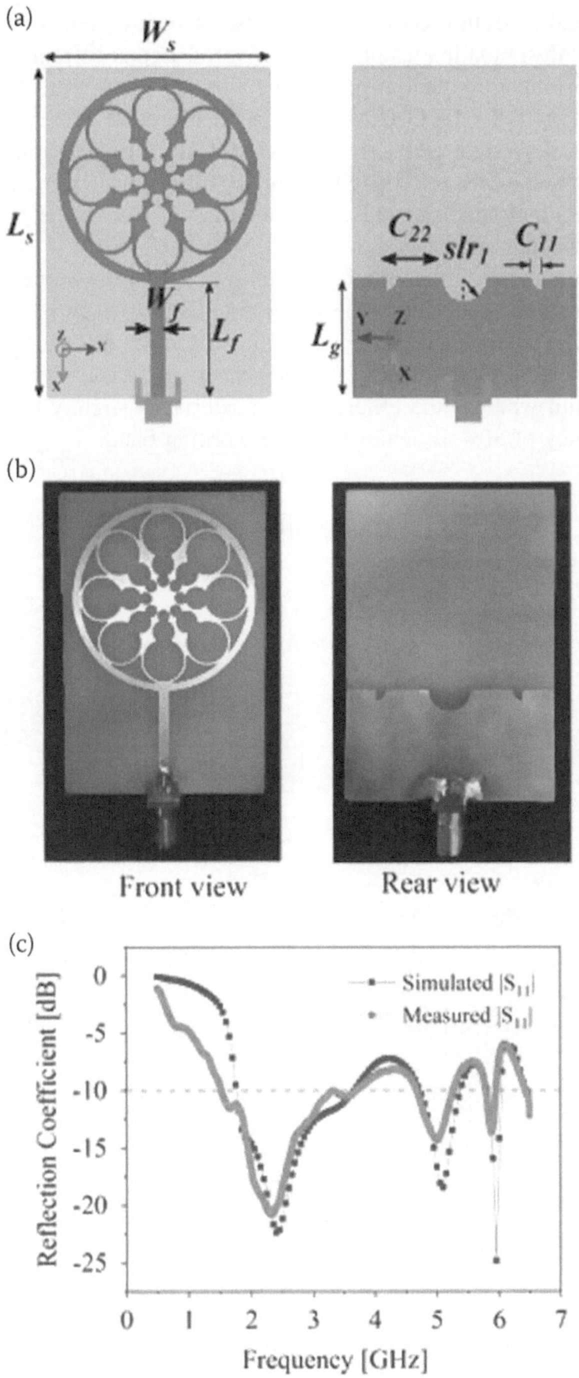

FIGURE 9.1 Wheel shape antenna [9], (a) simulation model, (b) fabricated model, (c) S_{11} parameter.

The calculated reflection coefficient characteristics are shown in Figure 9.2(a) and (b) along with simulation results and discovered a great correlation between the outcomes. The simulation outcome shows the working bands from 1.76–3.66 GHz, 4.72–5.38 GHz and 5.8–6.04 GHz, whereas the measured characteristics work from 1.51–3.69 GHz, 4.67–5.25 GHz, and 5.78–5.96 GHz. The determination of operating bands is based on the criteria that $|S_{11}|<-10$ dB. The variation in the reflection coefficient levels and smaller band deviations may be due to the fabrication conditions and losses in connectors.

The peak gain and radiation efficiency performance in the frequency domain of the proposed antenna is considerable in operating bands, which has been shown in Figure 9.2(c). The antenna gain improves from 1 dB to 4.3 dB in 1.5–3.6 GHz band with an improvement in frequency. At resonances of 5 GHz, 5.9 GHz bands the gain is maximum whereas the efficiency characteristics slightly decrease and the average efficiency of 80% is achieved in the working band.

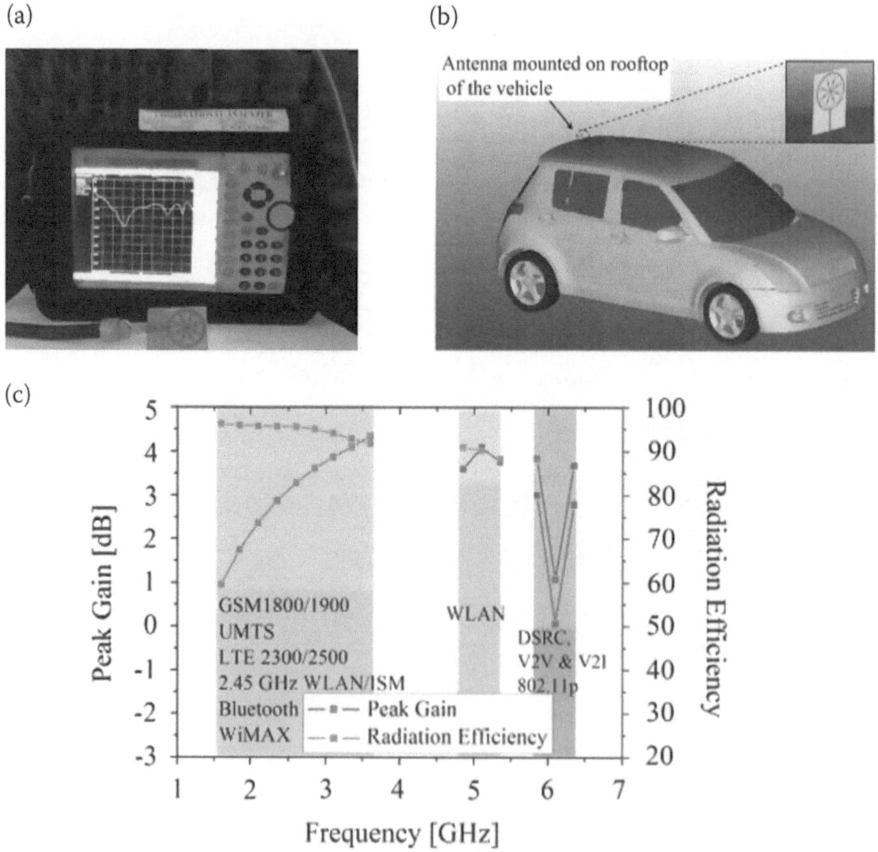

FIGURE 9.2 (a) S-parameter measurement setup using VNA, (b) roof top mounting application on a vehicle body, and (c) simulated peak gain and radiation efficiency vs frequency characteristics of the proposed antenna.

The obtained working bands and radiation performance shows that the antenna is well suited for potential vehicular applications such as ISM applications, UMTS, GSM 1800/1900, LTE 2300/2500, Bluetooth, WLAN 2.4 GHz/5GHz including the most notable DSRC band for V2V and V2I communications. The comparative performance characteristics of the antenna for its iterations are shown in Table 9.2.

The various parts of the vehicular design are allocated with various materials with their individual dielectric properties. The metallic surfaces are designed as a perfect electric conductor (PEC). When the antenna goes to the isolation state, the actual radiation pattern generated will be affected by the vehicular body. Dipole like radiation patterns is observed in the ideal case in the *XZ*-plane and *YZ*-plane with respect to the coordinate system of the car body. At 1.8 GHz, 2.4 GHz, and 3.5 GHz frequencies, dumbbells like elevation plane patterns are observed in azimuth plane and the patterns are omnidirectional in nature.

Four lobes patterns are observed at 5.25 GHz, 5.9 GHz frequencies in the XY-plane. Dipole like pattern with slight deviation has been observed in *YZ*-plane and omnidirectional patterns for Azimuth plane patterns are observed in *XY*-plane. The electrically large vehicular body changes the far-field characteristics. The region of maximum radiation lies between 30° to 60° in the elevation plane at 1.8 GHz, 2.4 GHz, 3.5 GHz, and with increasing the operating frequencies, this elevation plane coverage extends up to 80°. The fields that exist beneath the antenna feed portion (–45° to –90°) get a bounce back from the large PEC vehicular body. This contributing constructive pattern directs field patterns such that it covers 30°–80° elevation range.

It also provides extra beams, which can be shown in Table 9.3. The azimuth plane patterns of the installed antenna are in acceptable correlation with isolated

TABLE 9.2
Comparative Characteristics of Antenna Iterations

Antenna Itetration	Operating Bands [GHz]	Fractional Bandwidth (%)	Resonant Frequencies	S_{11} (dB)	Gain		Radiation Efficiency	
					Max	Avg	Max	Avg
Iteration-1 (without DGS)	2.06–2.66	12.711	2.3 GHz	−16.2981	3.3227	2.9002	95.93	95.68
	3.56–5.42	20.712	4.8 GHz	−24.2615	5.0690	4.4762	92.30	90.70
Iteration-2 (without DGS)	2.03–2.65	13.247	2.3 GHz	−17.7249	3.3227	2.9002	95.98	95.60
	3.7–5.33	18.05	4.65 GHz	−22.4617	5.0690	4.6150	92.29	91.61
	5.73–6.26	4.42	6.05 GHz	−12.2359	4.0923	4.0564	89.59	89.40
Iteration-3 (without DGS)	1.89–2.39	11.68	2.1 GHz	−37.2568	2.873	2.622	95.88	95.76
	3.68–5.19	17.02	4.65 GHz	−22.4817	5.1306	4.5947	91.96	90.98
	5.56–5.90	2.96	5.75 GHz	−15.1608	4.2108	3.7719	86.65	80.85
Proposed Antenna with DGS	1.76–3.66	35.05	2.4 GHz	−22.3793	4.34	3.306	95.95	94.65
	4.72–5.38	6.53	5.1 GHz	−18.4588	4.09	3.888	90.93	89.96
	5.80–6.04	2.02	5.95 GHz	−24.8515	3.843	3.843	80.08	80.08

performance. The visualized three-dimensional patterns are shown in Table 9.3(b), which represents the orientation of radiation pattern such that two dipole-like patterns are stacked with each other. However, the direction of maximum radiation can be shown in the horizontal planes (i.e., ±10°) and in 30°–80° elevation angles and yields ≥3 dB gain so that the provision of vehicle-to-vehicle (V2V), as well as vehicle-to-infrastructure (V2I) interactions, will be feasible.

9.3.2 Transparent and Conformal Wheel-Shaped Fractal Antenna for Vehicular Communication Applications

The prototyped antenna is manufactured on a transparent and flexible polyvinyl chloride material of size 55 mm × 40 mm × 3 mm. The proposed antenna is considered to work in Personal Communication Service (PCS-1900), GSM-1800/1900, Universal Mobile Telecommunications System (UMTS), Digital Communication System (DCS- 1800), Long-Term Evolution (LTE 2600), Scientific and Medical radio band (ISM 2. 4G), World Interoperability for Microwave Access (Wi-MAX), Industrial, Bluetooth, Wireless local area network (WLAN), IEEE802.11p protocol-based Vehicle-to-everything, Wireless Access in Vehicular Environments (WAVE), and Dedicated short-range communications (DSRC) communications bands.

The substrate is a significant element in the microstrip patch antennas that effectively holds the sensitive metallic layers and thus provides physical help and strength to the antenna. In this prototype, a thermoplastic polymer material named clear polyvinyl chloride (PVC) is utilized as a substrate in the flexible structure. Few plasticizers doped in this material to impart the softness and transparent color with a light blue tint that possesses a refractive index ranging from 1.54–1.56. Its effective properties such as corrosion-resistant, water-resistant, chemically inertness, and thermal and electrical insulation, weather-resistant, and self-extinguish toward flammability are affected to utilize this material as the substrate in the antenna fabrication.

The 3D electromagnetic investigation of the proposed antenna is carried out using the ANSYS HFSS simulation tool, which utilizes the finite element method (FEM) for investigating the antenna geometry and obtaining the fundamental characteristics such as the reflection coefficient and radiation patterns. The platform mounting analysis of the antenna is performed in the simulation tool ANSYS Savant, which is dependent on an asymptotic (high frequency) method of Shooting and Bouncing Rays (SBR).

The circular ring element in the iteration-1 with inner radius 'R_2' and outer radius 'R_1' placed on a material of relative permittivity 'ε_r' and the effective relative permittivity 'ε_e' is given by [8] is

$$\varepsilon_e = 0.5(\varepsilon_r + 1) + 0.5(\varepsilon_r - 1)\sqrt{1 + \frac{10t}{W}}, \tag{9.4}$$

where

TABLE 9.3(A)
Far-Field Performance of Installed Antenna on a Vehicular Body in Comparison with Isolated Performance

XZ-Plane	YZ-Plane	XY-Plane
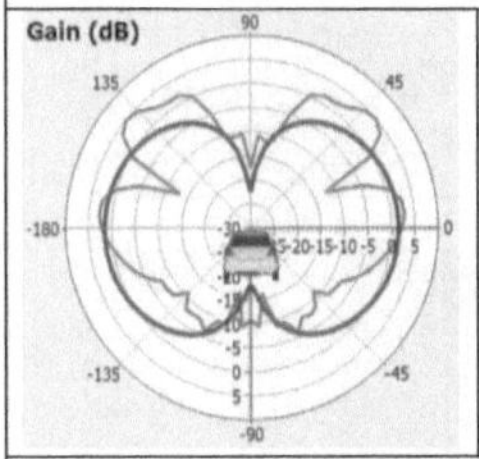	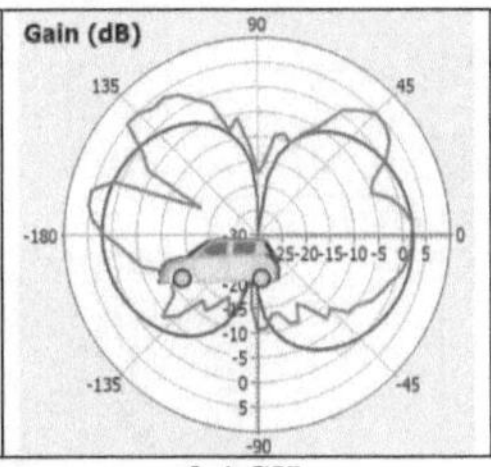	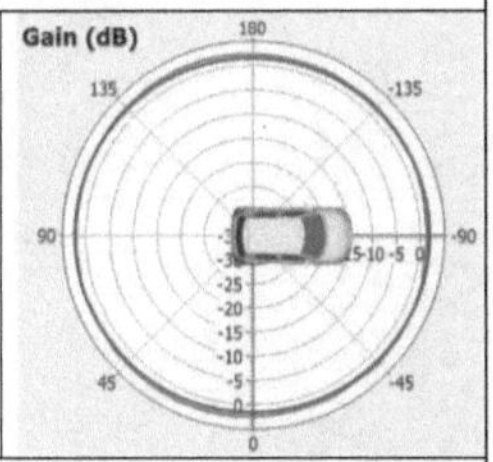
2.4 GHz		
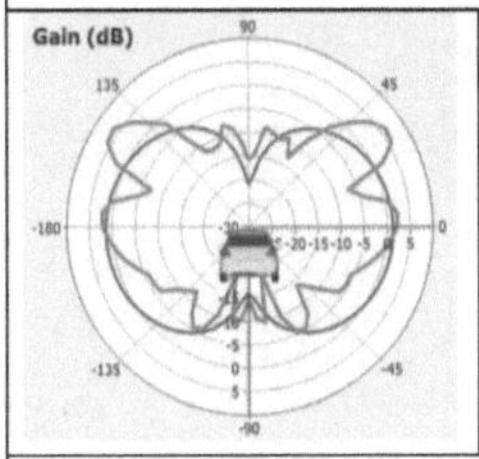	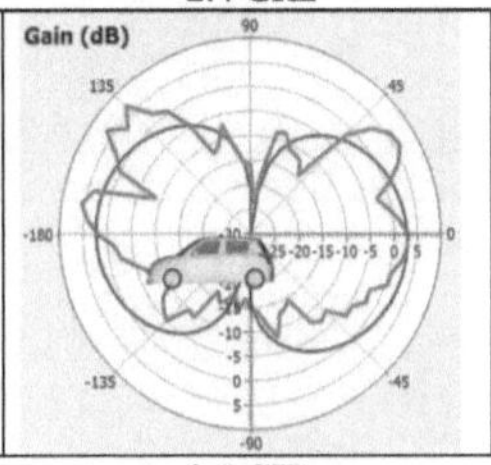	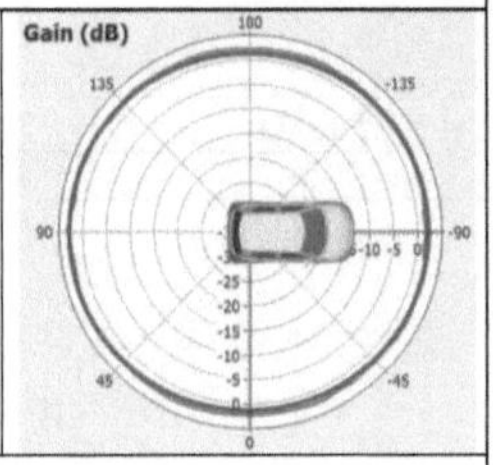
3.5 GHz		
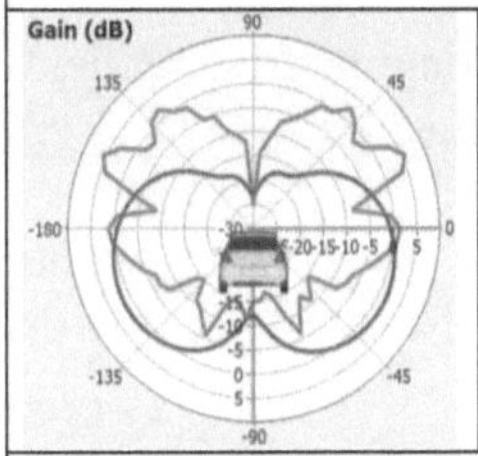	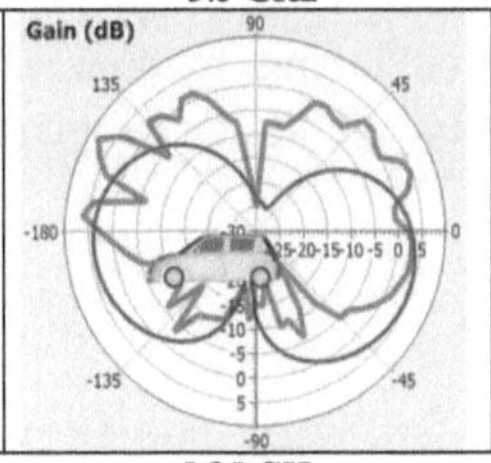	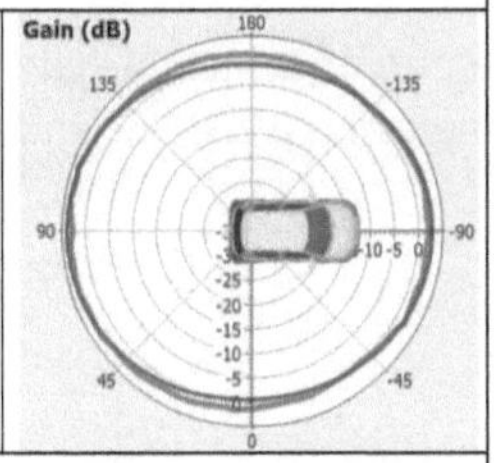
5.25 GHz		
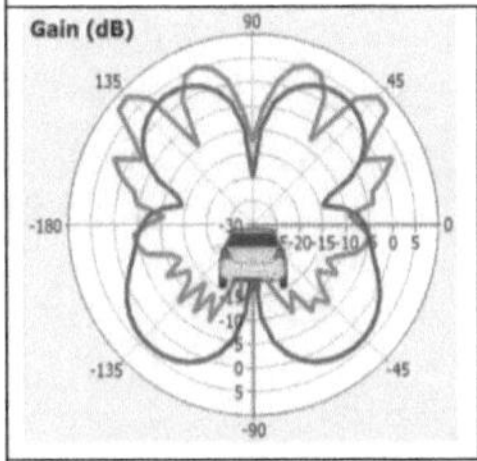	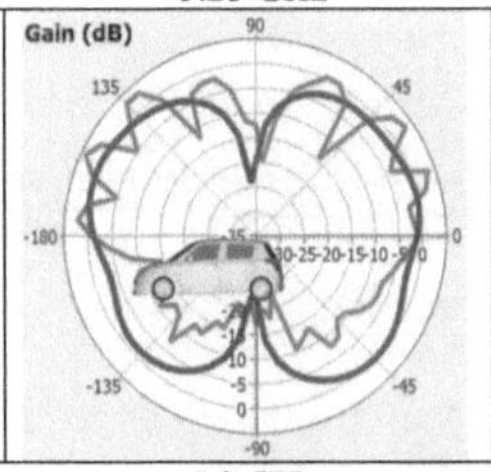	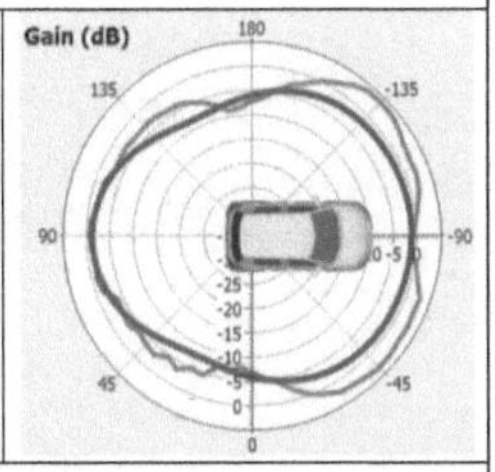
5.9 GHz		
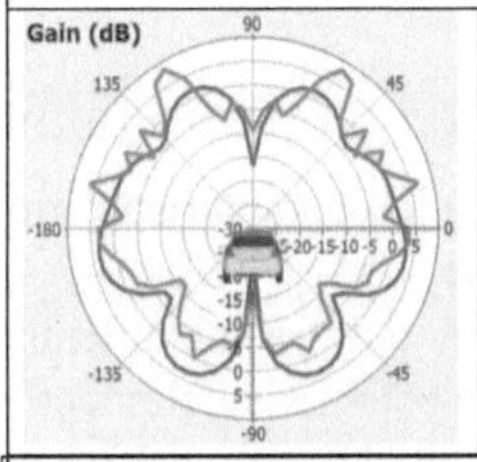	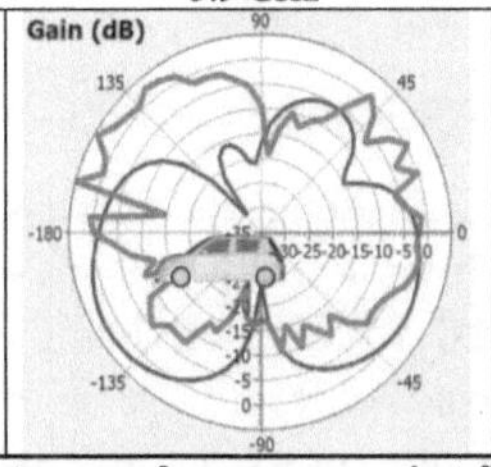	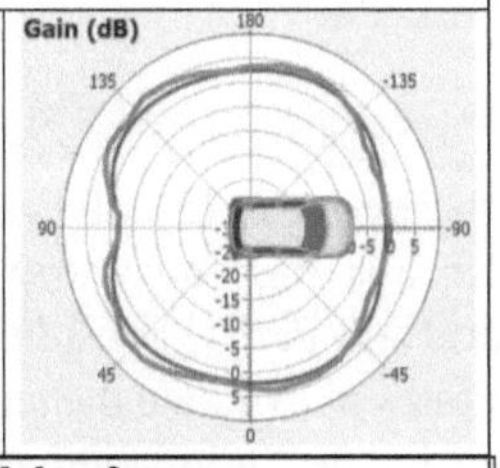
▬ isolated antenna performance ▬ installed performance		

TABLE 9.3(B)
3D Far-Field Performance of Installed Antenna on a Vehicular Body

At 1.8 GHz	At 2.4 GHz	At 3.5 GHz

At 5.25 GHz	At 5.9 GHz	

$$W = (R_1 - R_2). \tag{9.5}$$

Along with the curved edges of the circular ring, by considering the fringing fields into account, the effective inner and outer radii can be given as [8]

$$R_{1e} = R_1 + 0.5[W_e(f) - W], \tag{9.6}$$

$$R_{2e} = R_2 - 0.5[W_e(f) - W], \tag{9.7}$$

where

$$W_e(f) = \frac{W + [W_e(0) - W]}{1 + \left(\frac{f}{f_p}\right)^2}, \tag{9.8}$$

$$W_e(0) = 120\pi t / z_0\sqrt{\varepsilon_e}, \tag{9.9}$$

$$f_p = z_0/(2\mu_0 t). \tag{9.10}$$

μ_0 represents the permeability and z_0 represents the quasi-static characteristic impedance of a circular ring microstrip line of width $W = R_1–R_2$.

The structural similarity considered which is existing in the fractal antenna geometry facilitates to evaluate the surface current flow on the conductive elements. This goes to formulate the mathematical expressions for the model. The halfwave resonant length 'L_r' is computed using

$$L_r = c/(2f_r\sqrt{\varepsilon_e}), \tag{9.11}$$

where 'c' is velocity of light in free space (3×10^8 m/s), the frequency 'f_r' corresponding to the guided wavelength is 3.5 GHz (center frequency). The parameters of the antenna geometry can be mathematically mentioned with the dependency on halfwave resonant length 'L_r' as follows:

$$Lr = \begin{cases} 0.564\pi(R_1 + R_2) \\ 3.074\pi a1 \cdot sf\,[0.9 \cdot sf + 0.85] \\ 1.064(Lg + W_g/2) \\ 1.052 \cdot \pi(2a_1 + t_1) \end{cases}, \tag{9.12}$$

where the parameters 'L_g', 'W_g' are length and width of the ground plane respectively, 'a_1' is the radius of the inner circular cut made in iteration-4, and 'sf' is the scaling factor expressed as

$$sf = a_n/a_{n-1}, \tag{9.13}$$

where 'a_n', 'a_{n-1}' are the radii of nth and (n_1)th circular cuts, respectively.

After choosing the final layout of the antenna, which fulfills the coverage of essential vehicular working bands, the antenna characteristics are simulated for many bending angle iterations for the PVC substrate material. The formal bending angles mostly 30°, 60°, 90°, and 120° are examined for this analysis. The mathematical Eq. (9.14) is utilized to calculate the bending radius essential to fabricate the substrate in a flexible manner.

$$\text{Bending radius, } br = 0.5(Ls - fb) \times (360/\theta). \tag{9.14}$$

where 'f_b' represents the length of the feed line section on the flexible substrate, 'L_s' represents the length of the substrate, and 'θ' is the bending angle specified illustrated in Figure 9.3(a). The equivalent reflection coefficient and peak gain vs frequency characteristics are illustrated in Figure 9.3(b) and (c) respectively. It can be recognized from Figure 9.6(b) that all its bending iterations are working in the closest equivalent bands. As the bending angle (flexibility) rises, the frequency shift toward the right side of the spectrum is noticed at the starting band of the antenna and hence little bit reduction of full impedance bandwidth appeared, though the considerable matching can be view at resonant frequencies. The obtained bands and peak gain values for equivalent iterations are illustrated in Table 9.4. The performance is also differentiated with the planar CPW version of the antenna layout, which is developed on the FR-4 substrate. Compared to the FR-4 the PVC material-based antenna geometry indicates high gain and the maximum gain characteristics have been noticed for the flexible antenna iteration for the bending angle of 30 degrees. Later, with the improvement in the bending angle, the antenna gain is

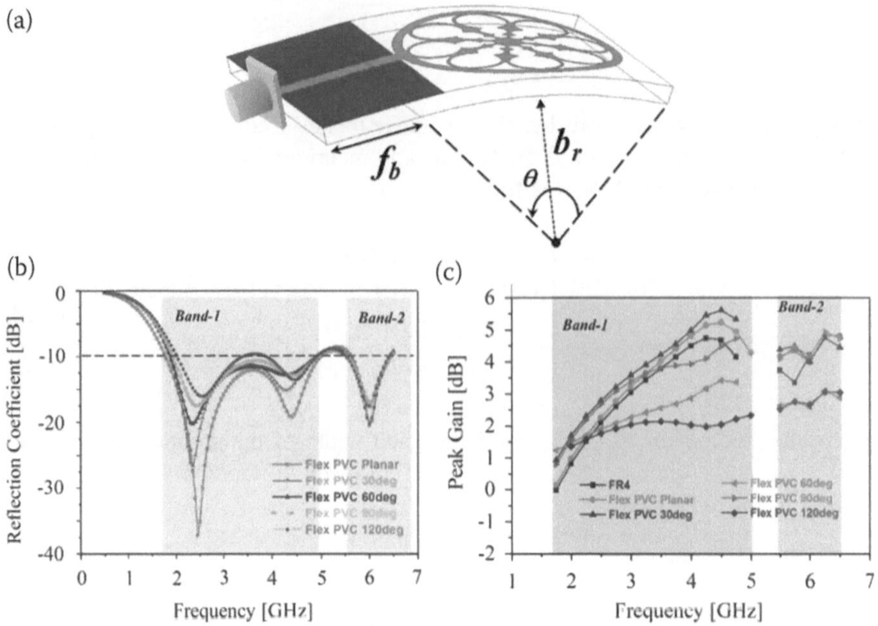

FIGURE 9.3 (a) Geometry of flexible antenna [9] characteristics of proposed flexible antenna at different bending iterations, (b) reflection coefficient vs frequency characteristics of proposed flexible antenna, and (c) simulated peak gain vs frequency.

noticed to be decreasing in both of its working bands, which can be viewed from Figure 9.3(c).

The manufactured transparent antenna with PVC material exhibits the working characteristics from 1.85 GHz–4.94 GHz and 5.38 GHz–6.5 GHz with low reflection losses observed at frequencies 2.57 GHz (−31.174 dB), 4.38 GHz (−21.79 dB), and 6.22 GHz (−17.08 dB), which are obtained under antenna testing.

The magnitudes of surface current density on the conductive layers of the designed antenna, which are derived utilizing the simulation for planar as well as flexible configurations, are illustrated in Figure 9.4. The plots are generated at three different frequencies: 2.4 GHz, 3.5 GHz, and 5.9 GHz.

From Figure 9.4, the surface currents are mostly denser at the portions on the main circular ring-shaped radiating layout. After noticing the nature of current flow at different instants, the surface currents obtained at the feeding layout are fed to the circular ring-shaped radiating layout and the edges of the ground plane are having a high concentration. Later, these currents assembled on a circular ring layout are splits to traverse in two different parallel directions, which are managed to converge at the top of the patch. At resonant frequencies, the currents also traverse on the fractal circular ring arms and thus initiate the essential resonant length to fabricate the antenna functioning. Lastly, these currents converged at the concentric region of the patch where all the eight fractal branches are converging.

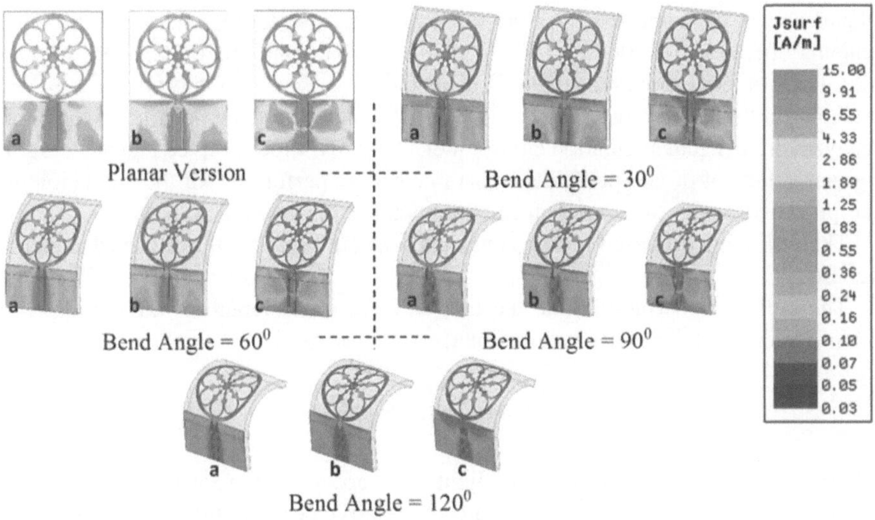

FIGURE 9.4 Magnitude of surface current density of proposed antenna structure for planar configuration and flexible iterations at (a) 2.4 GHz, (b) 3.5 GHz, and (c) 5.9 GHz.

The two-dimensional patterns are illustrated in *XY*, *YZ*, and *XZ* planes, which indicate the corresponding orientation of the vehicle as depicted in Figure 9.5. The dumbbell-shaped patterns observed for the free-standing antenna at 2.4 GHz and 3.5 GHz are somewhat altered after mounting on the vehicular rooftop. The orientation of the pattern appears like the stacking of two dipole antenna patterns, which is because of the reflections provided by the large perfectly electric conducting surface of the vehicular roof. The directions of maximum radiations in elevation plane patterns are slanted between 30°–60° elevation angular zones, which is equivalent to communicating with base-stations. At those two frequencies, the azimuth plane (*XY*-plane) patterns observed for freestanding antenna and when mounted on a car are following nearly omnidirectional and maintaining good correlation. The peak gain of the antenna with respect to frequency is illustrated in Figure 9.5(d) with the simulated one, measured with and without antenna placement on the vehicle. When the proposed antenna is utilized in conformal mode at 30 degrees bending angle, the maximum gain is observed as 5.9 dB.

9.3.3 Flexible LCP Based Conformal Fractal Antenna for Internet of Vehicles (IoV) Applications

A fractal structured flexible antenna is proposed for conformal vehicular communication applications. The proposed antenna is designed and fabricated on Rogers ULTRALAM 3850 liquid crystal polymer substrate of relative permittivity 2.9 and occupies a footprint of 40 × 30 × 0.1 mm^3. The fractal structure is obtained through etching several circular cuts in a recursive manner, which enhanced the impedance bandwidth of the antenna. The operating band characteristics with respect to the planar and flexible mode of the proposed antenna are studied and analyzed. The antenna operates in 2.30–7.89 GHz, 8.97–10.42 GHz, and 10.76-extended beyond 12 GHz. The conformal placement is also studied by placing the proposed flexible antenna on the side-wing mirror of the vehicle and the antenna characteristics are analyzed in a virtual simulation environment. The real-time received signal strength measurements of prototyped antenna parameters are performed for the validation of the proposed design for the conformal application on the vehicle body.

The following proposed design considerations are taken for the designed antenna geometry:

The radius 'R_p' which is like the half-wave resonant length of the circular radiator (traditional circular patch antenna)

$$R_p = c/4f_r\sqrt{(\varepsilon_r + 1)/2}, \tag{9.15}$$

where 'c' represents the velocity of light, 'f_r' represents the lowest resonant frequency of the antenna and 'ε_r' represents the relative permittivity of LCP substrate.

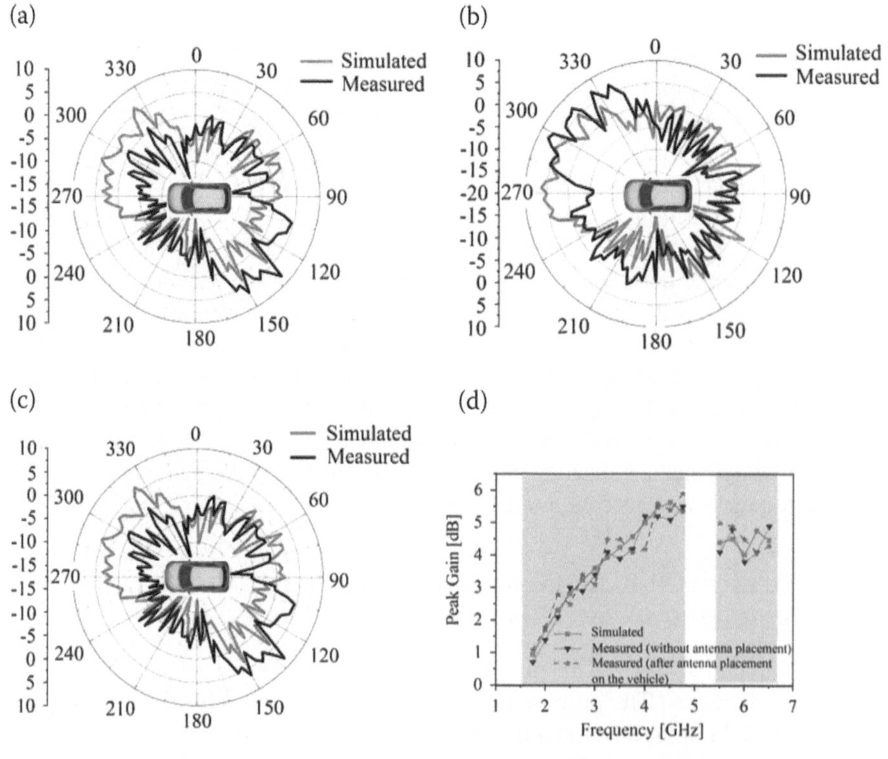

FIGURE 9.5 Simulated and measured radiation patterns of the proposed antenna when mounted conformal to the surface of the side-view mirror of a car obtained at (a) 2. 4 GHz (b) 3. 5 GHz (c) 5.9 GHz and (d) Simulated and measured gain of the proposed antenna.

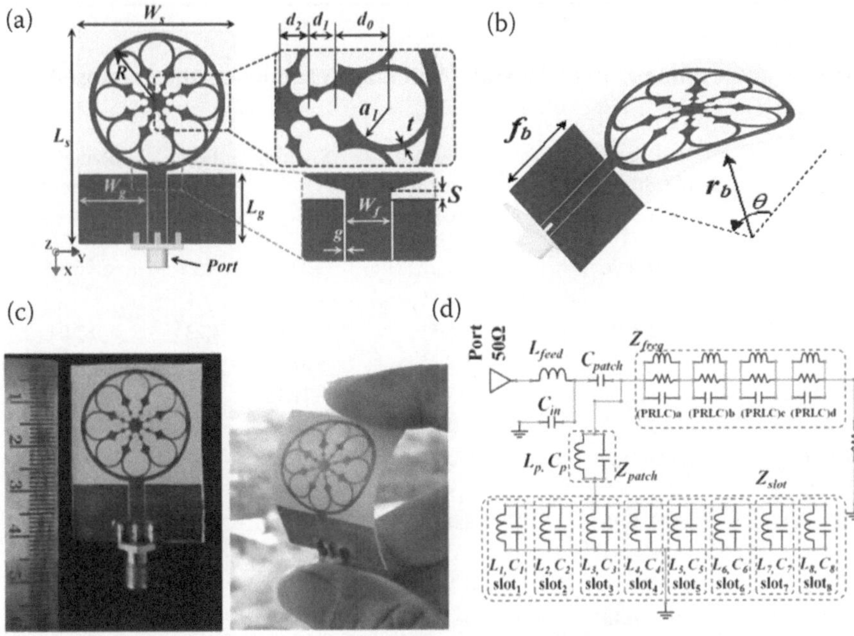

FIGURE 9.6 (a) Geometry of proposed antenna [10], (b) flexible version of antenna (c) fabricated antenna. Geometrical parameters are: $L_s = 40$, $W_s = 30$, $L_g = 13$, $W_g = 13$, $R = 13$, $g = 0.25$, $t = 0.5$, $S = 0.69$, $t_1 = 1$, $d0 = 4.2$, $d_1 = 2.1$, $d_2 = 2.43$, $W_f = 3.84$, $a1 = 3$ (in mm) (d) equivalent circuit.

The width of the circular ring 't_1', can be mathematically equivalent to the lowest resonant frequency 'f_r' as

$$f_r = 4.66 \times \exp\{-9.08 \times 10 - 4 \times \pi(R - t_1)^2\} - 0.47. \tag{9.16}$$

The radius of fractal circular ring is represented as 'a_1', and the subsequent circular rings are having the radii as mentioned below.

$$a_{(n+1)} = a_{(n)} \times S_f, \tag{9.17}$$

where 'S_f' represents the scale factor of value 0.52. The optimized dimensions of the designed antenna design are shown in Figure 9.6(a). The flexible antenna is illustrated in Figure 9.6(b) at bending radii 'r_b' with a bending angle of 'θ' and the manufactured flexible antenna is illustrated in Figure 9.6(c). The bending radius can be calculated by using

$$r_b = 0.5(L_s - f_b) \times (360/\pi\theta), \tag{9.18}$$

TABLE 9.4
Performance Comparison of Flexible Antenna Iterations of the Proposed Antenna

Antenna Iteration	Impedance Bandwidth [GHz] ($\lvert S_{11} \rvert \leq -10$ dB)	Resonant Frequencies [GHz]	S_{11} [dB]	Peak Gain [dB]	Radiation Efficiency
FR4	1.79–4.96	2.22	−22.37	4.78	96.88
		4.15	−17.03		
	5.42–6.07	5.81	−27.94	4.234	87.04
Flex PVC Planar	1.78–5.0	2.45	−37.15	4.767	98.84
		4.4	−19.15		
	5.63–6.42	6.0 GHz	−20.41	4.848	96.85
Flex PVC 30 deg	1.86–4.9	2.35 GHz	−26.74	5.632	99.27
		4.3 GHz	−15.19		
	5.56–6.45	6.0 GHz	−17.68	4.766	98.66
Flex PVC 60 deg	1.88–4.94	2.35 GHz	−20.17	3.44	99.36
		4.3 GHz	−12.82		
	5.53–6.43	6.0 GIIz	−17.39	3.110	98.64
Flex PVC 90 deg	1.93–4.94	2.45 GHz	−17.42	4.748	99.23
		4.4 GHz	−12.86		
	5.55–6.45	6.0 GHz	−17.08	4.974	98.64
Flex PVC 120 deg	2.02–3.40	2.55 GHz	−16.04	2.138	99.15
	3.88–4.90	4.45 GHz	−13.53	2.236	98.96
	5.53–6.41	6.0 GHz	−20.02	3.086	98.88

where 'θ' represents the bending angle, 'L_s' represents length of the substrate, and 'f_b' represents length of the feed line section on the flexible substrate as shown in Figure 9.6(b).

The antenna works in three bands at four resonant frequencies. The performance of the proposed antenna can be modeled by a similar circuit as illustrated in Figure 9.6(d) with a 50 Ω matched load [11]. The feed inductance and the static capacitance of the antenna are given by L_{feed}, C_{patch} individually. The impedance of the circular ring-shaped patch is represented by Z_p. The fractal slots incorporated in the antenna geometry can be noticed with a parallel combination of eight *L-C* resonator circuits whose equivalent impedance is Z_{slot} and the attained resonant frequencies can be realized through an impedance element Z_{freq}, which is established by the series combination of parallel RLC (PRLC$_n$) circuits. Here, $n = 4$ (the number of resonant frequencies). The working bands are almost stable at all bending angles and equivalent values are illustrated in Table 9.5. The peak gain at resonant frequencies is above 4 dB for all bending conditions, which can be observed from Table 9.5.

The antenna placement analysis is achieved in the ANSYS Savant simulation tool. The shooting and bouncing ray (SBR+) based simulation calculates

TABLE 9.5
Operating Band Characteristics of the Proposed Flexible Antenna

Bending Angle	Operating Bands [GHz]	f_r [GHz]	Peak Gain [dB]
Planar	2.3–7.89	3.11	5.043
	8.97–10.42	6.92	6.276
	10.76–ext	11.1	5.041
15°	2.33–7.91	3.16	5.254
	8.93–10.04	7.03	5.338
	10.56–ext	10.87	5.62
30°	2.35–7.90	3.1	5.342
	8.92–9.98	7.02	5.342
	10.81–ext	11.18	5.323
45°	2.35–7.89	3.15	5.302
	8.82–9.83	7.01	5.608
	10. 71–ext	10.99	4.798
60°	2.29–7.88	3.08	5.114
	8.77–9.78	7.02	5.653
	10.49–ext	10.7	4.785
75°	2.34–7.93	3.11	4.874
	8.82–9.93	7.02	6.20
	10.62–ext	10.86	4.869
90°	2.35–7.98	3.13	4.452
	8.82–9.90	7.03	6.057
	10.52–ext	10.73	4.761
105°	2.35–8.05	3.09	4.423
	8.91–10.04	7.11	6.017
	10.74–ext	11.01	4.832
120°	2.33–8.14	3.07	6.187
	8.90–10.07	7.13	6.165
	10.68–ext	11.02	4.725

the far-field solution for the antenna performance on electrically large layouts. The 2-Dimensional and 3-Dimensional radiation patterns that are achieved at working frequencies are illustrated in Figure 9.7. In all these cases, the XY-plane (azimuth) patterns are tilted with respect to the front direction of the vehicle, which is because of the conformal installation of the antenna over the curved surface of the wing mirror. At 45 degrees, maximum radiation is achieved.

The patterns in Figure 9.7 depict that the structure of the wing mirror opposes the backward radiation of the flexible antenna and hence the effect of radiation is minimum inside the vehicle.

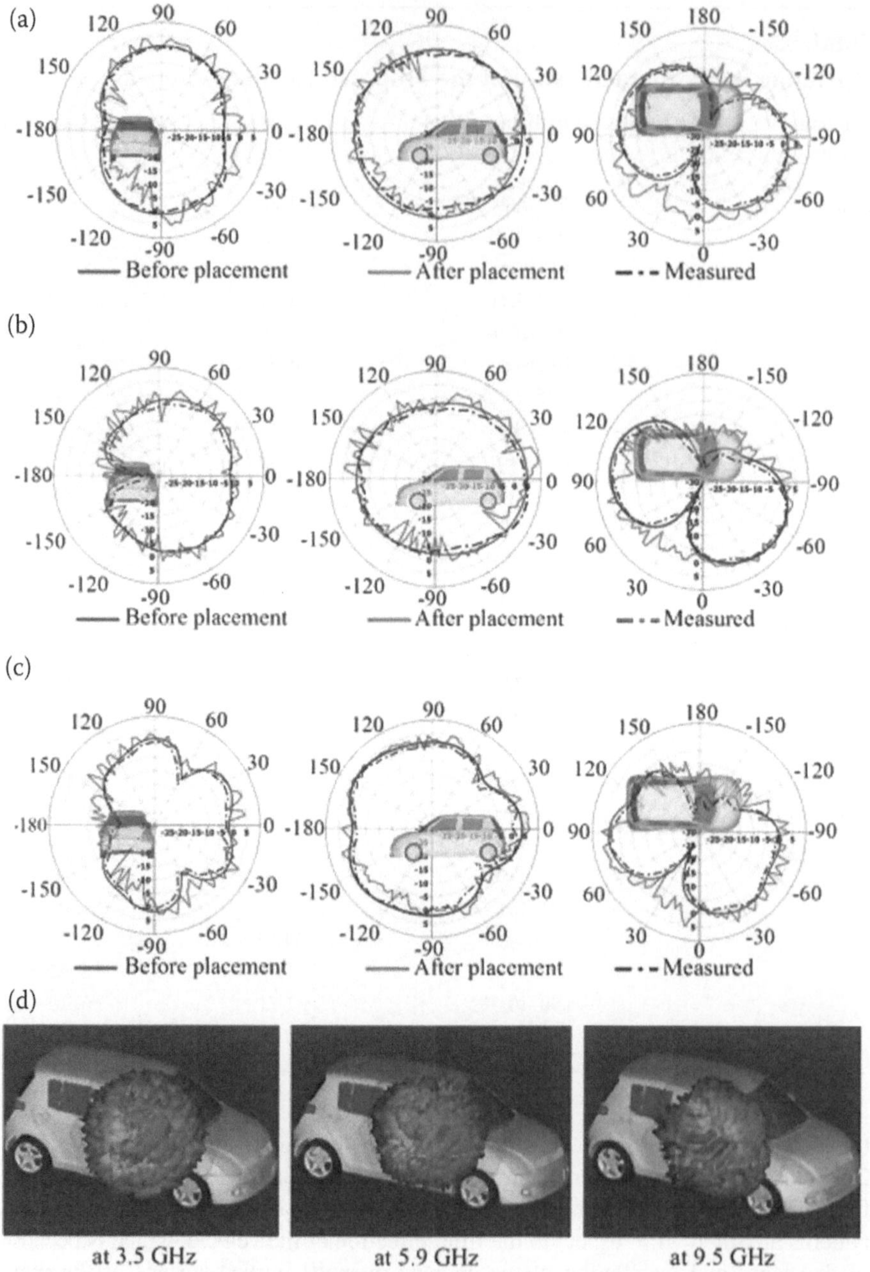

FIGURE 9.7 Simulated and measured Two-dimensional radiation patterns of the proposed antenna before and after placement on the vehicle body [*YZ*-plane (figure in the left), *XZ*-plane (figure in the middle), *XY*-plane (figure in the right)] (a) at 3.5 GHz (b) at 5.9 G GHz (c) at 9.5 GHz.(d) 3D far-field patterns of the proposed antenna with on-vehicle placement.

However, the radiation patterns obtained after the antenna placement are having the ripple nature due to the combined effect of the radiation from the secondary sources like windows, metallic structures, etc. of the vehicle. The maximum gain of the antenna before the placement is noted as 2.738 dB, 5.252 dB, and 4.525 dB. The antenna gain after placement on the vehicle is identified as increased to 7.612 dB, 9.398 dB, and 9.258 dB at operating frequencies 3.5 GHz, 5.9 GHz, and 9.5 GHz respectively.

After computing the far-field characteristics for the antenna placement on the vehicle, a practical scenario is considered as shown in Figure 9.8(a). It represents a road-traffic scenario in which two vehicles (Vehicle-1, Vehicle-2) are coming toward each other and in the middle, a junction is also introduced, where another vehicle (Vehicle-3) is coming across the straight path. These vehicles mutually exchange their information through their installed antenna modules. This scenario is practically implemented to observe the signal reception performance by installing the antenna on the wing mirror of the vehicle (Vehicle-1) and placing a high gain transmitting horn antenna (POWERLOG70180) fixed at a location as shown in Figure 9.8(b). The distance between these two is varied from 1 m to 10 m and the receiving signal power is recorded through the Combination analyzer (Anritsu MS2037C).

The received signal power values are measured with respect to variation in distance between the transmitting antenna and vehicle and plotted in Figure 9.8(c). It is observed that there is a decrease in the received signal power at 5 m due to reflections caused by another vehicle placed at that location.

9.3.4 Bandwidth Reconfigurable Antenna on LCP Substrate for Automotive Communication Applications

A compact and ultrathin frequency reconfigurable antenna is shown in this particular section in Figure 9.9(a) ($40 \times 30 \times 0.1$ mm^3) for vehicular based communication applications. The designed antenna contains a coplanar waveguide fed circular ring layout along with the angularly placed parasitic circular elements manufactured on a Rogers ULTRALAM-3850 LCP substrate of having $\varepsilon_r = 2.9$. The working bandwidth around the targeted resonant modes 2.4 GHz, 5.9 GHz, 9.5 GHz, and 12.5 GHz of the antenna are reconfigured by controlling the BAR64-03W PIN diode switching elements in the antenna layout. The fabricated antenna and its measurement setup are shown in Figure 9.9(b) and (c).

Later, this design focused on studying the radome housing effects and radome substrate permittivity effects on the performance of the designed reconfigurable antenna at various switching states and then far-field characteristics of the radome enclosed antenna after placing the antenna on the rooftop of the vehicle are also illustrated. A step later, the radome paint effects on the propagation of the signal in a real-time environment are experimentally analyzed by computing the received signal strength parameter and illustrated in this book. The gain of the proposed antenna varies between −0.51 dB and 6.12 dB for the antenna in stand-alone mode whereas it swings from 2.85 dB to 8.54 dB for the antenna installed with radome.

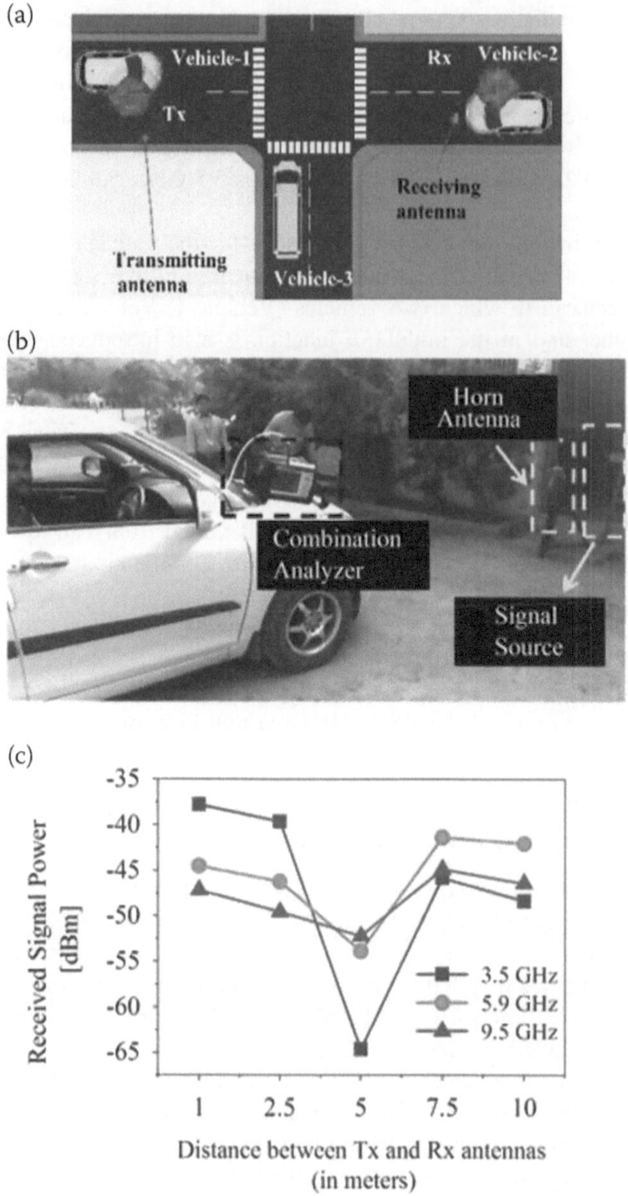

FIGURE 9.8 (a) Road-traffic scenario of vehicles (b) practical measurement setup of received power measurements (c) received signal power vs distance characteristics of proposed antenna (Tx power = 10 dBm).

Other than fundamental mode 'f_{r1}', the resonant frequencies f_{r2}, f_{r3}, and f_{r4} are related as:

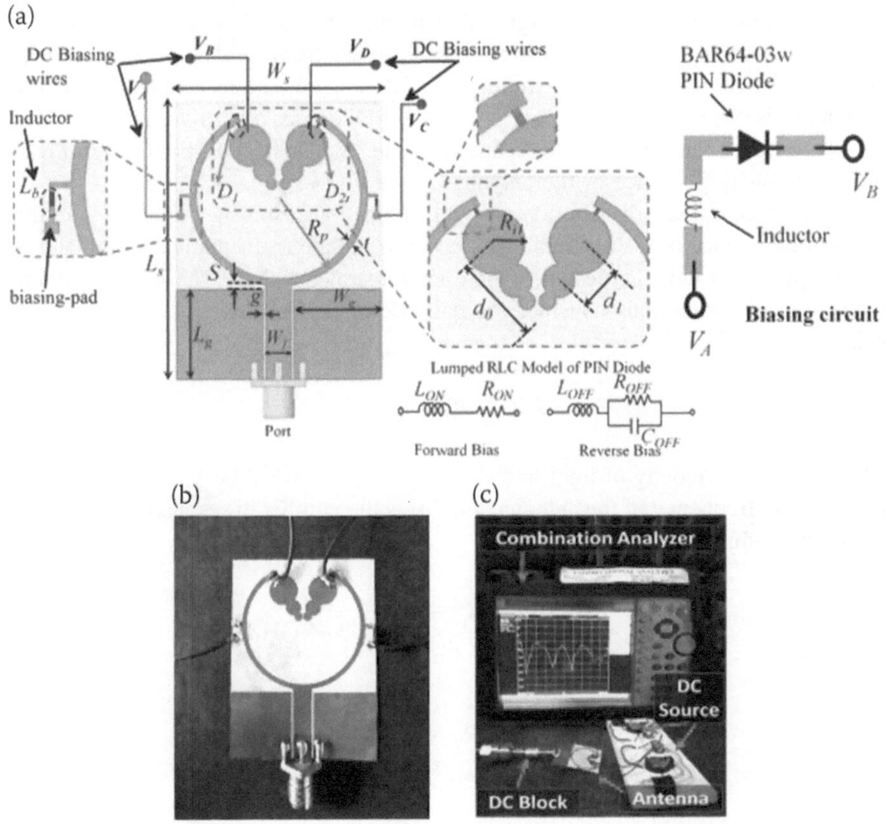

FIGURE 9.9 (a) Antenna geometry of the proposed reconfigurable antenna [12]. [Geometrical parameters: $L_s = 40$, $W_s = 30$, $L_g = 13$, $W_g = 12.73$, $R_p = 13$, $S = 0.69$, $t = 1$, $d_0 = 8.35$, $d_1 = 4.2$, $W_f = 3.84$, $g = 0.35$, $Ri_1 = 3$, all dimensions are in mm, $S_f = 0.52$], (b) Fabricated prototype of the reconfigurable antenna on LCP substrate, (c) Measurement setup for Reflection coefficient vs frequency characteristics of the proposed stand-alone reconfigurable antenna.

$$\left.\begin{aligned} f_{r2} &\approx 2.614 \times f_{r1} \\ f_{r3} &\approx 3.84 \times f_{r1} \\ f_{r4} &\approx 5.472 \times f_{r1} \end{aligned}\right\}. \tag{9.19}$$

The relation among the obtained resonant frequencies f_{r1}, f_{r2}, f_{r3}, and f_{r4} can be confirmed as a set of mathematical impressions as:

$$\left.\begin{aligned} f_{r2} &\approx 2.614 \times f_{r1} \\ f_{r3} &\approx 1.468 \times f_{r2} \\ f_{r4} &\approx 1.424 \times f_{r3} \end{aligned}\right\} \tag{9.20}$$

The peak gain has reached its high value in region-3 and low value in region-1 among all working regions, which can be observed from Table 9.6. Peak gain is noticed to be equivalent to the symmetrical switching conditions of the antenna and a decrease is observed when both or any of the PIN diodes are forward biased. This can be because of the insertion loss caused by the PIN diode at the contact terminals on the conducting surface of the antenna.

The surface current distribution patterns attained at resonant frequencies on the conducting surface of the antenna have led to formulate mathematical formulations for the proposed antenna design, which are illustrated below. These approximate mathematical expressions consider the half-wave resonant length 'L_{rn}' calculated using

$$L_{rn} = c/(2f_{rn}\sqrt{\varepsilon_r}), \tag{9.21}$$

where 'c' is the velocity of light in free space (3×10^8 m/s), the 'f_r' corresponds to the resonant frequency of the antenna, and 'n' is the number of resonant frequencies in an operating region.

$$L_{r1} \approx \begin{cases} 0.915\pi(2R_p - t) \\ 1.5L_g + W_g \end{cases} \quad \text{at 2.45 GHz,} \tag{9.22}$$

$$L_{r2} \approx \begin{cases} 0.187\pi(2R_p - t) \\ 0.537\pi\left(\frac{2R_p - t}{6} + R_{i1}(1 + S_f)\right) \\ 0.767[(0.5 \times Lg) + W_g] \end{cases} \quad \text{at 5.97 } GHz, \tag{9.23}$$

$$L_{r3} \approx \begin{cases} 0.223\pi\left(\frac{2R_p - t}{6} + 2R_{i1}(1 + S_f)\right) \\ 0.114\pi(2R_p - t) \\ 0.118\pi(2R_p - t) \\ 0.546[(0.334L_g + W_g)] \end{cases} \quad \text{at 9.45 GHz,} \tag{9.24}$$

$$L_{r4} \approx \begin{cases} 0.08\pi(2R_p - t) \\ 0.246\pi[0.23(2R_p - t) + R_{i1}] \end{cases} \quad \text{at 13.3 GHz.} \tag{9.25}$$

The radome affects the resonance in the X-band region (nearer to 9.0 GHz) to own a low value of reflection loss. In entire performance, it can be explained that the resonant frequencies are shifting low with the addition of the radome layout and a large ground plane. The reflection coefficient vs frequency of the antenna with shark-fin radome is finally illustrated in Figure 9.10 with simulated and measured characteristics. In the practical aspect, the large ground plane was prepared with an aluminum conducting foil layered on cardboard.

TABLE 9.6
Operating Band Characteristics of Stand-Alone Reconfigurable Antenna

Switching Case		Operating Bands [GHz]	%BW	f_r [GHz]	S_{11} [dB]	Peak Gain [dB]	Operating Bands [GHz]	%BW	f_r [GHz]	S_{11} [dB]	Peak Gain [dB]
D$_1$	D$_2$	Simulated					Measured				
OFF	OFF	2.16–3.11	36.05	2.55	–23.16	1.096	2.06–3.41	49.36	2.48	–25.86	1.01
		6.11–7.25	17.06	6.63	–19.71	1.963	6.03–7.72	24.58	6.58	–15.90	1.54
		9.18–10.22	10.72	9.63	–27.44	3.863	8.90–10.45	16.02	9.95	–21.05	3.67
		12.77–ext	--	13.7	–47.86	2.782	12.63–ext	--	13.89	–25.06	2.45
OFF	ON	2.09–2.92	33.13	2.45	–31.14	0.096	1.82–3.23	55.84	2.40	–25.21	0.03
		5.65–7.04	21.91	6.36	–25.42	2.915	5.55–7.27	26.83	6.34	–20.33	2.54
		8.76–10.04	13.61	9.4	–31.87	5.985	8.46–10.45	21.04	9.32	–22.21	5.48
		12.32–ext	--	13.46	–19.98	5.817	11.79–ext	--	12.72	–16.64	5.61
ON	OFF	2.08–2.89	32.59	2.42	–27.41	0.096	1.92–3.38	55.09	2.51	–20.60	0.03
		5.71–7.08	21.42	6.41	–26.60	2.915	5.26–7.30	32.48	6.61	–21.76	2.53
		8.81–10.10	13.64	9.46	–30.36	5.985	9.01–10.29	13.26	9.53	–22.71	5.45
		12.32–ext	--	13.43	–19.36	5.817	12.13–ext	--	14.33	–17.88	5.52
ON	ON	1.95–2.66	30.80	2.26	–31.25	_0.369	1.88–3.14	50.19	2.36	–28.30	_0.51
		5.4–6.67	21.04	5.98	–20.57	2.717	5.04–6.91	31.29	6.06	–13.42	2.52
		8.57–9.79	13.28	9.09	–37.61	6.393	8.36–10.13	19.14	9.13	–18.84	6.12
		12.08–14.36	17.24	12.82	–21.14	5.097	11.93–14.22	17.51	12.95	–18.36	4.92

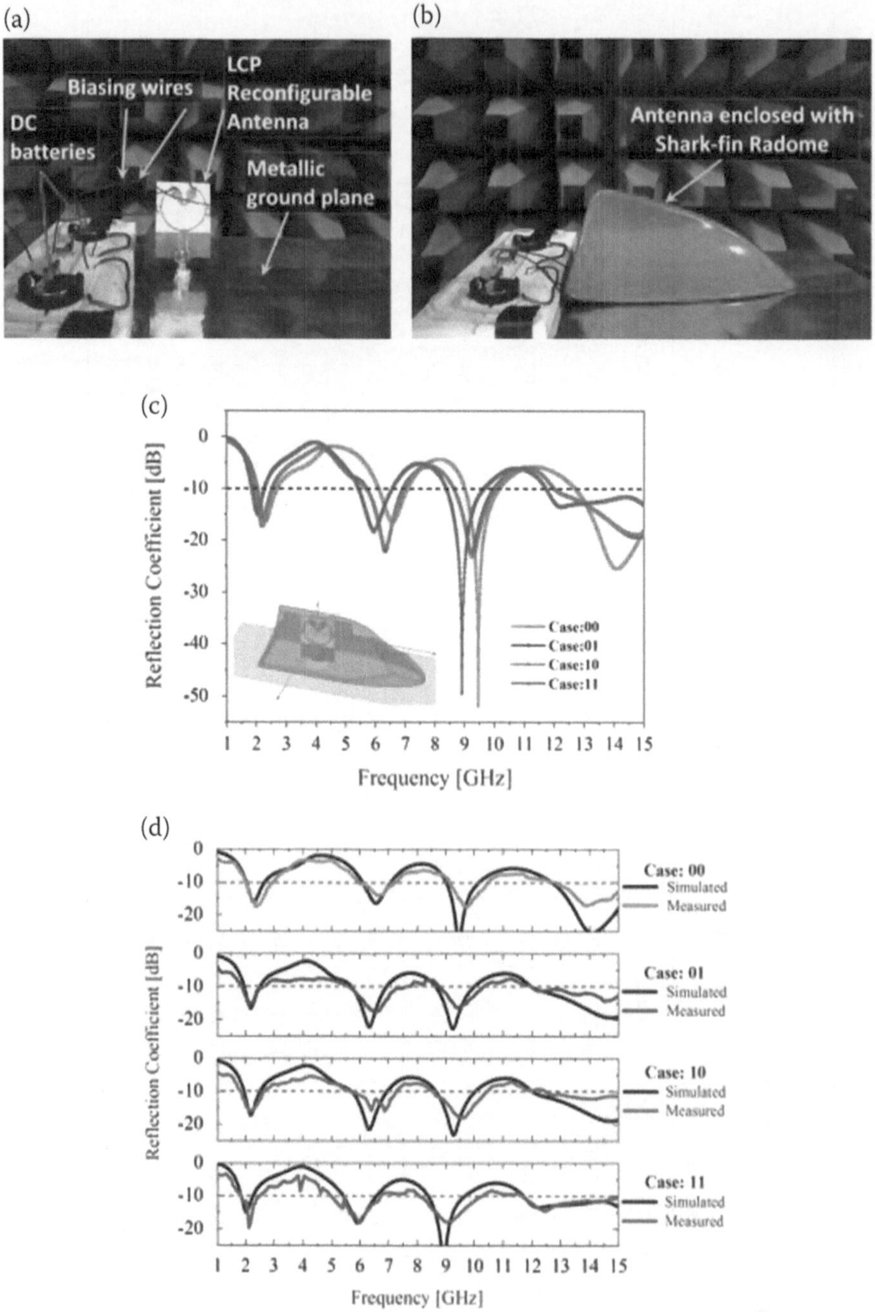

FIGURE 9.10 Measurement setup of the antenna (a) with large ground plane (b) enclosed with shark-fin radome and reflection coefficient vs frequency characteristics of the proposed reconfigurable antenna inside a radome (c) simulated (d) measured.

The differentiation in the simulated and measured results can be due to some manufacturing tolerances, soldering issues, permittivity tolerances, the insertion loss of the surface mount components such as inductors, and PIN diodes and biasing wires, and the antenna alignment problems with the radome. The gain is however similar for the stand-alone antenna when the symmetric switching states (01 & 10) are operated. But, due to the asymmetric layout of the radome, equivalent behavior can't be noticed (Table 9.7).

After examining the characteristics of radome material permittivity effects $\varepsilon_{radome} = 2$ gives effective performance compared to other radome materials. The vertcal orientation of the antenna is included with the shark-fin radome structure that is virtually placed on the rooftop of the vehicle body as depicted in Figure 9.11 and simulated using the ANSYS Savant EM tool. This will examine the current densities within the volume of the entire layout and solves the total far-field solution for the electrically large vehicular platform-mounted antenna performance.

The vehicle roof is similar to a ground plane and behaves as a mirror plane, which will inevitably affect the antenna performance. The radiation patterns are omnidirectional in XY and *YZ* planes of the stand-alone antenna at its first two working bands. The patterns attained after installation on the vehicle have perfectly matched with the patterns attained for the antenna with big ground plane and radome. However, in some directions, the differentiation may be because of the other parts of the vehicle. Total elevation plane patterns in the *XZ* plane are symmetrical along the 0–180° axis in the gain at which the radiation is zero (along 0–180° axis), which is attained as the second band is raised by the large ground plane of the vehicle and forms two sidelobes equivalent to the vertical axis in the *XZ* plane. At the third band, the A null along the vertical axis in the *XZ* plane is raised to flat response when one of the PIN diodes is turned ON, which is noticed in the third band. In the *YZ* plane, the patterns are symmetric but slightly changed around 50° to the curvature of the rooftop of the vehicle body. Many lobes can be observed in the radiation patterns at middle and higher working bands. The *YZ*-plane radiation pattern has been slightly tilted, and its side lobe level (SLL) also increases. Then, the appropriate increase in SLL can enlarge the coverage area in the elevation plane. The three-dimensional patterns presented in Figure 9.12(a) give a perfect presentation of the occurrence of radiation.

The paint consequences for the presentation of the antenna are inspected experimentally and examined in this chapter. Different paints, such as water-based paints, nitrocellulose paint (one of the oldest varieties), acrylic enamel (cheapest), acrylic lacquer paint (for smooth and glass-like appearance), urethane paints, and a lot of trending paints, are popularly accessible like pearlescent and metallic with dozens of shades appropriate for automotive applications. In this examination, metallic copper paint is considered to apply on the outside of the radome. It is prepared by blending fine grade copper powder of almost 4 grams blended with the 15 ml of acrylic enamel and mixed on. The paint is covered on the shark-fin radome and later curing it is placed on the proposed reconfigurable antenna with bias circuitry as illustrated in Figure 9.12(b–d).

In this analysis, the Keysight's EXG Microwave signal generator is utilized as a transmitting source that gives excitation to the AARONIA-AG POWERLOG70180 standard wideband horn antenna. The manufactured reconfigurable antenna is taken

TABLE 9.7
Operating Band Characteristics of Reconfigurable Antenna Enclosed with Vehicle Radome

Switching case		Simulated						Measured					
D_1	D_2	Operating Bands [GHz]	%BW	f_r [GHz]	S_{11} [dB]	% Change in f_r	Peak Gain [dB]	Operating Bands [GHz]	%BW	f_r [GHz]	S_{11} [dB]	% Change in f_r	Peak Gain [dB]
0	0	2.05–2.69	27	2.31	−16.31	−9.41	3.280	2.03–2.97	37.6	2.35	−17.53	−5.24	2.97
		6.10–7.08	14.87	6.57	−16.43	−0.9	5.474	5.97–7.23	19.09	6.64	−14.13	0.91	5.23
		9.02–10.17	11.98	9.46	−51.79	−1.76	6.748	9.14–10.48	13.65	9.68	−17.59	−2.71	6.82
		12.7–ext	—	14.08	−25.42	2.77	9.195	12.81–ext	—	13.86	−17.09	−0.21	8.05
0	1	1.96–2.52	25	2.2	−16.61	−10.21	3.685	1.89–2.62	32.37	2.11	−15.51	−12.08	3.12
		5.71–6.93	19.3	6.33	−22.10	−0.47	5.218	5.63–7.57	29.39	6.55	−17.47	3.31	4.98
		8.65–9.99	14.37	9.25	−22.78	−1.59	6.945	8.78–10.31	16.02	9.40	−16.41	0.85	6.57
		12.01–ext	—	14.74	−19.32	9.51	9.104	11.99–ext	—	14.48	−14.13	13.83	8.54
1	0	1.95–2.53	25.89	2.2	−17.28	−9.09	3.061	1.82–2.66	37.5	2.12	−15.91	−15.53	2.85
		5.71–6.95	19.58	6.33	−21.47	−1.24	5.364	5.67–7.26	24.59	6.87	−15.96	3.93	4.92
		8.66–10.03	14.66	9.27	−23.15	43.49	6.603	8.55–10.44	19.9	9.48	−18.08	−0.52	6.28
		12.02–ext	—	14.73	−18.91	9.67	8.464	12.47–ext	—	15.09	−12.01	5.30	8.14
1	1	1.84–2.28	21.35	2.03	−14.95	−10.17	3.475	1.85–2.51	30.27	2.12	−19.74	−10.16	3.17
		5.42–6.58	19.33	5.94	−18.22	−0.66	4.723	4.97–6.73	30.08	5.96	−18.19	−1.65	4.38
		8.38–9.67	14.29	8.92	−49.41	−1.87	6.837	8.08–10.25	23.67	9.01	−17.86	−1.31	6.44
		11.71–ext	—	12.24	−13.34	−4.52	6.624	11.63–ext	—	12.45	−14.57	−3.86	6.39

Note: 0-OFF or reverse bias, 1-ON or forward bias; in the % change in 'f_r', the negative sign (−) represents the frequency shift toward lower side and vice versa.

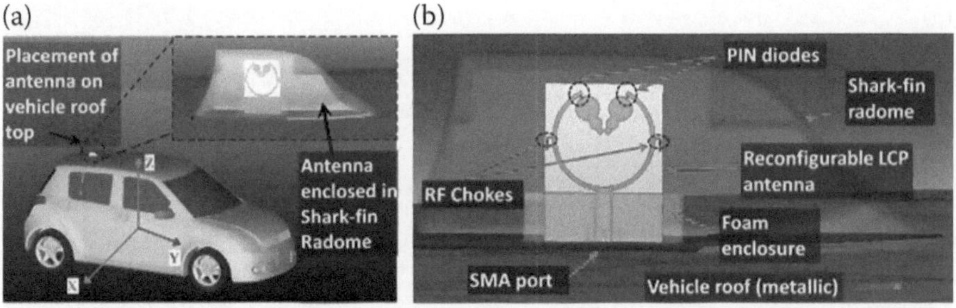

FIGURE 9.11 (a) Final configuration of antenna placement on vehicle (b) geometrical illustration.

as the receiving antenna. The combination analyzer is set to the Spectrum Analyzer mode and associated with the receiving antenna to record the received power (in dBm). The received power values are calculated for the three cases: (i) antenna with a radome (ii) antenna with painted radome (iii) antenna without radome. In the analysis, two frequencies 2. 4 GHz and 5. 9 GHz are excited to horn antenna through the signal generator set to many input power-levels 0 dBm, 5 dBm, 10 dBm, 15 dBm, and 19 dBm. This process is carried out by varying distances 'd' between the transmitting and receiving antennas where d = 0.5 m, 1 m, 1.5 m, and 2 m.

The signal reception is high at 5.9 GHz when the antenna is without radome and d= 0.5 m and 1 m, whereas at d = 1.5 m and 2 m, the antenna without radome has great signal reception capacity at 2.4 GHz. However, the received signal amplitude 'P_r' increases with the increase in the transmitting power 'P_i'. The low amplitude level of the transmitted signal that the proposed antenna received is –49.82 dBm, –52.85 dBm, –58.72 dBm, –60.21 dBm at 0.5 m, 1 m, 1.5 m, 2 m respectively whereas the maximum amplitude is recorded as -18.41 dBm, –24.73 dBm, –28.52 dBm, and –27.82 dBm respectively. The signal reception performance of the antenna with painted radome is better than the normal one at 5.9 GHz for d = 1 m, 1.5 m. A similar case is observed at 2.4 GHz for d = 1.5 m and 2 m. The small reduction in received signal amplitude can be because of the surface problems while painting and the stacking of the present dielectric layer of paint on the existing layer. The reduction in received signal amplitude may be affected by the reflections at the air/dielectric interface.

A frequency bandwidth reconfigurable ultrathin antenna examined in this section is made of parasitic circular components and an open-ended circular ring. The bandwidth reconfigurability is attained by controlling the biasing conditions of the incorporated PIN diode. This reconfigurable bandwidth can help a vehicular band communication system to cover the necessary bandwidth through switching. The proposed antenna is appropriate for the LTE (1.9–2.025 GHz), GSM-1900 (1850–1990 MHz), UMTS (1920–2170 MHz), V2I IEEE 802.11p (5.850–5.925 GHz), V2V, Mobile Satellite Service (forward link: 1980–2010 MHz, reverse link: 1980–2010 MHz), DSRC, and WLAN 802.11a/n (5.710–5.835 GHz). The proposed layout is experimentally validated with a shark-fin radome enclosure and a large ground plane. The examination is also conducted on dielectric permittivity effects of the radome on antenna performance yields suggests the radome with relative

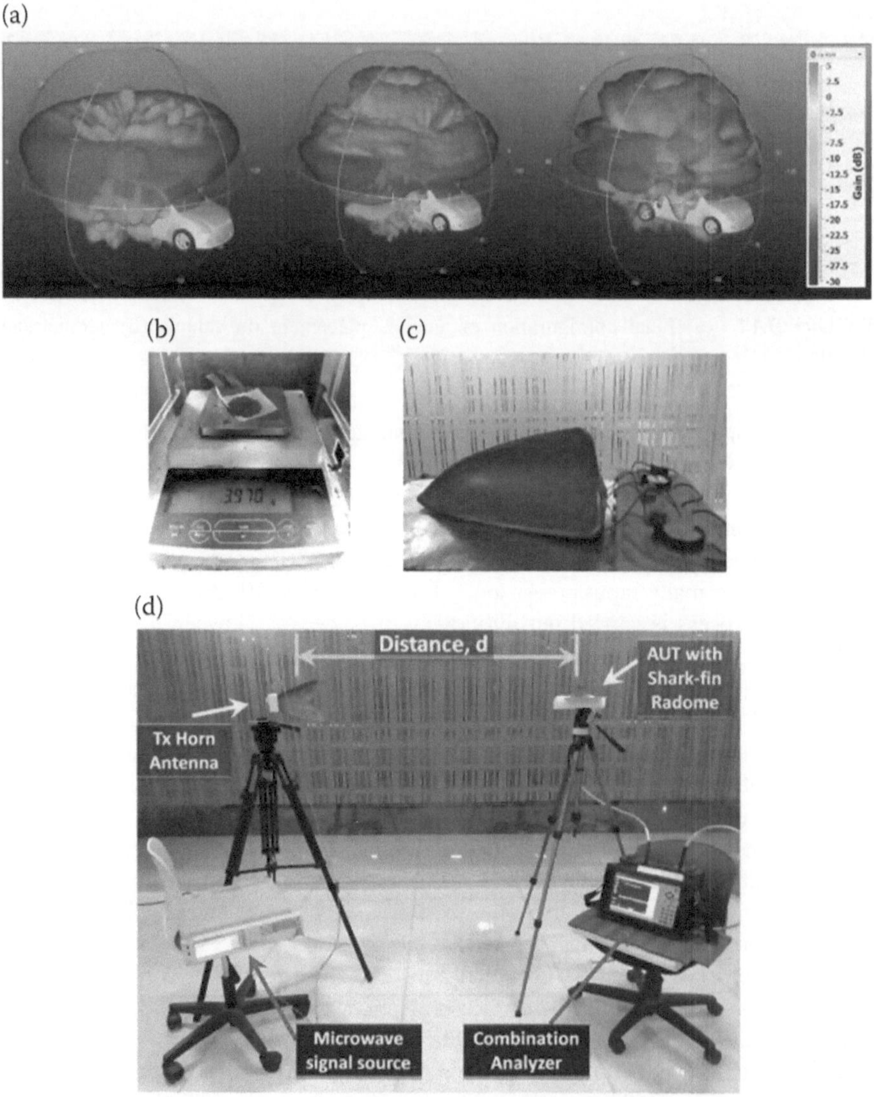

FIGURE 9.12 (a) 3D-far-field characteristics of the antenna after installed on a vehicle (virtual simulation environment), (b) fine copper powder weighing in electronic balance, (c) painted shark-fin radome, (d) experimental setup.

permittivity 2–2.1 for modern practical applications for the designed antenna. The relative permittivity of more than 2.1 causes the existing working bands to change more or improve reflection loss. The measured results of paint impacts portray that the impact of radome paint has little change on the exhibition of the antenna and can make the designed antenna reasonable for bandwidth reconfigurable vehicular band communication applications.

9.4 CONCLUSION

The conclusion deals with the summary of the entire research work. In this research, endeavors have been made for designing a miniaturized multiband and reconfigurable antenna for vehicular communication band applications. Various antenna structures have been designed, simulated, fabricated, and functionally verified experimentally, to develop the final design. These designs are also verified for antenna placement at various locations on the vehicle body like roof-top, side-view mirror, and on the windshield glass. For the final antenna, i.e., a reconfigurable one, the operating band and radiation characteristics are analyzed for antenna placement on the vehicle with radome enclosure effects and paint effects on the antenna performance. The reconfigurable feature is attained through a step-by-step approach that initially includes the design of a multiband antenna and then converting to switchable design through the incorporation of switching elements (PIN diodes). The design evaluations have been performed for the planar and flexible configurations of the antenna.

REFERENCES

[1]. Wang, C.X., Cheng, X., & Laurenson, D.I. Vehicle-to-vehicle channel modeling and measurements: recent advances and future challenges. *IEEE Commun Mag.* 2009; 47(11):96–103.

[2]. Kumar, M., & Nath, V. Introducing multiband and wideband microstrip patch antennas using fractal geometries: development in last decade. *Wireless Pers Commun.* 2018; 98: 2079–2105. https://doi.org/10.1007/s11277-017-4965-x

[3]. Madhav, B.T.P., Krishnam Naidu Yedla, G.S., Kumar, K.V.V., Rahul, R., & Srikanth, V. Fractal aperture EBG ground structured dual band planar slot antenna. *Int J Appl Eng. Res.* 2014; 9(5): 515–524.

[4]. Allen, B., Ignatenko, M., & Filipovic, D. S. Low profile vehicular antenna for wideband high frequency communications. *IEEE Int Symp Antennas Propag (APSURSI).* 2016; 115–116. 10.1109/APS.2016.7695766.

[5]. Lopez, D.G., Ignatenko, M., & Filipovic, D.S. Low-profile tri-band inverted-F antenna for vehicular applications in HF and VHF bands. *IEEE Trans Antennas Propag.* 2015; 63(11):4632–4639.

[6]. Mohammod, Ali, Guangli, Y., Hwang, H.-S., Tuangsit, S. Design and analysis of an R-shaped dual-band planar inverted-F antenna for vehicular applications. *IEEE Trans Vehicular Technol.* 2004; 53(1): 29–37.

[7]. Madhav, B.T.P., & Anilkumar, T. Design and study of multiband planar wheel-like fractal antenna for vehicular communication applications. *Microw Opt Technol Lett.* 2018; 60(8):1985–1993.

[8]. James, J.R., and Peter S, Hall, eds. Handbook of microstrip antennas. *IET.* 1989; 28;169.

[9]. Madhav, B.T.P., Anilkumar, T., & Kotamraju, S.K. Transparent and conformal wheel-shaped fractal antenna for vehicular communication applications. *AEU-Int J Electron Commun.* 2018; 91:1–10.

[10]. Anilkumar, T., Madhav, B.T.P., Hawanika, Y.S., M, Venkateswara Rao, & B, Prudhvi Nadh. Flexible liquid crystal polymer based conformal fractal antenna for internet of vehicles (IoV) applications. *Int J Microw Opt Technol.* 2019; 14(6):423–430.

[11]. Majumdar, Anurima, Sisir, K. D., & Annapurna, D. Ultra wide band CPW fed patch antenna with fractal elements and DGS for wireless applications. *Prog Electromagn Res*. 2019; 94:131–144.

[12]. Anilkumar, T., Madhav, B.T.P., Venkateswara Rao, M. & Prudhvi Nadh, B. Bandwidth reconfigurable antenna on a liquid crystal polymer substrate for automotive communication applications. *AEU-Int J Electron Commun*. 2020; 117:1–13.

Part V

Importance and Uses of Microstrip Antenna in IoT

10 Importance and Uses of Microstrip Antenna in IoT

A. Birwal
University of Delhi, New Delhi, India

10.1 INTRODUCTION TO IOT

The next step in the computer age is out of the conventional laptop or desktop world. In the paradigm of the Internet of Things (IoT), many of the objects or things that surround us will be available in one form or the other on the network. This network can be of any form, either wired or wireless, or sometimes a combination of both. Wireless sensor network (WSN) technologies will rise in the near future to meet this new challenging demand, where communication and information systems are invisibly embedded or hidden in the environment surrounding us. This leads to the generation of huge amounts of data that must be stored, analyzed, and represented in a seamless way, which is not only efficient but also easy to interpret. For such ubiquitous computing, cloud computing capabilities can provide the best virtual infrastructure that integrates everything from sensing devices, data storage devices, data analytics tools, data visualization platforms, to delivery to clients end interface [1]. An important feature of IoT is intelligent communication with existing wireless networks and background software-based services. The evolution of growing ubiquitous data and communication networks involved is already apparent with the increased presence of internet access availability provided by Wi-Fi, 4G, WLAN, WiMAX, and upcoming 5G wireless networks [2]. Nevertheless, to implement the concept of the IoT successfully and effectively, the computational criteria would have to become faster than conventional mobile computing models and evolve in a way to develop a linkage between real objects or things and integrating their information into our computing environment.

A revolutionary transformation of the modern Web as a network of interacting things that not only harvests knowledge from the atmosphere (sensing) and communicates with the physical universe (actuation/control/command), but also makes use of established Internet protocols to include information sharing, storage, software and networking infrastructure.

Driven by the proliferation of devices based wireless technologies as shown in Figure 10.1, as well as embedded wireless sensor and actuator devices, as can be seen the IoT has come out as a latest development to bring revolution in the existing

DOI: 10.1201/9781003187325-10

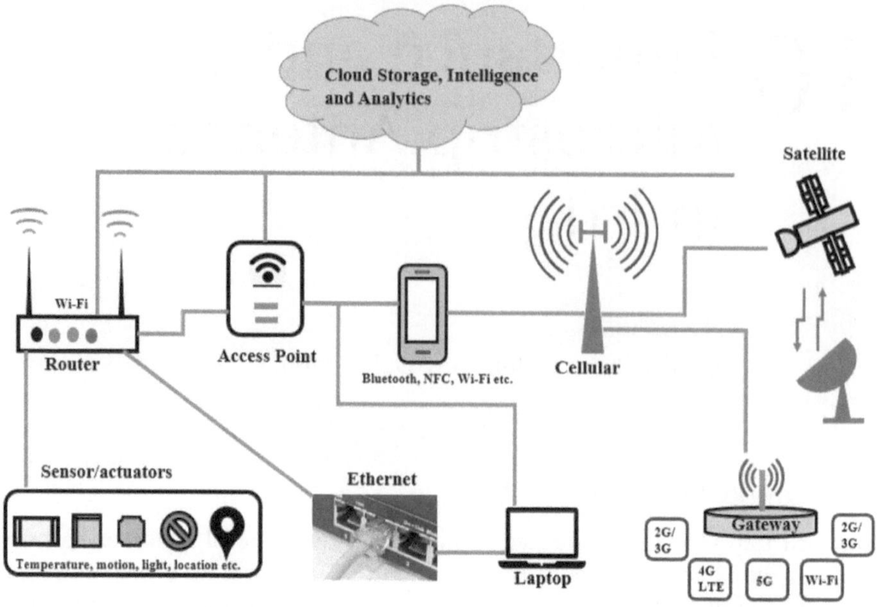

FIGURE 10.1 IoT architecture showing interconnected wireless devices or things.

static internet to a truly connected digital world. Online transition contributed to unimaginable size and speed of people-to-people communications. The next innovation is the communication of things to build an intelligent connected environment. The number of connected things on the world exceeded the number of people, i.e., the world population [3] in 2011. As can be seen from Figure 10.2, there are currently 35 billion connected devices and this is expected to be 75 billion

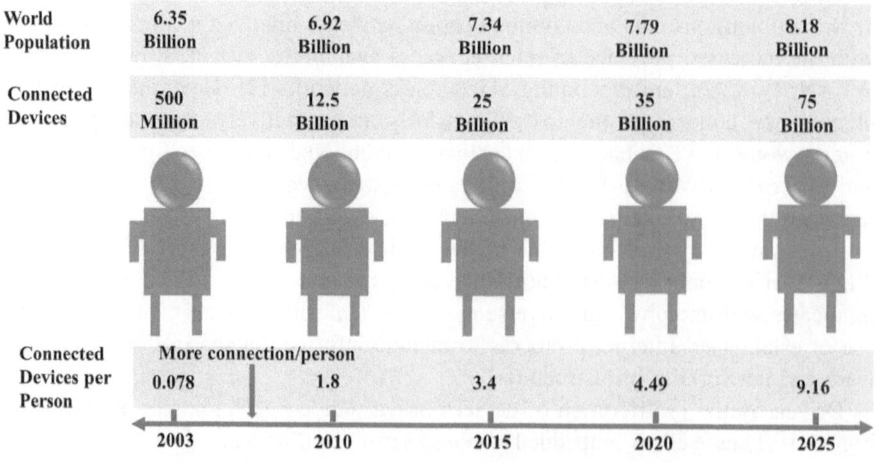

FIGURE 10.2 Estimated number of connected IoT devices per person by 2025.

[4,5] by the end of 2025. Many such IoT devices are considered to be based on low-cost, wireless technologies, with limited computational and storage capabilities. As the demand of IoT systems increases gradually concerns regarding the stability, security, and reliability of the user data from such IoT devices is becoming important.

The vast IoT environment puts together countless end products that use different kinds of radio modules with onboard sensors and actuators, battery packs, Micro-Electro-Mechanical Systems (MEMS), but above all of these devices, antennas are the most important and critical component. The entire performance of a communication system depends on the type of antenna being used. The design and choice of antenna depends on the type application and its operating frequency. Another important aspect of antenna design for IoT systems is the availability of space, as we know the IoT systems are usually compact systems, so designing an efficient antenna within a limited space is the biggest challenge. Antennas are the critical component of the majority of wireless networks, which provides higher throughput between various network typologies. Some of the most common types of antenna used in IoT systems are simple wire antenna, whip antenna, and PCB antenna, which are usually less complex antenna types and provide omni-directional radiation pattern. So many development kits for IoT applications utilize these kinds of antennas for connectivity with GPS, GSM, Wi-Fi, Bluetooth, and other wireless standards. Many specific applications such as Body Area Networks (BAN) and other wearable electronics often include low profile antennas that conform to the device's surface [6]. This convinces the need to develop antenna based on IoT device to fulfil the application specific requirements. IoT antennas may either be classified based on their operational bands or depend on the types of application. For example, ISM bands frequencies which are unlicensed are more prevalent in the development of IoT. Some specific applications require licensed band antennas like military applications. Among all IoT application, no antenna can fulfil all the requirements or can be termed as a perfect antenna. Some antenna preferred to be installed at outside of IoT module while some are integrated with the module itself. Each of the antenna types has associated advantages and disadvantages. Microstrip line-based antenna is the most widely used antenna today as they are based on planar transmission lines [7–10], which has many design benefits such as light weight, low profile, and conforming to any surface.

10.2 DESIGN CHALLENGES OF ANTENNAS FOR IOT APPLICATIONS

With an increasing demand of compact IoT systems, System-in-Package (SiP) and System-on-Chip (SoC) are designed and optimized for available radio frequency (RF) interfaces, exposure to wireless technologies is now easier than ever, but one thing that still requires special attention is the antenna, which plays a critical role in overall system design. As we know, link budgets are vital in providing efficient communications, and maybe the single most essential aspect of designing wireless interface. The selection of antennas and, more importantly, the way they are designed within the system will play a major role to influence the link budget. Due to this, a detailed understanding of the antenna design instructions is essential at the design phase itself. The first point to understand is that antennas are passive in nature, so for them it doesn't really matter what the type of energy (protocols or signals) is that they are carrying; what matters for them is the operating frequency and signal power. An antenna can operate in transmit and receive mode and it should behave in exactly a similar way in both the cases; this also known as reciprocity theory. It also implies that an antenna doesn't really care if the system is a transmitter type, or a receiver type, or both, the design of antenna would remain the same in all the circumstances. An antenna when we describe as a part of the IoT system is either classified as embedded, indicating that it is mounted completely on the PCB board and connected via microstrip line, or cabled connected, indicating that it is connected via a coaxial cable to the PCB board [11]. Cabled connected antennas are mostly installed within the enclosure; however, such antennas can also be placed outside the enclosure unit or, in some situations, outside the building. It is necessary to understand several other criteria as part of the IoT antenna design, such as the required data rate, operating frequencies, and the required range of operation and power levels for wireless connection. Most of these requirements would be standard in a variety of technologies, but they are important parameters for an antenna designer. Range is indeed the most fundamental parameter that can help to decide the most suitable protocol for an application. This can cover small, medium, or long distances, covering a few centimeters to over several kilometers. Range or distance is often directly linked to data rate or throughput, and this may always be a major deciding factor. However, the two rely primarily on the transmission power. Many protocols may accept very low throughput both over long distances at low power, while higher throughput is usually limited to shorter range and require more power to operate. It is necessary to note that the wireless protocol is determined by the type of RF electronic circuitry used, while the antenna merely transmits and receives the RF signals fed by the connected electronics, independent of the used protocol. Embedded or internal antennas are usually made up of ceramic, plastic, metals, and other costly substrate which design low profile and light weight mountable antenna. These features make it easy to find a most suitable antenna design for any application. Many other aspects like the PCB size, its shape, and the mount location of the antenna on the PCB board will also have an effect on the antenna efficiency [12]. So, to design an efficient antenna for IoT system with high efficiency, an antenna designer should observe various important factors such as antenna size, antenna shape, and placement location.

10.3 ANTENNA DESIGN CONSIDERATION FOR IOT SYSTEMS

As a greater number of IoT devices are connected wirelessly to the internet gateway, electronic engineers face several challenges for example how to design radio transmitters installed to existing equipment space, as well as how to develop and produce devices with incredibly small sizes. In addition, they are working to fulfil the challenges of consumers on IoT products compliant with ergonomic design, relevant usability, and environmental harmony.

Size requirement is among the most significant factors when IoT modules are considered, besides this radio communication and price are often widely considered. Usually, engineers prefer compact IoT parts, excellent RF efficiency, and low prices. Often IoT devices do not include all of the above listed features, so IoT module designers have to face such challenges. As the electronics industry is continuously dependent on new silicon processing technologies, the silicon chips have become progressively smaller in recent years. Space issues have been successfully resolved for IoT module design by incorporating MCU (micro-programmed control unit) including RF front end within the SoC (system on chip). However, the rapid development in SoC design has not solved the problem regarding the physical design of the RF transmitter, i.e., the antenna element, which is typically leave it to the organization for custom antenna design, or instruct them to select an easy-to-use module with an integrated antenna. Antenna space is also another challenge to build small IoT devices with high performance and secure wireless communication.

It is a known fact that an antenna has to face multi-dimensional complexity when both efficiency and size are considered simultaneously. Usually to make a cheaper module the cost of BOM (bill of material) is low, so most of the time it is required that the antenna to be used for IoT design should be based on PCB tracing approach. But PCB antennas consume a remarkable size that typically covers $25 \times 15\ mm^2$, resulting the overall size increase of IoT product. The PCB antennas also have another drawback when printed to the module PCB, i.e., their performance is vulnerable to decay as a result of shielded enclosing material and needed a special consideration in the final product assembly phase ito achieve the optimum working environment. The space in IoT module design reserved for RF components should include the required clearance range with no part or trace included within this range. In addition, some space between antenna and edge of the shield should be retained. For IoT devices comparable to button size, the efficiency of the antenna is highly impacted. Usually the antenna efficiency decreases when we try to make the IoT device compact.

10.4 DESIGN OF MICROSTRIP ANTENNA FOR IOT APPLICATIONS

Microstrip patch antennas are a family of planar antennas based on planar transmission lines that have been most popularly used antenna design technology used by antenna designer over the last five decades. They have been used for many IoT applications involving wireless connectivity. The concept behind microstrip patch antennas originated from the printed circuit board technology, which is used not only for conventional planar transmission lines and circuit design but also as a radiating element. Figure 10.3 show the physical structure of the microstrip patch antenna, which can be calculated by using the design equations given in [13]. It consists of a top metallic patch supported by a dielectric substrate material over a metallic ground plane. The top patch can take any shape such as rectangle, circle, triangle, etc.

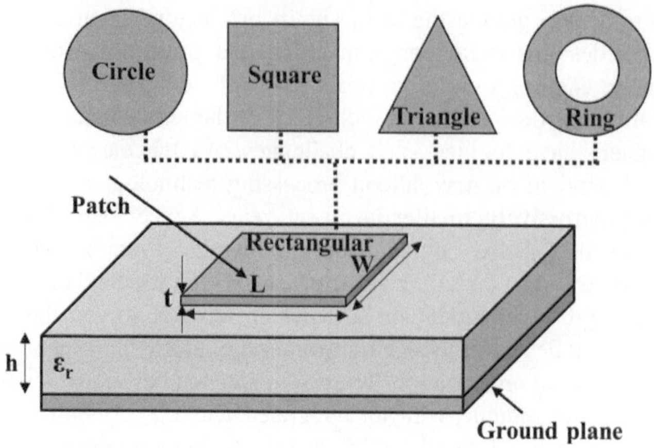

FIGURE 10.3 Physical structure of the rectangular microstrip patch antenna and collection of other possible patch shapes.

There are multiple techniques to feed the patch antenna viz coaxial feed, microstrip line feed, inset feed, proximity coupled feed, and aperture coupled feed [14,15]. Electromagnetic energy from the feeding line is first directed or coupled and concentrated mostly in the region below the patch, which serves as a resonant cavity with open circuits at two edges. Some of this energy is flowing out of the cavity and radiating into space, contributing to the antenna radiation. Microstrip antennas have many advantages compared to others such as low profile, conformal, rugged, and suitable for PCB manufacturing, easy integration with other circuit elements, easy to design for dual polarization, and can support multi frequency operation. Considering these examples, the microstrip patch antenna has gained much popularity among antenna designers for designing antennas for commercial IoT applications, mobile, and other wireless communication systems. Some of the commonly used microstrip antenna used for IoT applications are given below:

10.4.1 PCB Antenna

PCB antenna is also commonly used for IoT applications as shown in Figure. 10.4. They are usually the metallic trace of some form like meander or loop. Since these antennas are compact and don't provide a complete 50 Ω matching, a separate LC based matching network is usually required at the input end to match the complex impedance. Some of the common challenges encountered in the design of PCB antenna are the proximity of the nearby electronic component, metal or plastic enclosure, and ground plane variation changes the input impedance of antenna.

10.4.2 External Microstrip Antenna

External microstrip patch antenna is the antenna connected to the IoT module using a coaxial cable. Such antenna is usually placed outside the IoT module at some

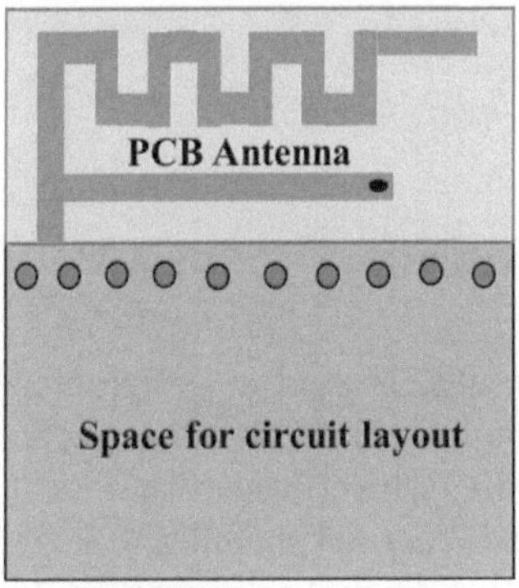

FIGURE 10.4 Schematic of PCB antenna.

distance apart as shown in Figure 10.5, which show a co-planar waveguide (CPW) feed circular microstrip patch antenna. External antenna provides more stable radiation pattern and coverage as it is free from unwanted radiation coming out from the connected circuit PCB. These antennas are usually omni-directional antenna and thus capable to receive/transmit signal from any direction.

10.4.3 Chip Antenna

Chip antennas (CA) are a specific type of microstrip antenna recognized for its low footprint as shown in Figure 10.6. They are more generally integrated into printed circuit boards at some outer location to avoid interference with the other circuit

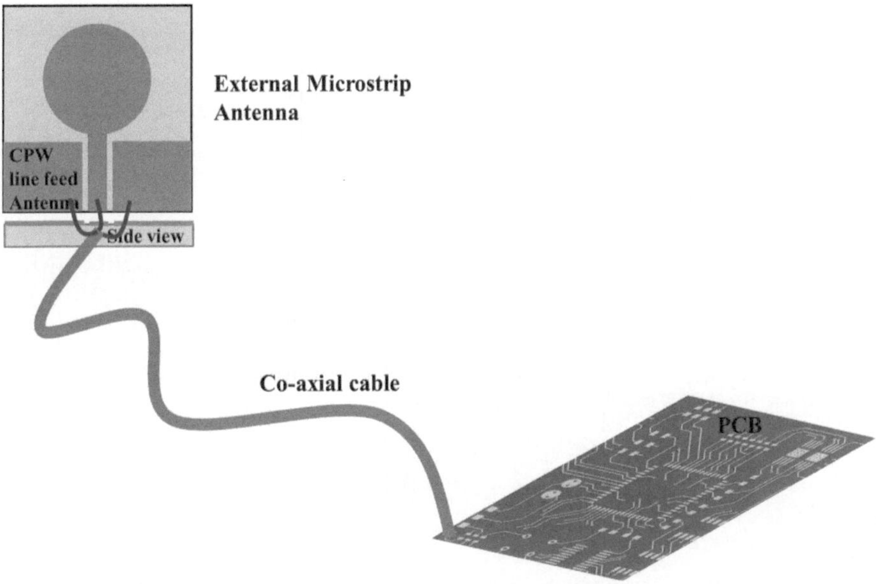

FIGURE 10.5 CPW-fed external antenna connected to PCB using a coaxial cable.

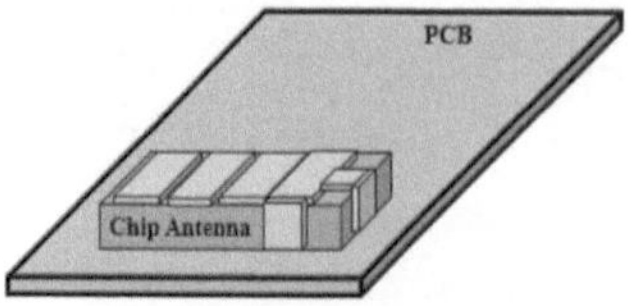

FIGURE 10.6 Schematic of chip antenna.

elements. The high permittivity materials such as ceramic is used to design chip antenna. They act as a standalone radiating element to radiate electromagnetic waves at high frequencies. They occupy a limited space, which makes them ideal for compact devices like cell phones, Wi-Fi routers, and other IoT modules. CA are usually cheaper in bulk production and a best alternative when it is impractical to use large-size antenna.

10.4.4 Wire Antenna

Wire antennas can be used in almost any design, as among the lowest-cost antenna with highest design flexibility. For optimal designs, these antennas typically require electromagnetic simulation at the design frequency. They can also demand custom designs depending on the housing or casing of the IoT module. As frequency decreases, wire antennas often increase in size, which can contribute some associated challenges in reproducibility and reliability. Wire antennas can be designed in half or quarter wave configuration as shown in Figure 10.7, which depends on the design requirement. An additional advantage of a wire antenna is that they are usually cost-effective. Almost any type of wire antenna can be designed at any frequency with very little lead time.

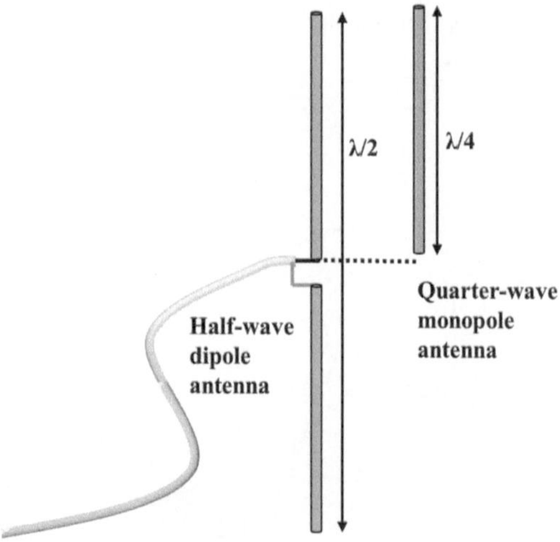

FIGURE 10.7 Schematic of wire antenna.

Most of the time it is required that the IoT module support complex multiple wireless transceivers. Multiple wire antenna can be used to address such problems.

10.4.5 Radio Frequency Identification Tag (RFID) Antenna

RFID system mainly consists of three parts, a reader antenna used for scanning, an RFID tag antenna that excites the chip that contains all the product related information. RFID technology can become a popular opportunity in the IoT system, where each tiny object or thing can be uniquely identifiable by just attaching a RFID tag antenna along with the data chip. RFID tag antenna can be designed in any shape and size depending on the operating frequency. The main operation of a RFID tag is to receive and transmit radio waves using the antenna element connected to the tag. The RFID tag antenna can able to generate enough power using these radio waves without the use of any external battery, and such tags are also known as passive tags.

A schematic diagram of a RFID system showing the tag antenna and reader antenna is shown in Figure 10.8. A reader control unit has access to all the data stored in all the tags further to any computer or any other IoT gateway.

10.5 COMPARISON OF VARIOUS ANTENNA TOPOLOGIES

As the IoT modules continue to shrink in size and adding more wireless technologies, it is becoming an extremely big challenge to create room for antennas. With the goal of computing, connecting, and sensing all interconnected things, IoT modules are developing many applications for the consumer and industry. To reduce the size of such devices while retaining high performance under low-power

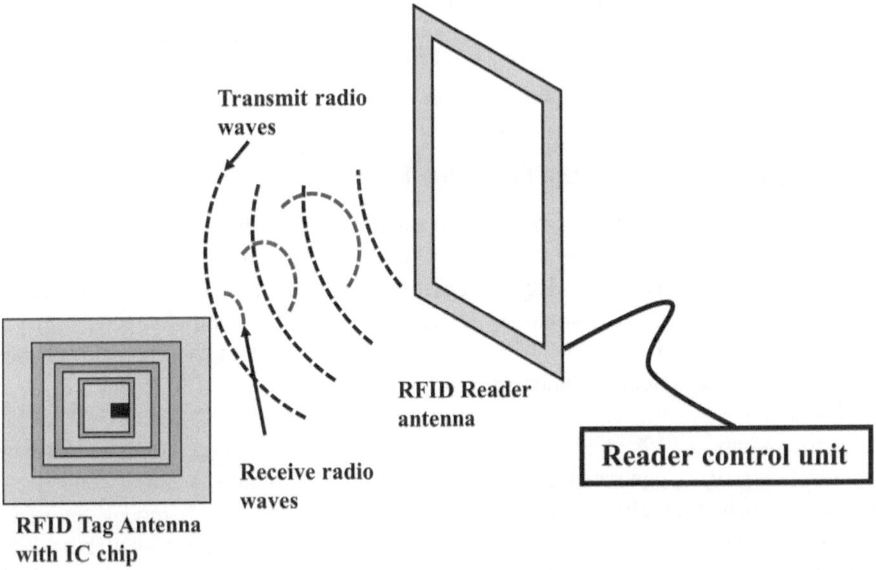

FIGURE 10.8 Schematic of RFID system and components involved in it.

conditions, a number of smart strategies are required. One of the commonly used techniques to minimize the antenna size is the folding technique, which not only minimizes the design effort and complexity but also lowers the design cost. Table 10.1 describes the commonly used IoT topologies used for antenna with some advantages and disadvantages.

10.6 CURRENT TRENDS IN THE DESIGN OF ANTENNAS FOR IOT APPLICATIONS

This section is devoted to various types of IoT antenna designed in the recent past, and this will help the antenna designer to select the right antenna suitable for desired IoT applications.

10.6.1 Low Profile PCB Multi-Band Antenna

Multi-band IoT antenna is useful where a single antenna can be used for many applications at operating at different frequency bands. Figure 10.9 shows the geometry of the multi-band PCB antenna [9] with major dimension are $L = 49$ mm, $W = 39$ mm, $L_1 = 17$ mm, $L_2 = 10$ mm, $W_1 = 4$ mm, $W_2 = 5.2$ mm, $G = 0.57$ mm, $R_1 = 16$ mm and $R_2 = 18$ mm. The multiband antenna consists of two monopole antenna with two folded microstrip arms to form a low-profile and compact structure. The antenna supports GSM (900/1800 MHz), LTE (1900/2100/2300/2500 MHz), and WLAN (2450 MHz) bands covering most of the IoT applications.

The antenna occupies a dimension of 49 mm × 39 mm where and printed on FR4 substrate having relative permittivity of 4.4 and thickness 1.5 mm. The antenna is simple in design and a potential candidate for internal antenna to be used in IoT device. This antenna provides almost omni-directional radiation pattern which distributed uniformly in all direction. The simulated (S_{11}<−10 dB) impedance bandwidth is 38% at higher LTE and WLAN band (1700–2500 MHz) and 12% at lower GSM band (850–960 MHz) with an average gain of around 2–3 dBi in these bands.

10.6.2 CPW-Fed Circular Polarized External IoT Antenna

Sometime it is not possible to integrate the antenna along with the circuit board, and external antennas are used in such circumstances. In [10], a CPW-fed circular polarized antenna is presented for location based IoT applications. The antenna designed to operate in required GPS band, i.e., L1 (1575.42 MHz) and simultaneously cover other bands such as L2 (1227.6 MHz) and L5 (1176.45 MHz). A schematic of the external CPW-fed antenna connected to a GPS receiver is shown in Figure 10.10.

The antenna provides a right-hand circular polarization (RHCP) with a 3 dB axial ratio bandwidth of 11% (1.15–1.29 GHz) and 18% (1.5–1.8 GHz). Here the circular polarization (CP) is generated by inserting slots in the ground structure, which provides the quadrature phase necessary to generate two orthogonal modes for generating CP. The antenna can be integrated with the commercially available GPS receiver to get the correct location coordinates. The measured gain value is found to be in between 1.5 to 3.5 dBi with almost onmi-directional radiation pattern.

TABLE 10.1
A Comparison of Various Antenna Topologies used for IoT

Antenna Design Type	Advantages	Disadvantages
PCB	• Such antennas are low cost antenna. • Occupies less area ~25 × 15 mm^2 • Many different geometries already reported. • Many simulation software available. • Matching network can also be designed on the same PCB, if required. • Compact size at higher frequencies.	• At lower frequencies, size becomes large. • Antenna efficiency reduced. • Separate clearance area required. • Not possible to redesign. • Multiple antenna design not possible.
Chip	• Such antennas are compact compared to PCB. • Low frequency can be realized using multilayer approach. • Easily reproducible with less test cycle. • Easily tunable and mountable to any surface.	• Less performance compared to PCB antenna. • Redesign and tuning not possible. • Usually costly compared to PCB.
Wire	• Simple design and high performance. • Consume very less footprint on PCB. • Simple simulation for any frequency. • Flexibility in tuning. • Provides satisfactory performance and usually cheaper.	• It is an external antenna which consume more space. • Very large size at lower frequency.
External microstrip antenna	• Such antenna can be designed on any substrate. • Light weight and low profile. • Can be designed for narrowband, wideband and ultra-wideband. • Conformal to any host surface. • Can be designed on textile material. • Multi-input-multi-output (MIMO) configuration for diversity applications.	• Suffers from dielectric losses and conductor losses. • Offers lower gain. • Lower power handling capability. • Spurious radiation.

10.6.3 Wearable Antenna for Wireless Body Area Networks

Wearable antenna used in wireless body area networks (WBANs) has found many IoT applications that can be useful to communicate with both in-body and off-body devices nearby. One of the primary requirements of such antenna is to have flexible design based on textile material using planar microstrip line conformal to any surface [16]. While designing wearable antenna, a designer should concentrate on

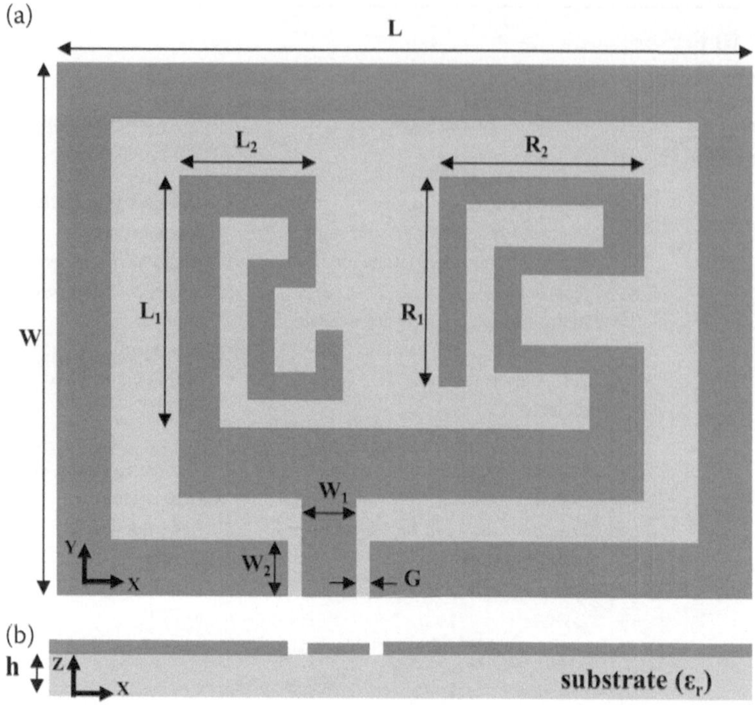

FIGURE 10.9 Geometry of the proposed antenna (a) top view (b) side view.

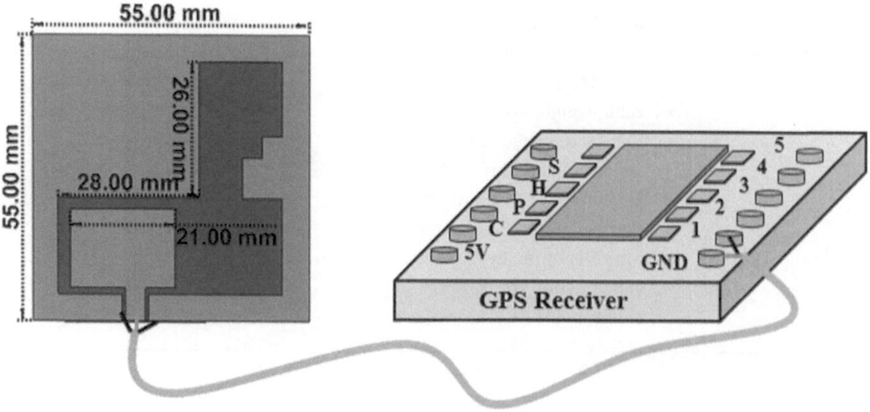

FIGURE 10.10 CPW-fed external antenna connected to a gps receiver via coaxial cable.

various antenna characteristics and their response to the nearby human body under various conditions such as bending, rotation, tilting, etc. One such design is reported in [17], where a button type antenna is proposed for wearable applications. The reported antenna is a dual-band antenna, which covers 2.45 GHz and 5.8 GHz band

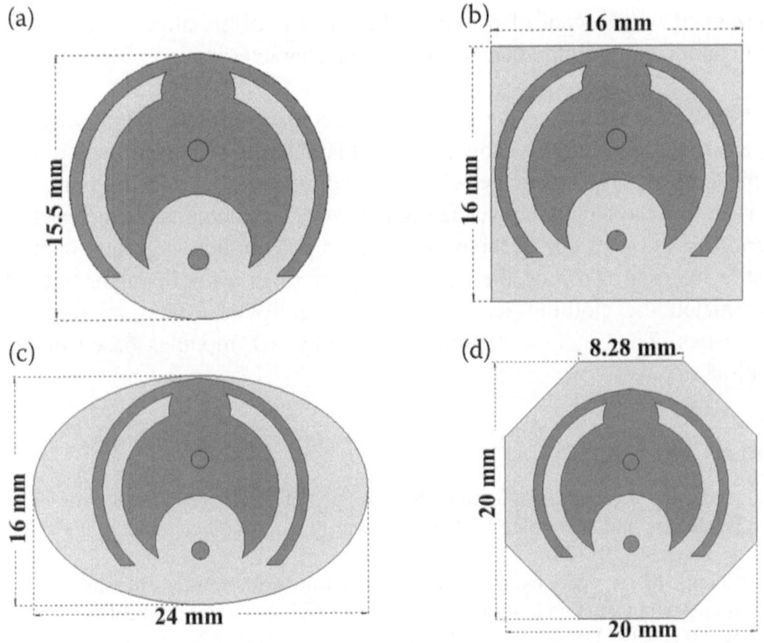

FIGURE 10.11 Different button geometries: (a) circular, (b) square, (c) ellipse, and (d) octagon.

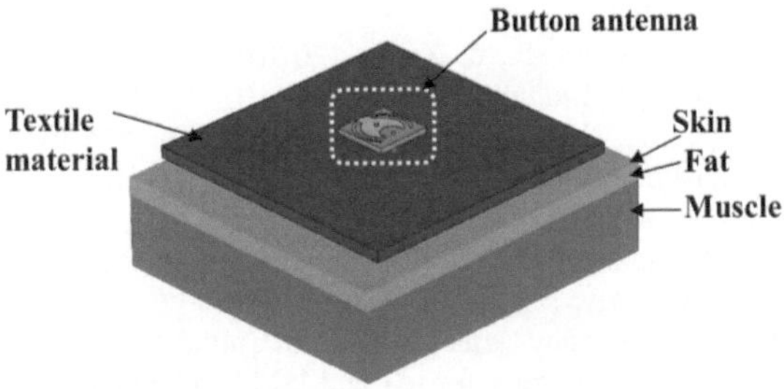

FIGURE 10.12 Square button antenna placed on large textile material above bio-tissue.

useful for on-body and off-body communications. Here on-body devices can be other wearable or implantable devices whereas off-body devices can be a nearby wireless system such as Wi-Fi. Figure 10.11 shows the different button geometries designed on a substrate with permittivity of 2.2 and thickness 1.5 mm, that can be used as a button in wearable shirt or coat having dimension equivalent to that of a button. This button patch is placed on top of a large ground plane that can be made using conductive garment or flexible substrate.

The antenna provides a narrow band resonance at 2.45 GHz with a gain of 2.3 dBi while wideband resonance between 4.6–6.4 GHz covering the desired

5.8 GHz band with a gain of 5.1 dBi. The antenna offers almost omnidirectional radiation pattern in both the bands and a stable characteristic considering the tilting effects.

To monitor the effects on the antenna characteristics due to the presence of bio-tissue, a simulation study is shown where this antenna is placed on top of a bio-tissue made up of three layers, i.e., skin, fat, and muscle as shown in Figure 10.12. It is observed that the bio-tissue has negligible impact on antenna characteristics such as return loss and gain due to the presence of large conductive ground plane below the textile material. To hide the circular copper rings a textile superstrate can be used to match the clothing style, which has negligible effects on the radiation characteristics. The antenna is suitable for many IoT modules based on WBAN applications.

REFERENCES

[1]. Ande, R., Adebisi, B., Hammoudeh, M., & Saleem, J. (2020). Internet of Things: Evolution and technologies from a security perspective. *Sustainable Cities and Society*, *54*, 101728.

[2]. Chung, M. A., & Chang, W. H. (2020). Low-cost, low-profile and miniaturized single-plane antenna design for an Internet of Thing device applications operating in 5G, 4G, V2X, DSRC, WiFi 6 band, WLAN, and WiMAX communication systems. *Microwave and Optical Technology Letters*, *62*(4), 1765–1773.

[3]. World Population Projections. https://www.worldometers.info/world-population/world-population-projections/. [Accessed 27May2020].

[4]. G. D. Maayan, The IoT rundown for 2020: Stats, risks, and solutions. https://securitytoday.com/Articles/2020/01/13/The-IoT-Rundown-for-2020.aspx?Page=2. [Accessed 27June2020].

[5]. Jurcut, A. D., Ranaweera, P., & Xu, L. (2020). Introduction to IoT security. *IoT Security: Advances in Authentication*, 27–64. 10.1002/9781119527978.ch2.

[6]. Boyuan, M., Pan, J., Wang, E., & Yang, D. (2020). Wristwatch-style wearable dielectric resonator antennas for applications on limps. *IEEE Access*, *8*, 59837–59844.

[7]. Elijah, A. A., & Mokayef, M. (2020). Miniature microstrip antenna for IoT application. *Materials Today: Proceedings*, *29*, 43–47.

[8]. Olan-Nuñez, K. N., Murphy-Arteaga, R. S., & Colmn-Beltrán, E. (2020). Miniature patch and slot microstrip arrays for IoT and ISM band applications. *IEEE Access*, *8*, 102846–102854.

[9]. Birwal, A., Singh, S., & Kanaujia, B. K. (2020). Smart compact-folded microstrip antenna for GSM, LTE, and WLAN applications. In *Smart Systems and IoT: Innovations in Computing* (pp. 475–481). Singapore: Springer.

[10]. Birwal, A., Singh, S., Kanaujia, B. K., & Kumar, S. (2019). Broadband CPW-fed circularly polarized antenna for IoT-based navigation system. *International Journal of Microwave and Wireless Technologies*, *11*(8), 835–843.

[11]. Qu, L., Xu, Y., Piao, H., & Kim, H. (2019). Antenna performance improvement by utilizing a small parasitic slot in stacked PCBs for IoT devices. *Journal of Electromagnetic Waves and Applications*, *33*(17), 2287–2295.

[12]. Singh, P. P., Goswami, P. K., Sharma, S. K., & Goswami, G. (2020). Frequency reconfigurable multiband antenna for IoT applications in WLAN, Wi-Max, and C-Band. *Progress in Electromagnetics Research*, *102*, 149–162.

[13]. Balanis, C. A. (2016). *Antenna Theory: Analysis and Design*. New York: John Wiley & Sons.
[14]. Fang, D. G. (2017). *Antenna Theory and Microstrip Antennas*. Boca Raton: CRC Press.
[15]. Hedayati, M. K., Abdipour, A., Shirazi, R. S., Ammann, M. J., John, M., Cetintepe, C., & Staszewski, R. B. (2019). Challenges in on-chip antenna design and integration with RF receiver front-end circuitry in nanoscale CMOS for 5G communication systems. *IEEE Access*, *7*, 43190–43204.
[16]. Jiang, Z. H., Brocker, D. E., Sieber, P. E., & Werner, D. H. (2014). A compact, low-profile metasurface-enabled antenna for wearable medical body-area network devices. *IEEE Transactions on Antennas and Propagation*, *62*(8), 4021–4030.
[17]. Birwal, A., Singh, S., & Kanaujia, B. K. (2019, December). Dual-band circular-button antenna for WBAN applications. In *2019 IEEE Indian Conference on Antennas and Propogation (InCAP)* (pp. 1–4). New York: IEEE.

11 Importance and Uses of Microstrip Antenna in IoT

Neha Parmar
Core IoT Technologies, Andhra Pradesh, India

11.1 INTRODUCTION TO IOT

1. Internet of things (IoT) has a broad and diverse ecosystem that includes a wide range of different connectivity types. These types are mainly divided into 5 stages that are sensors/controllers, gateway devices, communications network, software for analyzing and translating data, and end application services. These end application services are used for smart and connected cities, connected vehicles, Industrial IoT, smart home, smart health care, smart agriculture, medical instruments connectivity, etc. The technologies used to communicate the end devices for providing services are Cloud computing, low-power wide area technologies, big data analytics, network function virtualization (NFV) and software defined networking (SDN), 5G, and edge computing [1].
   The end devices utilize sensors/actuators, Micro-Electro-Mechanical Systems (MEMS) devices, batteries/cells, energy harvest techniques, radio modules, and antennas. The choice and design of the antenna varies depending on the transmission signal strength, frequency range of transmitted signal, and the space available [2] (Figure 11.1 and Table 11.1).
2. IoT networks and devices support licensed and unlicensed sub—6 GHz bands using bandwidth less than 5 MHz, Long-Term Evolution (LTE)—A uses up to 20 MHz with career aggregation, and mm-Wave spectrum utilizes 5G. To guide uplinks and downlinks between gateways and end devices there is a requirement of IoT network. IoT network used for this purpose are mesh, star, or point to point technologies. The communication link is accomplished by Omni directional antenna structures such as Chip, PCB, Whip, rubber duck, patch, and wire antenna [2].

The antennas are usually used in those IoT devices that have GPS, Bluetooth, Wi-Fi, and other communication capabilities. The signals transmitted by satellites are

DOI: 10.1201/9781003187325-11

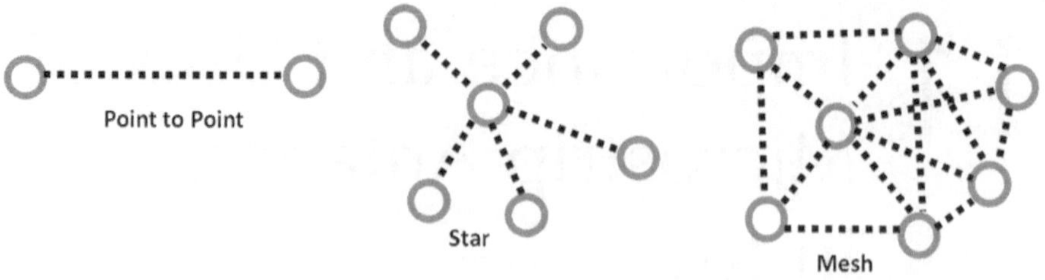

FIGURE 11.1 To relay data between node-node or gateway-node, star and mesh topologies are used. Mesh topologies are limited to short distance and will always be on standby. Star topology based devices can go into sleep mode [2].

TABLE 11.1
IoT Technologies and Their Frequencies [2]

Applications	IOT Technologies	Frequency
General (Smart home, smart building, etc.)	Zigbee	915 MHz, 2.4 GHz
	Z wave	2.4 GHz
	Bluetooth	2.4 GHz
	WiFi	2.4 GHz, 3.6 GHz,4.9 GHz, 5 GHz, and 5.9 GHz
	GPS	1575.42 MHz, 1227.6 MHz, 1176.45 MHz
IIoT	Wireless HART	2.4 GHz
	ISA 100.11a	2.4 GHz
LPWAN (Smart city, smart farming, etc.)	LORA	915 MHz for the U.S. 868 MHz for Europe, and 433 MHz for Asia
	Sigfox	868 MHz, 902 MHz
Medical	MBAN (IEEE 802.15.6)	2360 to 2400 MHz
	WBAN (IEEE 802.15.6)	2.4 GHz, 800 MHz, 900 MHz, 400 MHz
Avionics	WAIC	4200 MHz to 4400 MHz

either right handed circular polarization (RHCP) or left handed circular polarization (LHCP). The patch antennas are also designed for dual polarization, and it can be reconfigured by using PIN diodes or RF MEMS switches. Moreover, antenna is designed for various configurations like High/Low Bandwidth, Reconfigurability of frequency, polarization, bandwidths, operating frequencies, polarizations, radiation patterns, smart beam width and pattern, ultra wide band of 1–3 GHz, circular polarized radiation pattern, miniaturization, multiband operated frequency bands, low cost antennas, etc.

Antenna Selection

There some antennas that have some of these configurations such as parabolic antenna, lens antenna, yagi uda antenna, microstrip antenna, etc.

Parabolic antenna has high gain, high directivity, narrow beam widths and it is used in high frequency wave of 100 MHz to 1 GHz.

Lens antennas are mostly used in microwave relay communication and have less side lobes and back lobes. These antennas are difficult to manufacture with a complex structure, and they have lower efficiency and high manufacturing cost.

Microstrip antennas operated at microwave frequencies, have small size, minimum excitation, come with various shapes (such as rectangular, triangular, square, other polygon), lower fabrication cost, supports multiple frequency bands, supports dual polarization types, light in weight, and robust while mounting on rigid surfaces. Besides these advantages, antenna fabrication in integrated circuit technology allows higher accuracy, which was difficult for traditional antennas to achieve. Microstrip antennas are also easily etched on PCB and provide easy access for design troubleshooting. Hence it is easily integrate with MICs and MMICs.

IoT uses NBIoT, LTE Cat1, Cat M1, EC-GSM IoT, Zigbee, Google thread, LoRa, SigFox, and Bluetooth LE, etc. All the radiation properties used in these IOT devices were only suited to microstrip antenna.

The following are the listed use of microstrip patch antenna for different applications in IoT devices:

1. The narrowband internet of things (NB-IOT) is mostly used in Low Power Wide Area (LPWA). It uses minimum bandwidth for uplink and downlink of 180 kHz. The area coverage by NBIOT is 1 to 8 km for urban areas and up to 25 km in rural areas. The NBIOT devices will have long operation, low manufacturing coverage, improved efficiency, and wide area coverage [3]. The requirements of narrow bandwidth IoT devices are perfectly fulfilled by patch antennas. The inverse relation between relative permittivity of substrate and the bandwidth is used. The higher value of substrate permittivity causes narrow bandwidth and also allows shrinking of antenna and high impedance of antenna [4]. The relation between the length of patch antenna with permittivity and bandwidth scaling with length and permittivity can be represented in equation form, which is written below:

$$L = \frac{1}{2f_c\sqrt{\varepsilon_0 \varepsilon_r \mu_0}} \tag{11.1}$$

$$B \propto \frac{\varepsilon_r - 1}{\varepsilon_r^{\,2}} \frac{W}{L} h \tag{11.2}$$

NB-IOT has application in wireless sensor networks, smart metering, smart cities, and e-Health and tracking. The use of narrow band information has low cost, low power consumption, and longer battery life.

2. Reconfigurable antennas are used for faster and accurate communication with different antenna modules and different methodologies, without changing the model structure of antenna switching between various bands in possible by reconfigurable antennas. The reconfigurability provides functioning of antenna in multi bands using switches [5]. The only way to introduce multiband functionality is by using microstrip antennas. To attain the multiband frequencies the multi length rectangular slots are made.
3. Low profile: The IoT systems provide a huge number of application in communication various devices worldwide, which also require securing data and identifying the suspicious devices. To track these devices and systems, low profile antennas are required. The antennas can be in the shape of Tie patches, L shape, U shape, etc., for the low profile multiband operations. Low profile antennas have compact size as their dimensions are reduced to small values and have short distance communication. However, the performances of antennas are not compromised [6].
4. High bandwidth: The IoT is a fast growing network of interconnected devices, sensors, and actuators. Wireless sensor technology plays an important role in improving device connection and efficiency of IoT applications. Antennas for IoT devices also plays an important role for the explosive development in various fields, including tracking, advertising, security building, smart cities, etc. Thus, there is high demand of antennas that have multi frequency and multi function operations, and they are easily fabricated and mounted on devices. Microstrip antennas have narrow bandwidth and it will limit the information transfer process, which will correspondingly increase the communication time. The bandwidth can be enhanced by using parasitic patch with other patch fed with power [7].

11.2 DESIGN OF MICROSTRIP ANTENNA FOR IOT APPLICATIONS

The designing of the microstrip antenna is dependent on the application and the transmission frequency, signal strength, and available space to deploy on the IoT devices. In this section we will look over the different designing of antennas for different applications and conditions.

The IoT sensors/devices are connected through cellular, satellite, Wi-Fi, Bluetooth, RFID, NFC, LPWAN, and Ethernet. For better connectivity low power consumption, large range and high bandwidth is best suited, but unfortunately this combination is not possible. Instead segments are made of various connectivity options and there are three groups.

High power consumption, long range and high bandwidth: It takes a lot of power to send lot of data wirelessly over a large distance. The cellular and satellite communication are the best chosen options for this group.

Low power consumption, small range and high bandwidth: When data transmission is large to send, then the distance of communication is reduced to consume less power. Wi-Fi, Bluetooth, and Ethernet provided this connectivity with small amounts of power.

Low power consumption, large range and low bandwidth: If the power consumption is to be reduced with a long range distance, then the only way to do so is to reduce the data quantity. Low power wide area networks (LPWANs) are used for this type of group of communication.

These three groups can be achieved by the microstrip patch antenna. A dielectric material separates the radiating patch and ground plane. There are two types of radiating patch and both have different radiation patterns: rectangular patch and circular patch.

Rectangular patch antennas are linearly polarized and show good results for Return loss, VSWR, Gain, and it has better suppression for side lobe level. Circular patch antenna shows better results for bandwidth, radiation pattern, impedance matching, and side lobes. According to the application requirement, the patch shape is decided. We will go through the latest applications for IoT device and its antenna design.

11.2.1 Patch Antenna Designing Having Circular Polarization

This microstrip antenna has different feeding methods and among them aperture coupling method is widely used because as the frequency rises, it cannot be physically used. It has advantages of presenting higher bandwidth and isolating radiation structure to the feed. Sometimes circular polarized antenna is used to improve reliability of communication and to overcome multipath fading problem. To obtain circular polarized a crossed slot is configured, which is formed by single microstrip feed placed diagonally to the cross. This is simple to design and has good polarization bandwidth [7] (Figure 11.2).

The quadrangular patch has crossed slot, and crossed slot is 45° shift with respect to feedline of microstrip. The slot width is set to be 10% of corresponding slot length. Slot length defines the coupling between feedline and patch, and thus it is

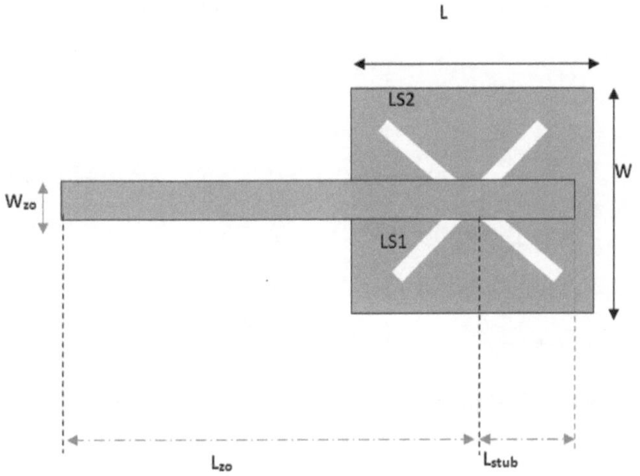

FIGURE 11.2 Design of microstrip patch antenna for circular polarization [7].

concluded that operating frequency is dependent on slot length. The ratio between slot lengths has an impact on axial ratio and impedance matching is affected by slot length [7].

$$\text{Slot length ratio, } K_S = \frac{L_{S1}}{L_{S2}}, \tag{11.3}$$

$$L_{S1} = \frac{2L_a}{K_S + 1}, \tag{11.4}$$

$$L_{S2} = \frac{2L_a K_S}{K_S + 1}, \tag{11.5}$$

where L_a is the average slot lengths, L_{s1} and L_{s2} are length of each slot. The values for aperture length and slot ratio are set and then all design parameters are calculated using the above equations. For this designing aperture average is set to 3.65 mm and slot ratio for good polarization is found to be 1.1 and 1.2. Two dielectric substrates are separated by a ground plane. The crossed slots are cut in the ground plane and patch element is the upper element. The feeding lines are present in the bottom substrate. The antenna was designed for 17 GHz band and for large bandwidth. Using Eqs. 11.1–(11.3) and these values other dimensions of crossed slot, patch, and feedline. The final dimension of antenna are $W = 4$ mm, $L= 4$ mm, $L_a = 3.65$ mm, $K_s = 1.15$ mm, $W_{z0} = 0.55$ mm, $L_{z0} = 12$ mm, $L_{stub} = 2.1$ mm.

The software design and hardware design of patch antenna for 17 GHz is shown in Figure 11.3. The substrate is made by Rogers RO4725JXR, which has substrate thickness of 0.78 mm, dielectric constant of 2.55, and loss tangent of 0.0026 at frequency of 10 GHz.

The designed antenna is simulated in high precision software such as Ansys HFSS, CST microwave studio, and the results are obtained.

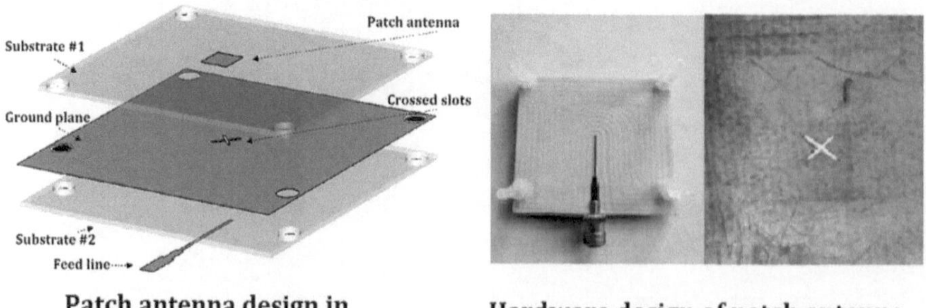

Patch antenna design in simulation software

Hardware design of patch antenna

FIGURE 11.3 Patch antenna design for circular polarization at 17 GHz and the hardware design of patch antenna with crossed slots [7].

The S_{11} was found to be less than −10 dB and −30 dB for the 17 GHz. The measured bandwidth is 2.2 GHz and gain of antenna is 5.8 dBi. For good circular polarization, axial ratio was considered less than 3dB, and it was found that the antenna was circular polarized for 350 MHz band centered at 17 GHz. The obtained bandwidth is more than 2 GHz and the size of antenna is also reduced, and thus they are important aspects for antenna to be integrated with IoT devices or sensors [7].

The use of circular polarized is more stable and has reliable links when orientation and direction of other antenna of IoT device are constantly changing. Thus, circularly polarized antennas are used where orientation of other device is unknown. These types of antennas have applications in mobile Wi-Fi, Bluetooth and ISM applications, Autonomous Vehicles, Unmanned Aerial Vehicle (UAV), and Robotics.

The above microstrip patch designing was more focused to produce a circular polarization and aperture fed coupling. For other application requirements, let us move to the next section.

11.2.2 Microstrip Patch Antenna With High Directivity and Diversity for IoT Devices

As we know, microstrip patch antennas consisted of a thin metallic strip. The height of substrate from the ground plane is dependent on the wavelength. Various types of substrate can be used to design microstrip antenna but the performance of antenna is mainly dependent on dielectric constant of substrate, and hence it is to be considered properly. Array antenna at gateway improves the radio link such as that transmission power is reduced by antenna factor.

Antenna array of microstrip patch is used to achieve higher gain (directivity) and to give diversity. These two parameters are used in MIMO and it also increases communication reliability. The selection of feeding method is really important as it affects bandwidth, size of patch, return loss, and smith chart. There are four most popular feeding methods present: coaxial probe, microstrip line, aperture coupling, and proximity coupling. Among all these, microstrip line feed is chosen as it is easy to fabricate, easy for impedance matching, and has ease in modelling [8].

The two patches used for antenna array have substrate of dielectric constant 4.3 and 2.4 GHz operated frequency with thickness of 1.5 mm. Figure 11.4 shows a single microstrip antenna that would be used to make an array of antenna. The two array antennas are separated by a separation of distance L.

Using two patches microstrip antenna as shown in Figure 11.4, the results were obtained of magnitude and phase of S11 as shown in Figure 11.5, antenna parameter values over frequency ranges. The impedance is $z' = 0.928 - j0.064$ represented by yellow circle and $r = 0.050\angle - 136.406°$ of S11 parameter in 2.4 GHz.

The resulted gain is 6.9 dBi at 2.4 GHz. By observing result of gain, it can be concluded that effective output power is increased by focusing on radiated signals. The obtained efficiency is 72.2% and power radiated is 0.002 watts at 2.4 GHz. The directivity of the designed antenna array is 8.328 dBi.

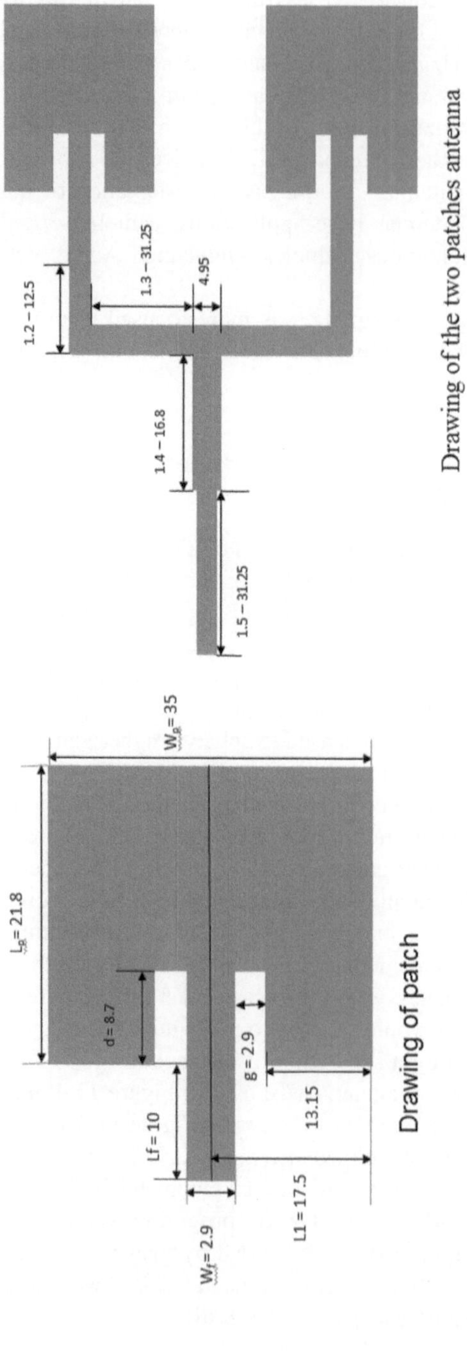

FIGURE 11.4 (a) Design of microstrip antenna for high directivity communication (b) Designing of two patches antenna array [8].

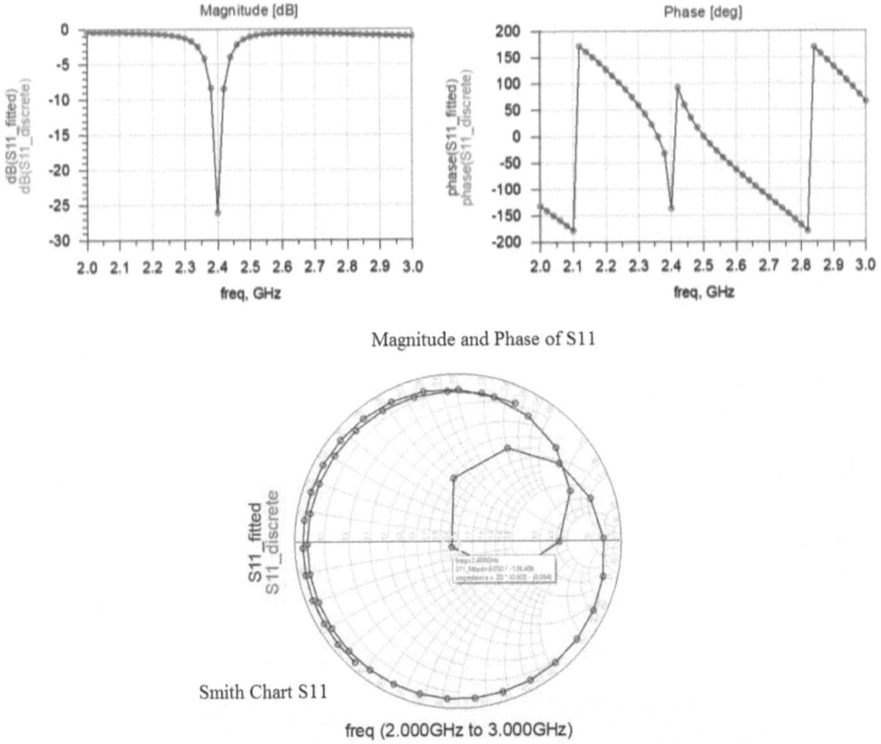

FIGURE 11.5 S11 parameters results in (a) magnitude and phase (b) smith Chart [8].

The radiation pattern obtained from the energy source is –40dBi at 2.4 GHz, which is also shown in Figure 11.6.

Array antennas are used for wearable technologies, mm-Wave mobile communications, 5G communication, wireless sensor networks, UHF, ISM bands, and cognitive radio. The obtained results from microstrip patch antenna array are favorable to use for IoT applications.

11.2.3 Multiband Microstrip Patch Antenna for IoT Devices

There are various methods to design a microstrip patch antenna by varying the feeding techniques, dielectric materials, number of substrate layer, modifying the ground plane structure, reshaping the geometry of patch antenna, etc. In all these ways, two ways for designing a antenna for multiband purpose are discussed in this section. Wireless communication capability is increased due to various portable devices in industry. There are many embedded sensors to these devices. Hence, terrestrial network has lack of connected network problem to all devices and sensors. For this purpose, satellite communication is established. The IoT devices can be WiMAX, WLAN, Bluetooth, etc.

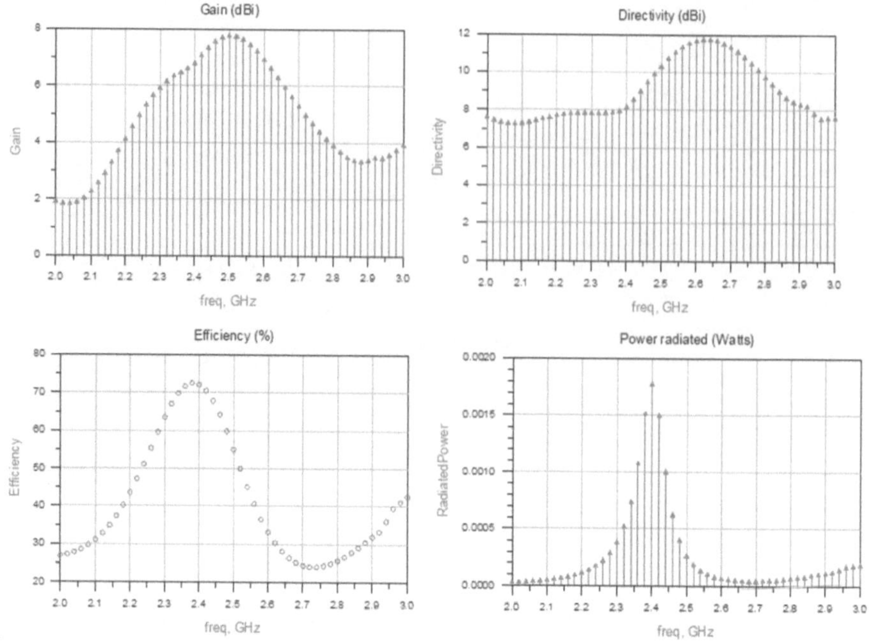

Antenna parameters versus frequency

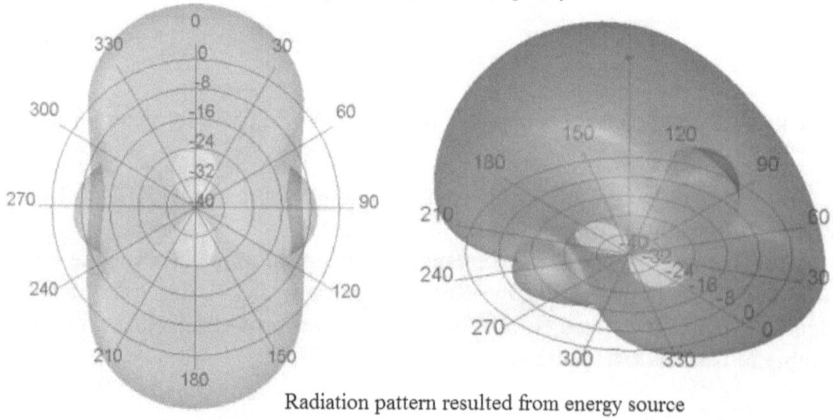

Radiation pattern resulted from energy source

FIGURE 11.6 (a) Antenna parameters such as gain, directivity, efficiency, and power radiated vs frequencies and (b) radiation pattern at 2.4 GHz [8].

a. **Multiband microstrip patch antenna with a rectangular shape patch**

The microstrip patches can be rectangular, square, triangular, etc. When the width and length of antenna is reduced, the antenna gets miniaturized and resonant frequency also goes higher. The substrate material plays an important role in designing aspect, and a larger substrate will result in large bandwidth desired for IoT devices. Thus for substrate FR4 is used that has dielectric constant of 4.4. The height of substrate of 1.6 mm and 0.002 values is chosen for loss tangent. Microstrip strip feed line is used as feeding technique as it has easy fabrication, good radiation by feed, and good matching of impedance. Let us discuss the designing process by finding the values of dimension of antenna patch, feedline, and effective constant [9].

The width of patch in microstrip antenna is calculated by [10,11] (Figure 11.7).

$$W_p = \frac{c}{2f_r\sqrt{\frac{\varepsilon_{r+1}}{2}}} \tag{11.6}$$

where f_r: resonating frequency

C: speed of light in air = 3 × 108 m/sec

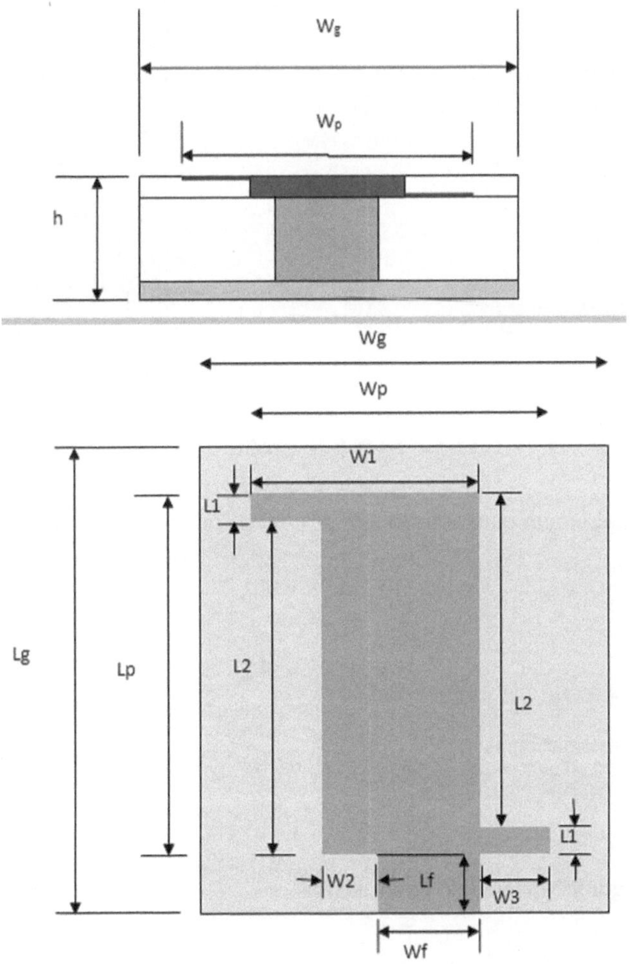

FIGURE 11.7 (a) Side view (b) Top view of antenna design for multiband [9].

ε_r: dielectric constant = 4.4

Effective dielectric constant, ε_{eff} of an antenna is given by

$$\left.\begin{aligned} Z_o &= Z_f 2\pi\sqrt{\varepsilon_{\text{eff}}}\ln\left(8\frac{h}{W_p}+\frac{W_p}{h}\right) \\ \varepsilon_{\text{eff}} &= \frac{\varepsilon_r+1}{2}+\frac{\varepsilon_r-1}{2}\left[\left(1+12\frac{h}{Wp}\right)^{-1/2}\right] \end{aligned}\right\}\frac{W_p}{h}>1, \tag{11.7}$$

$$\left.\begin{aligned} Z_o &= \frac{Z_f}{\sqrt{\varepsilon_{\text{eff}}}\left(1.393+\frac{W_p}{h}+\frac{2}{3}\ln\left(\frac{W_p}{h}+1.44\right)\right)} \\ \varepsilon_{\text{eff}} &= \frac{\varepsilon_r+1}{2}+\frac{\varepsilon_r-1}{2}\left[\left(1+12\frac{h}{Wp}\right)^{-1/2}\right] \end{aligned}\right\}\frac{W_p}{h}<1. \tag{11.8}$$

To find patch length L_p, first it is to find effective length (L_{eff}) and extended length of antenna, which are as follows [12]:

$$L_{\text{eff}} = \frac{c}{2f_o\sqrt{\varepsilon_{\text{eff}}}}, \tag{11.9}$$

$$\Delta L = 0.412h\frac{(\varepsilon_{\text{ef}}+0.3)\left(\frac{W_p}{h}+0.264\right)}{(\varepsilon_{\text{eff}}+0.3)\left(\frac{W_p}{h}+0.8\right)}, \tag{11.10}$$

$$L_p = L_{\text{eff}} - 2\Delta L. \tag{11.11}$$

The width and length equations of the ground plane are as follows:

$$L_g = 6h + L_p, \tag{11.12}$$

$$W_g = 6h + W_p. \tag{11.13}$$

Feed length estimation is

$$Lf = \lambda_g/4, \tag{11.14}$$

where λ_g is guide wavelength expressed as

$$\lambda_g = \frac{\lambda}{\sqrt{\varepsilon_{\text{eff}}}}. \tag{11.15}$$

$$\text{And}\lambda\text{canbeexpressedas}\lambda = \frac{c}{f_o}. \quad (11.16)$$

Results: After determining the design parameters of microstrip antenna, the bandwidth, return loss, frequency of multiband, gain, VSWR, etc. can be determined by simulating the designed structure on simulating software Ansys HFSS. The calculated design parameters are shown in Table 11.2.

The simulated results for three frequencies bands 5.8 GHz, 6.76 GHz, and 8.4 GHz have return loss of −36.67 dB, −27.22 dB, and −40.83 dB respectively. The bandwidth also is an important parameter that represents the satisfactory radiate or receives energy over a range of frequencies. The obtained bandwidth is 170 MHz, 335 MHz, and 560 MHz with return loss of 10dB for 5.8 GHz, 6.76 GHz, and 8.4 GHz critical frequencies respectively. The VSWR has to be less than 2 for better matching with transmission line or feedline and more power transfer. VSWR value can be determined by the following equation:

$$\text{VSWR} = \frac{1+\Gamma}{1-\Gamma}. \quad (11.17)$$

Thus the calculated VSWR are 1.029, 1.090, and 1.021 for the three frequencies of 5.8 GHz, 6.75 GHz, and 8.4 GHz respectively.

The radiation gain obtained is 3.95 dB, 5.05 dB, and 11.17 dB for the tri bands 5.8 GHz, 6.76 GHz, and 8.4 GHz respectively. Thus the high gain value shows that high efficiency of antenna and the quality effectiveness of antenna to convert input power into radio waves for required direction. The resonating frequency lies in C-band and X-band for the designed microstrip antenna. The other way to design the microstrip antenna for multiband application is discussed in the following section.

b. **Microstrip circular patch antenna for multiband IoT applications**

Introduction and Design: For designing patch antenna a icosidodecahedron shaped patch is made. This unique shape results in seven frequency band under UHF and SHF

TABLE 11.2
Calculated Antenna Parameters Value [9]

Parameter	Value (mm)	Parameter	Value (mm)
Ground width (Wg)	38.429	Ground length (Lg)	46.86
Patch width (Wp)	28.83	Patch length (Lp)	37.26
Feed width (Wf)	10.01	Feed length (Lf)	4.8
W1	23.92	L1	2.26
W2	4.99	L2	35
W3	9.41		

frequency bands. The material used for substrate is FR4, which has dielectric constant of 4.4 and loss tangent taken as 0.02. As we know various benefits of choosing microstrip feedline such as inexpensive to manufacture, easy to fabricate, has polarization diversity, light weight, and good radiation pattern. The planar projection of icosidodecahedron model is introduced as this shape has the largest peripheral length as compared to other simple polygons. Using this shape, the surface current circulation will be boosted, and this will lead to multiband response of antenna [13] (Figure 11.8).

The ground plane has introduced slots in the structure that are said to be defected ground structure (DGS). The DGS structure will disturb the shield current distribution and accordingly nature of inductance and capacitance will change. To enhance the bandwidth of antenna, rectangular patches were introduced in the icosidodecahedron structure.

Results: When proposed design is simulated, then the reflection coefficient values as good value for bandwidth of 140 MHz (1.47–1.61 GHz), 350 MHz (2.59–2.94 GHz), 120 MHz (3.57–3.69GHz), 140 MHz (3.94–4.08 GHz), 240 MHz (4.49–4.73 GHz), 200 MHz (5.2–5.4 GHz), 250 MHz (6.35–6.6 GHz) was found.

The antenna has gain of 23.785 W/Sr, 10.769 W/Sr, 5.1068 W/Sr, 1.0253 W/Sr, 3.0271 W/Sr, 1.3695 W/Sr, 3.3601 W/Sr at 1.54 GHz, 2.73 GHz, 3.63 GHz, 4 GHz, 4.61 GHz, 5.32 GHz, 6.44 GHz frequencies respectively. Designed antenna covers

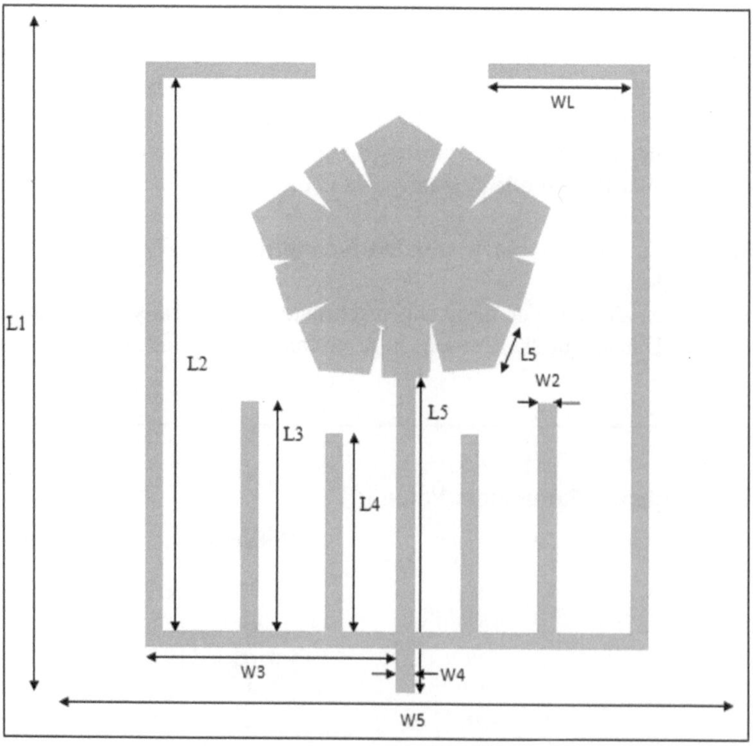

FIGURE 11.8 Icosidodecahedron shaped patch antenna design [13].

frequency ranges of SHF, Microwave, and UHF bands, which makes antenna suitable for WLAN and Wi-Fi.

Applications: Different communication technologies operated at different band are possible by multiband antenna. The multiband antennas are usually used in Wireless and Mobile Computing, Network and Communications, IOT applications in WSNs, WLAN, and LTE Bands, etc.

11.2.4 Miniaturized Antennas

Introduction: The communication established between IoT devices is totally different from other communication types as amount of information transfer is limited in IoT. Only information collected by sensor that is usually physical parameters of devices, geolocation data sensed is transferred. Thus antenna is to be designed by keeping these points in mind (Figure 11.9).

Design: We know the information transfer is limited; hence, bandwidth would be very low. Thus we can have our focus on other different aspects such as radiation efficiency, low cost antennas, miniaturization, etc. Antenna's VSWR bandwidth is related to quality factor; both are inversely proportional to each other.

$$Q_{lb} = \eta_r \left(\frac{1}{ka} + \frac{1}{(ka)^3} \right), \tag{11.18}$$

where $k = 2\pi/\lambda$, antenna is imagined to be in a sphere that has radius of 'a' and η_r is antenna radiation efficiency. (11.18) implies that when antenna radiation efficiency is reduced by reducing antenna dimension then operating bandwidth also gets reduced [14]. As the bandwidth requirement in IoT applications is less thus strong miniaturization can be obtained. For the metals whose conductivity values are finite, η_r value gets deceased with the dimensions of antenna, which results in limiting miniaturization. When miniaturization is allowed then sensitivity of the antenna is increased. Small variation in surrounding environment will cause a change of operating frequency band. Thus the known elements of antenna such as electronic circuit, battery, and casing must be taken into consideration while designing antenna and the unknown elements effects that are present in the environment have to be nullified to restore the antenna performance. Then antenna design is based on Inverted F Antenna (IFA), with some modification for dual band operation. It was

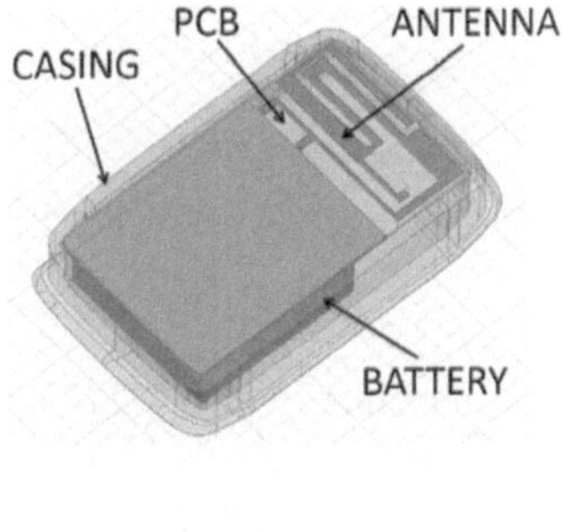

FIGURE 11.9 The antenna model and surrounding components attached to it [14].

also observed that when human body or a metallic surface is placed near to antenna then reflection coefficient of antenna is changed and accordingly the impedance matching is also changed. It was concluded that device under surroundings has a strong impact on antenna parameters values.

Results: To restore antenna performance in the presence of environmental changes a battery and casing are used, which shifts the antenna operable frequencies to lower values. Initially the reflection coefficient values are −14 dB at 0.91 GHz and when battery and casing are used the reflection coefficient is −11 dB at 0.87 GHz.

Applications: That connection between installed device is ensured by integration of numerous antennas in different devices. This integration in small devices requires miniaturization of various antennas. The miniaturized antennas are used in UHF-band Zigbee antenna, UHF-band Zigbee antenna devices, WLAN Applications, wearable antennas, ISM band small devices, UHF-band Zigbee antenna for the M2M/IoT communication, radio-frequency identification (RFID), global navigation satellite system (GNSS), and mobile devices, etc.

11.2.5 Reconfigurable Antenna

Introduction: The application of IoT is increasing every day and thus connectivity of device to device is much needed. Miniaturized antenna can be used for multiple device connectivity but radiation efficiency is reduced in it. So reconfigurable antennas can be used for this purpose and for switching between frequencies electrical switches such as PIN diode, varactor diodes, MEMS switches, and digital tunable capacitors are used. The switches have capacitance values and it impacts on radiation efficiency. Thus digitally tunable antennas are proposed for reconfigurability of patch antenna [15] (Figure 11.10).

Design: The patch antenna is in circular shape with a diameter of 58.8 mm and substrate made of Rogers RO 5880 which has dielectric constant of 2.2 and 0.787 mm is thickness of substrate. To achieve reconfigurability of frequencies one way is to use a slot of dimension of 30×2.4 mm^2 and other way is to use vias. Vias will tune antenna with matching and antenna will operate at lower frequencies without any alteration in dimensions.

Pe64907 Digitally Tunable Capacitor (DTC) is used to achieve the capacitance as per the requirement. This DTC can results in 32 capacitor values and they can be selected by 5 bit binary value. The model of DTC has inductive effect of 0.7 nH and capacitance and resistance values are variable, which are expressed by (11.19) and (11.20) [16].

$$\mathrm{Cs} = 0.056 \times \mathrm{state} + 0.38(\mathrm{pF}), \tag{11.19}$$

$$\mathrm{Rs} = \frac{20}{\mathrm{state} + \frac{20}{\mathrm{state} + 0.7}} + 0.7(\Omega), \tag{11.20}$$

where state defines the decimal value of corresponding 5 bit binary value. STM32 microcontroller is used to activate and control DTC operations. Let us take an

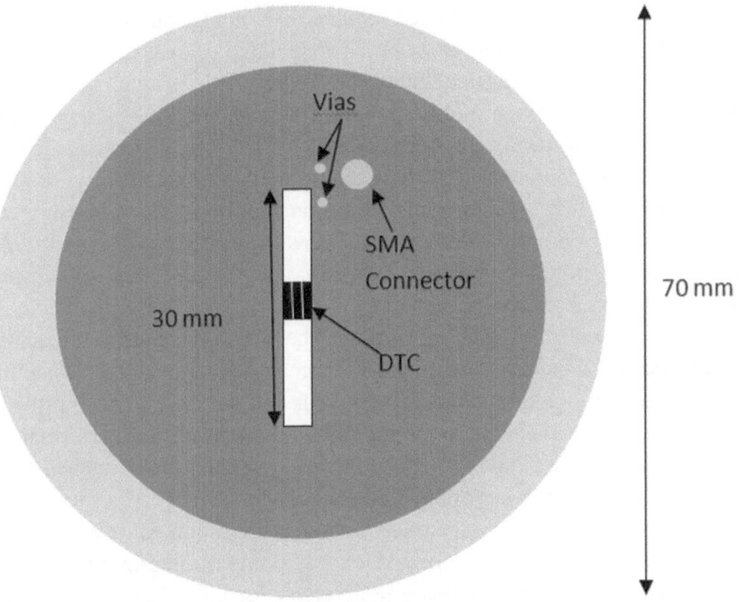

FIGURE 11.10 Circular patch antenna with DTC [15].

example to show the values of capacitance and resistance. If the first state, i.e., state 0 is '0000' and second sequence is '10100'. So using (11.18) and (11.19), state 0 has capacitance of 0.38 nH, resistance of 0.7 Ω and 0.35 nH inductance. While in state 20 (10100), capacitance is 1.5 pF, inductance of 0.7 nH and resistance of 1.65 Ω. For the state 0, antenna operated from 1.57 GHz to 1.58 GHz and 2.4 GHz to 2.41 GHz. When DTC switched to state 20, antenna operated between 864 MHz to 870 MHz and the frequency band of 2.4 GHz to 2.41 GHz [15].

At 1.58 GHz and 2.4 GHz, antenna produces broadside radiation pattern, which is useful in GPS and Bluetooth. At state 0 the reflection coefficient is −25 dB and at state 20 reflection coefficient is −15 dB. The designed antenna has radiation efficiency of 93% for 2.4 GHz frequency, patch diameter is equivalent to $\lambda/2$ at this frequency and 48% for 1.58 GHz with patch diameter of $\lambda/4$ at this frequency. The maximum gain of the antenna is 0.2 dB at 868 MHz, 1.8 dB, and 5.7 dB at frequency of 1.58 GHz and 2.4 GHz respectively. This antenna operates at the frequency bands of LORA, Bluetooth, and GPS.

Application: Reconfigurable antennas are used for cognitive radio systems, MIMO Systems, Biomedical Application, Wi-Fi, 3G Advanced, WiMAX, and WLAN wireless applications, ultra-low power IoT devices, etc.

11.3 DESIGN CHALLENGES OF ANTENNA FOR IOT APPLICATIONS

For any IoT product or device, the main crucial component is its antenna. But choosing the desired antenna for the particular device is a key design challenge. Designing and creating effective antenna performance requires examining various factors including size of antenna, shape, radiation efficiency, bandwidth, cost, etc. [17].

Size of antenna: A general device can contain 13–18 antennas for different purposes. For example, when mobile phones are considered it requires 13 antennas for cellular communication, Bluetooth, GPS, Wi-Fi, etc. However, IoT devices require one or two antennas for best communication possibility, e.g., fitness monitors, smartwatches, moisture and vibration monitors, etc. Choosing the right size of an antenna is one of the design challenges for IoT devices. Usually $\lambda/4$ or $\lambda/2$ length is found optimum to be used. Antenna length less than $\lambda/4$ drops signal strength and shrinks bandwidth.

Space for antenna on IoT devices surface: The other challenge is to decide which type of antennas to be used for a particular device and communication system. As we know IoT devices are getting smaller every year and we have to design the antenna according to the space available to mount without compromising performance of the antenna. Currently space available to the antenna is smaller than 2 inches × 2 inches.

Performance efficiency: Another challenge is to have good efficiency performance for higher bandwidth. Since many antennas are mounted on a single antenna, it is difficult to maintain the efficiency. Along with this is integrating many antennas in small space on IoT device and having diversity for LTE. Integration of antennas will also be difficult to isolate the radiation when close physically because most antenna radiation patterns are omni-direction.

If the LTE and GPS antennas are placed physically close, then radiation interference will occur. This issue can be controlled by observing current distribution and designing of GPS and LTE with a narrow band or by isolating two antennas.

Coverage of all bands: Usually the demand of antennas operable at maximum frequency bands is preferred by the industrialist, and designing and controlling such complex design is difficult. At low frequencies antennas are dependent on ground plane and efficiency is poor, and small effective aperture and bandwidth is also not that optimum. It is difficult to design antennas with small ground plane with the properties of covering worldwide bands. These are the main challenges faced by the researchers and industrialist in designing the microstrip antennas for IoT applications.

Compact antenna structures: The antenna dimension is reduced with increase in frequency and small handheld devices. Some devices uses various antennas to communicate multiple servers or devices at a same time. Thus antennas are to be fit in a small area available in the devices; many applications are increased, which requires compactness of antenna structure [18].

The other challenges faced by antenna designers are to have wide bandwidth, wide impedance, and high isolation of signals from surrounding components. There are also challenges in current path flow and impedance matching. The current path flow issues occur due to ground plane or slot antennas.

11.4 CURRENT TRENDS IN THE DESIGN OF ANTENNAS FOR IOT APPLICATIONS

There has been extensive research in IoT domains and it is going to increase as the connectivity will increase. The recent trends are focused in smart cities, smart cranes, and smart alarming calamity situations, smart grids, etc. Most of the nations have implemented these technologies into their day to day life. As the advancement in this field is increasing, the IoT standards, technologies, and interoperability are changing rapidly. The booming technology is *Smart Environment* with the inclusion of computing, sensors, data, robotics, and artificial intelligence with security and privacy. It covers transportation, utilities, home/offices, healthcare, police stations, school/colleges, and much more [19]. With the change in technologies, antennas are transforming to reciprocate the market requirements. As the sensor and IoT devices are becoming more compact and powerful, advancements in antenna are also required. Designing of future antennas is based on the adaptation with new technologies and increase in data transfer rate with high performance and high integration.

Instead of mounting antennas to devices, new research is focused on *Implanting Antennas onto Costumes or Placing Them on Skin*. So antenna parameters near human body add new research fields in antenna designing for IoT. While designing these types of antennas, some factors are to be considered: performance, matching, and selection of appropriate fabric. At high frequencies antennas perform with fading in radiation and wave propagation varies around the human body [20].

The fifth generation and IoT are widely used to improve society and human well being. There are different applications present and will also come in the future that require multiple wireless networks. So there is extensive requirement of better spectral efficiency by enabling multiple paths transmission using fewer power levels, and it is possible with *Multiple Input Multiple Output (MIMO) antennas* [21].

The applications of wireless are progressing quickly with features like portable wires and minimal efforts. With this working with wires is reduced and advancement of high transmission capacity and high radiation proficiency is possible. All of these are present in ultrawide band (UWB) applications. In recent years planar monopole receiving wire was extensively used for its capability of wide impedance data transfer with less efforts and a simple design structure. The mentioned advantages make *UWB Application Based Antennas* to be designed [22].

As the 5G technology is building, there will be an evolution in IoT devices, antennas, technologies, etc. Societies will be highly connected with great economic values and security. It is also estimated that 5G has wider frequency bands and larger bandwidth per frequency than the previous generations. Short range communication has Bluetooth and RFID while log range communication requires wired connections. For a larger geographical area, wired power supply will be difficult to implement. Hence there is a need of wireless power transmission system for the development of fast and reliable communication. For these requirements small sized *Rectenna Microstrip Array Antenna* is used to provide all the good radiation qualities [23].

The demand for seamless coverage, higher frequency ranges, higher data rate, low latency, quality of multimedia transmission, and higher reliability is required as the IoT is becoming in demand in every field of business and industries. To adapt these many applications microstrip patch antenna is the best candidate and its antenna size can reduce as frequency is increasing. It is lightweight, easy to manufacture, and has low cost of designing and fabrication. The increase in use of IoT is small devices enforces to design *Patch Antennas to Be Miniaturized* [24].

REFERENCES

1. IoT Technologies, https://stlpartners.com/research/the-iot-ecosystem-and-four-leading-operators-strategies/, accessed on Jul 2020.
2. End Devices and its Frequency Bands in IoT, https://www.5gtechnologyworld.com/specifying-antennas-for-various-iot-applications/, accessed on Jul 2020.
3. Jiming Chen, Kang Hu, Qi Wang, Yuyi Sun, Zhiguo Shi, Shibo He, "Narrow-Band Internet of Things: Implementations and Applications," *IEEE Internet of Things Journal*, 2017, doi:10.1109/JIOT.2017.2764475.
4. Antenna Theory for Microstrip Patch Antenna, http://www.antenna-theory.com/antennas/patches/patch4.php, accessed on Jul 2020.
5. Allam Vamseekrishna, Boddapati Taraka Phani Madhav, Tirunagari Anilkumar, and Lakkam Siva Shanker Reddy, "An IoT Controlled Octahedron Frequency Reconfigurable Multiband Antenna for Microwave Sensing Applications," *IEEE Sensors Letters,* 99:1–1, 2019, doi:10.1109/LSENS.2019.2943772.
6. Annamalai Selvarajan, M. Kumaresan, Prasanna Venkatesan G K D, "A low profile higher band IoT antenna for security applications," AIP Conference Proceedings 2039(1):020048, 2018, doi:10.1063/1.5079007.
7. T. Varum, M. Duarte, J.N. Matos, P. Pinho, "Microstrip Antenna for IoT/WLAN Applications in Smart Homes at 17 GHz," *12th European Conference on Antennas and Propagation* (*EuCAP* 2018), doi:10.1049/cp.2018.0475.
8. Yehuda Giay, Basuki R Alam, "Design and Analysis 2.4 GHz Microstrip Patch Antenna Array for IoT Applications using Feeding Method," *2018 International Symposium on Electronics and Smart Devices (ISESD) Conference*, October 2018, doi:10.1109/ISESD.2018.8605455.
9. Nischal Sanil, Pasumarthy Ankith Naga Venkat, Mohammed Riyaz, "Design and Performance Analysis of Multiband Microstrip Antennas for IoT Applications Via Satellite Communication," *Conference: 2018 Second International Conference on Green Computing and Internet of Things (ICGCIoT)*, August 2018, doi:10.1109/ICGCIoT.2018.8753037.
10. Nag, Vibha Raj, and G. U. R. P. A. D. A. M. Singh, "Design and Analysis of Dual Band Microstrip Patch Antenna With Microstrip Feed Line and Slot for Multiband Application in Wireless Communication," *International Journal of Computer Science and Information Technology and Security (IJCSITS)* 2.6:1266–1270, 2012.
11. Balur, Nazahat Jahan, and Sukanya Kulkarni, "Design of Multiband Microstrip Antenna," *International Journal of Advanced Research in Computer Science and Electronics Engineering (IJARCSEE)* 1(10): 82, 2012.
12. Internet of Things, https://www.rakon.com.pdf, accessed on July 2020.
13. Soumyadeep Das, T. Shanmuganantham, "Design of Multiband Microstrip Patch Antenna for IoT Applications," *2017 IEEE International Conference on Circuits and Systems (ICCS)*, pp. 20–21, December 2017, doi:10.1109/ICCS1.2017.8325968.
14. Fabien Ferrero, Leonardo Lizzi, C. Danchesi, S. Boudaud, "Environmental Sensitivity of Miniature Antennas for IoT Devices," *2016 IEEE International Symposium on Antennas and Propagation & USNC/URSI National Radio Science Meeting*, June 2016, doi:10.1109/APS.2016.7696581.
15. F. A. Asadallah, J. Costantine, Y. Tawk, L. Lizzi, F. Ferrero, C. G. Christodoulou, "A Digitally Tuned Reconfigurable Patch Antenna for IoT Devices," *2017 IEEE International Symposium on Antennas and Propagation & USNC/URSI National Radio Science Meeting*, July 2017, doi:10.1109/APUSNCURSINRSM.2017.8072501.
16. Peregine Semiconductor, Inc. "PE64907," http://www.psemi.com/products/digitally-tunablecapacitors-dtc/pe64907, accessed on July 2020.

17. Challenges in Antenna in IoT, https://www.designnews.com/challenge-mobile-phone-and-iot-antennas
18. Naveen Kumar, Dr. Hardeep Singh Saini, Shivangi Verma, Rajesh Kumar, Leena Mahajan, "A Small Microstrip Patch Antenna for Future 5G Applications," *IEEE International Conference on Reliability, Infocom Technologies and Optimization*, September 2016, doi:10.1109/ICRITO.2016.7784999.
19. Kinza Shafique, Bilal A Khawaja, Farah Sabir, Muhammed Mustaqim, Sameer Qazi, Internet of Things (IoT) for Next-Generation Smart Systems: A Review of Current Challenges, Future Trends and Prospects for Emerging 5G-IoT Scenarios, *IEEE Access*, November 2017, doi:10.1109/ACCESS.2020.2970118.
20. Sajjad Hussain, "Current Trends in Antenna Designing for Body Centric Wireless Communication," *IJSER Journal*, 2012, ISSN: 2229-5518
21. Wensong Wang, Zhenyu Zhao, Zhongyuan Fang, Quqin Sun, Xinqin Liao, Kye Yak See, and Yuanjin Zheng, "Compact Broadband Four-Port MIMO Antenna for 5G and IoT Applications," *IEEE*, December 2019, doi:10.1109/APMC46564.2019.9038745.
22. Inderpreet Kaur, Anil Kumar Singh, Jagriti Makhija, Sakshi Gupta, Krishna Kumar Singh, "Designing and Analysis of Microstrip Patch Antenna For UWB Applications," *IEEE Conference*, February 2019.
23. Christina Gnanamani, Aishwarya Suriyakumar "RF Energy Harvesting Using 2 × 4 Circular Microstrip Array Antenna for 5G-IoT Zero Power Wireless Sensors," *IEEE Conference*, October 2019, doi:10.1109/ICRAMET47453.2019.8980434.
24. Amiya B. Sahoo, Ninaad Patnaik, Smarak Behera, B. B. Mangaraj, Aditya Ravi "Design of a Miniaturized Circular Microstrip Patch Antenna for 5G Applications," *IEEE,* February 2020, doi:10.1109/ic-ETITE47903.2020.374.

Part VI

Ultra-Wide-Band Antenna Design for Wearable Applications

12 Design of an Edge-Fed Rectangular Patch Antenna for WBAN Applications and Analysis of Its Performance for M-ary Modulation Schemes

Qudsia Rubani, Sindhu Hak, Azan Khan, and Shuchismita Pani
Department of Electronics and Communication Engineering, ASET, Amity University, Noida, Utter Pradesh, India

12.1 INTRODUCTION

The recent progression in wireless technology along with expanding requirement to get data from any system that requires monitoring has permitted the advancement of Wireless Sensor Network (WSN). Sensors in WSN have detecting and processing capabilities, therefore they can be utilized as a part of an extensive variety of survivability applications, for example, health monitoring, environmental monitoring, fire monitoring, intelligent transport systems (ITS), and numerous other applications [1]. It is generally trusted that WSNs would be an ideal part of the next generation networks to give adaptable mobile connectivity and deployment. WSNs have existed in numerous fields that need abundant access in both non-real-time and real-time data, and the most accurate information is provided for the application in need. WSNs generally are comprised of nodes that require low power and are of low cost that do not need any existing infrastructures to function and can communicate untethered short distances. The nodes are also capable of sensing, communication, and data processing [2]. However, in a broader sense, the exploration of WSNs does not manage the difficulties related to human body monitoring. Human body monitoring has different constraints in comparison to environmental monitoring. Hence a subtype of WSNs has been developed by researchers: the Wireless

DOI: 10.1201/9781003187325-12

Body Area Network (WBANs). There has been an impressive increment in the interest of WBAN applications, for example, military services, personal entertainment, healthcare systems, and ambient intelligence [3]. It can likewise incorporate into embedded biosensor devices. Indeed, the development of WBANs ought to follow the regularly expanding advancement in the medicinal area; its principal aim is to confirm steady unavoidable and intensive care of patients at home or at work. A WBAN is an accumulation of small nodes, outfitted with biomedical sensors, wireless communication devices, and motion detectors. These nodes gather imperative body signs to process this gathered data, and it is then transmitted to a central unit. WBAN nodes have numerous benefits, like omnipresence connectivity, interoperability, and mobility. The system architecture of a WBAN is shown in Figure 12.1 [4].

The antenna has a critical part in achieving the task of a biosensor. As the biosensor containing the antenna can be set on a human body, it is a challenging task to implement an antenna to fulfill the basic prerequisites, for example, Omnidirectional pattern, linear phase, low dispersion, ultra-wide bandwidth, and linear gain, and considerably additionally difficult when the biosensor with the antenna is used on the human body for WBAN applications. It is realized that the performance of antenna altogether is influenced by body tissues because of the electrical qualities of the body, for example, radiation pattern distraction, resonant frequency shift, and changes in input impedance of antenna [5].

In every wireless sensor network modulation schemes are also equally important when it comes to transmitting the information. Many M-ary modulation schemes are there, which direct the transmission of data from the sender node to the receiver

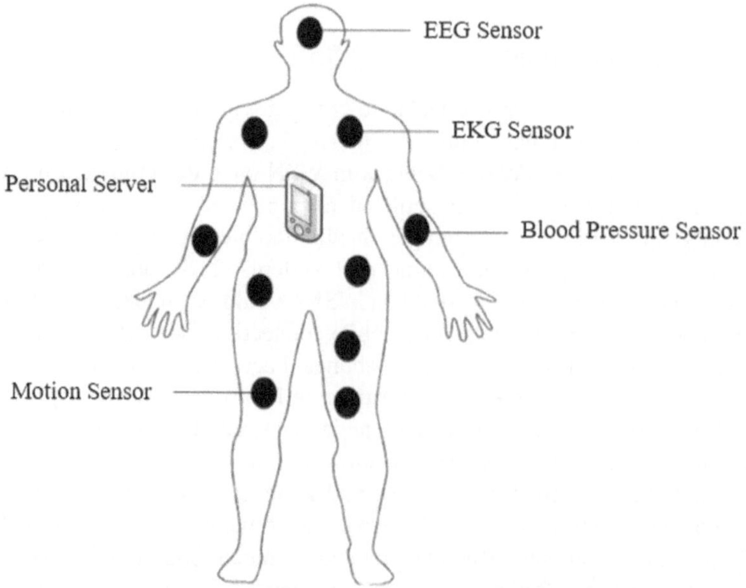

FIGURE 12.1 Architecture of a WBAN.

node. These schemes usually consider three main parameters in the transmission of data, which are as frequency, time, and phase. Some of the simple modulation schemes are M-ary FSK, i.e., Frequency Shift Keying, M-ary PSK, i.e., Phase Shift Keying, ASK, i.e., Amplitude shift keying, etc. Among all the available modulation schemes, we are mainly emphasizing M-ary FSK and M-ary PSK, because these modulation schemes have better bandwidth efficiency [6]. So in any WSN, the choice of the modulation scheme is a critical point that helps in increasing the life span, efficiency, and reliability of a sensor node [7].

In this work, an Edge-Fed Rectangular patch antenna for Wireless Body Area Network application has been proposed. The proposed antenna operates at 5.18 GHz and offers a bandwidth of 188.3 MHz. The designed antenna simulation results are presented and are explained in detail. A soft High-Frequency Structure Simulator, HFSS 3D Modeler, which is commercially available, has been used for simulating the antenna.

Specific Absorption Rate (SAR) is an important parameter for ensuring that the antenna designed is harmless to the human body. SAR determines the amount of RF energy absorbed by the human tissues [8]. Higher value of absorption means the higher the risk to human health [9]. As per the American National Standards Institute, the SAR value should be less than 1.6W/kg over a 1g tissue [10].

The antenna performance is further investigated for wireless body area networks by studying the effect of modulation schemes used in WBAN mainly M-ary Frequency Shift Keying and M-ary Phase Shift Keying.

The main contributions of this work are:

- An Edge-Fed Rectangular patch antenna for WBAN application has been designed, and parameters like VSWR, Gain, and Radiation pattern for the antenna have been analyzed.
- The compatibility of an antenna with the tissues of the human body is investigated by analyzing the Specific Absorption Rate (SAR).
- The performance of the antenna has been analyzed by investigating the effect of M-ary FSK and M-ary PSK on the amount of power transmitted by the antenna.

The rest of the paper is divided into sections as follows. Section 12.2 defines the system design and the proposed antenna design. Section 12.3 discusses the simulation results of the antenna. In Section 12.4, the performance of the antenna is discussed as per the WBAN applications. Finally, the concluding remarks are presented in Section 12.5.

12.2 SYSTEM AND ANTENNA DESIGN

The performance of the designed antenna with regards to the modulation schemes used in WBAN, mainly M-ary PSK and M-ary FSK, has been analyzed. Also, the design, geometry, and dimensions of the Edge-Fed Rectangular patch antenna are presented.

12.2.1 System Design

The choice of the modulation scheme is an essential point. This requires balancing of several factors: the desirable and required data and symbol rates, the expected channel characteristics, the relationship between target Bit Error Rate and radiated power, and the implementation complexity. Time spent in sleep mode by the transceiver is maximized by minimizing the transmit times. For transceiver/modulation, the higher the data rate, the lesser the time expected to transmit a certain measure of data, thusly minimizing the energy utilization. Another essential perception is that the power utilized by the modulation schemes is determined less by data rate and more by the symbol rate.

For the system, it is assumed that WBAN with homogeneous nodes used to monitor various parameters such as blood pressure, glucose level, electrocardiogram (ECG), electromyogram (EMG), and electroencephalogram (EEG) signals have been deployed on a human body. Consider a WBAN comprised of two biosensor nodes A and B as shown in Figure 12.2. These nodes comprise microcontroller, memory, battery, antenna, etc. Both nodes are separated by distance d and consist of separate antennas with antenna gain as G_A and G_B for A and B respectively. Since the nodes are homogenous, the values of G_A and G_B are equal. P_{tx} is the power transmitted and P_{revd} is the power received by nodes.

According to [11], energy per bit for M-ARY modulation schemes is given as

$$\frac{E_b}{N_o} = \text{SNR}.\ \frac{1}{R} = \frac{P_{revd}}{N_0}.\ \frac{1}{R} = \frac{1}{N_o.\ R}\frac{P_{tx}.\ G_A.\ G_B \lambda^2}{(4\pi)^2.\ d_o^{\gamma}.\ L}\left(\frac{d_o}{d}\right)^{\gamma}. \tag{12.1}$$

The above equation gives the required relationship between $\frac{E_b}{N_o}$ and the received power P_{revd}, where γ is a Path Loss exponent, L is packet length, R is data rate, and $\frac{E_b}{N_o}$ is the ratio of energy consumed by each bit and the level of noise in the system which has a definite value for all the M-ary modulations schemes available.

Table 12.1 illustrates $\frac{E_b}{N_o}$ (dB) required in the channel for M-ary FSK and M-ary PSK [11].

The value of $\frac{E_b}{N_o}$ for M-ary PSK rises with the rise in modulation number however for M-ary FSK, $\frac{E_b}{N_o}$ decreases with the increase in modulation number.

The proposed antenna has been analyzed for M-ary modulation schemes under various condition, such as varying the value of distance (d) and $\frac{E_b}{N_o}$ which differs for different modulation schemes. Then with the help of these values, P_{tx} is calculated from equation (12.1).

12.2.2 Antenna Design

The WBAN antenna is designed using FR4 epoxy as a substrate with dielectric constant (ε_r) and resonant frequency (f_r). The antenna is designed considering the

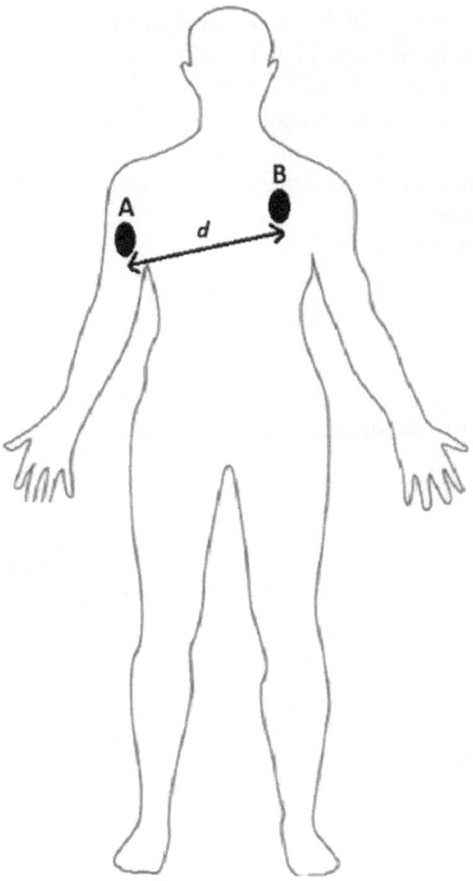

FIGURE 12.2 System model representation.

TABLE 12.1
Variation of $\frac{E_b}{N_o}$ with Changing Values of M for M-ary PSK and M-Ary FSK Modulation Schemes

M	M-ary PSK: $\frac{E_b}{N_o}$	M-ary FSK: $\frac{E_b}{N_o}$
2	10.5	13.5
4	11	10.8
8	14	9.3
16	18.5	8.2
32	23.4	7.5
64	28.5	6.9

network lifetime and energy efficiency. So, to make the antenna energy efficient and to increase the network lifetime, higher gain has to be achieved. Size is also an important parameter to be considered for the WBAN antenna. Biosensors are small devices used for WBAN, so the antenna being the part of the biosensor has to be as compact as possible.

Parameters for the Rectangular Microstrip patch antenna are calculated using the formulas [12] given below:

Patch width is given as

$$W = \frac{c}{2f_r\sqrt{\frac{\varepsilon_r+1}{2}}}. \tag{12.2}$$

The effective dielectric constant is given by

$$\varepsilon_{\text{eff}} = \frac{\varepsilon_r+1}{2} + \frac{\varepsilon_r-1}{2}\left[1+12\frac{h}{W}\right]^{\frac{-1}{2}}. \tag{12.3}$$

The increment in length, which is caused by the fringing field, is calculated as

$$\Delta L = \frac{h}{\sqrt{\varepsilon_{\text{eff}}}} \tag{12.4}$$

Patch Length is given by

$$L = \frac{c}{2f_r\sqrt{\varepsilon_{\text{eff}}}} - 2(\Delta L), \tag{12.5}$$

where

ε_r = Relative dielectric constant of the substrate

f_r = desired resonant frequency

h = dielectric thickness

The antenna dimensions are listed in Table 12.2 and are calculated with the help of equations (12.2–12.5). ε_r in equation (12.2) is taken as the relative dielectric constant of FR4 epoxy substrate which is 4.4. The dimensions calculated give the patch size as $17.56 \times 12.56 \times 1.6\ \text{mm}^3$ and the feed length is $8.294 \times 0.723\ \text{mm}^3$.

Figure 12.3 demonstrates the antenna geometry of the designed WBAN antenna to be deployed on the biosensor.

The designed Edge-Fed Rectangular Patch Antenna geometry is shown in Figure 12.3 while its schematic diagram is shown in Figure 12.4, where dimensions of the antenna are mentioned. The antenna is designed using the parameters listed in Table 12.2 and antenna geometry in Figure 12.3.

The values for G_A and G_B used in equation (12.1) are obtained from the designed antenna by simulating it in HFSS. Then with the help of these values, P_{tx} is calculated and antenna performance is studied for M-ary FSK and M-ary PSK.

TABLE 12.2
Dimensions of the Designed Antenna

Parameters	Value (mm)
Patch width (*W*)	12.56
Patch length (*L*)	17.56
Substrate height	1.6
Edge feed width	3.059
Edge feed length	14.896
Feed width	0.723
Feed length	8.294

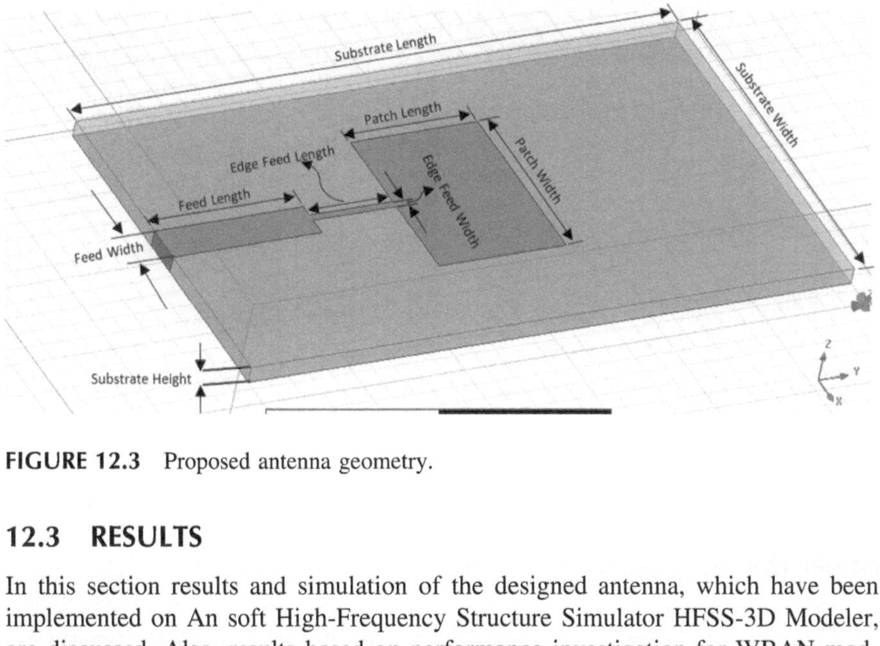

FIGURE 12.3 Proposed antenna geometry.

12.3 RESULTS

In this section results and simulation of the designed antenna, which have been implemented on An soft High-Frequency Structure Simulator HFSS-3D Modeler, are discussed. Also, results based on performance investigation for WBAN modulation schemes (M-ary PSK, M-ary FSK) with regards to power transmission for the designed antenna which were achieved with the help of MATLAB 2015 have been discussed in this section.

12.3.1 S_{11} Parameter

The desired value of S_{11} parameter for good impedance matching is S_{11} <−10dB. S_{11} of the antenna is displayed in Figure 12.5, where it can be seen that the resonant frequency attained by the proposed antenna is 5.18 GHz and shows a return loss of −23.84, which is clearly <−10 db. Therefore the designed antenna offers a good

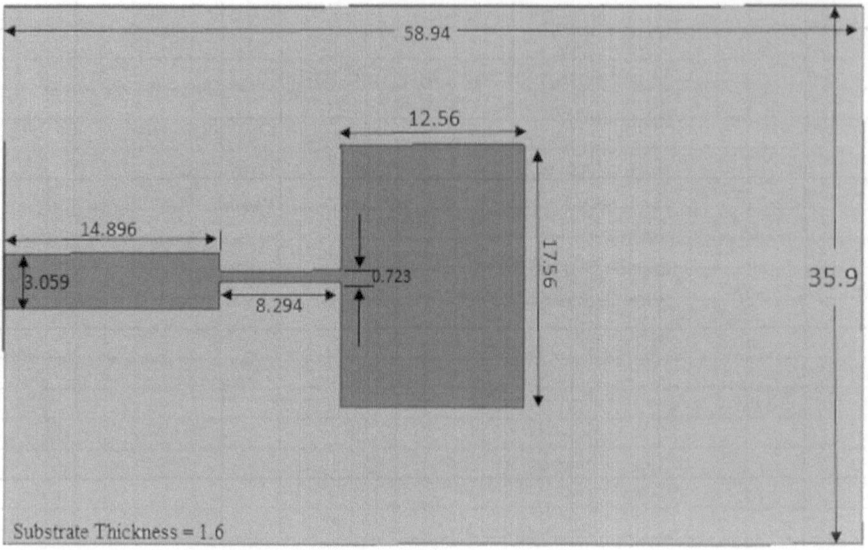

FIGURE 12.4 Designed antenna dimensions.

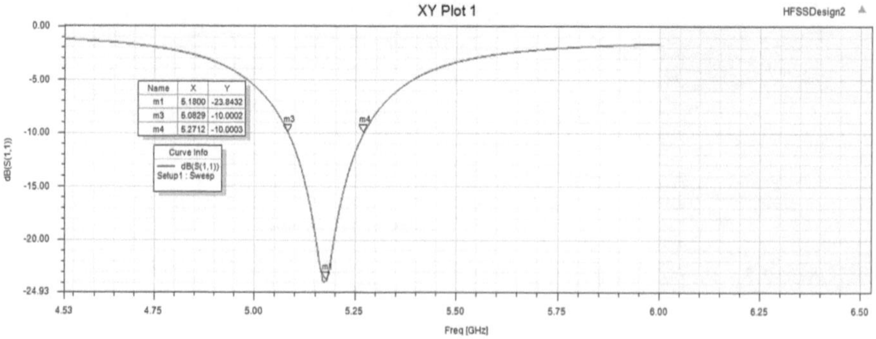

FIGURE 12.5 *S*11 parameter of the antenna.

impedance matching. The bandwidth offered by the designed antenna is 188.3MHz. Other parameters like efficiency, directivity, etc. are tabulated in Table 12.3.

The parameter S_{11} gives data about the transfer of power to the antenna from the generator. This parameter is also known as the reflection coefficient and it shows the connection between the wave reflected by the antenna and the incident wave on the antenna.

12.3.2 Voltage Standing Wave Ratio (VSWR)

This parameter is important for determining the impedance matching of the antenna with the transmission line, to which it is connected [13]. VSWR value of antenna,

TABLE 12.3
Computed Antenna Parameters of the Proposed Antenna

Quantity	Value
Peak directivity	8.57
Radiated power	0.00031104 W
Radiation efficiency	0.72832
Accepted power	0.00042707 W
Front to back ratio	30.384

for body area networks, especially, should be less than 2. The designed antenna in this work shows a VSWR value of only 1.17, indicated in Figure 12.6, which is undoubtedly good for this particular area of application.

12.3.3 Antenna Gain

Antenna gain is specifically identified with its directivity. It is the relation between the radiation intensity from the antenna and from an isotropic antenna which accepts the same transmitted power contrasted with the antenna under examination. Figure 12.7 shows the peak gain is 5. 1 dB for the antenna.

Antenna gain is specifically identified with its directivity. The directivity is characterized as the connection between the radiation intensity from an antenna toward every path at a given distance and the radiation intensity that would transmit a similar distance, contrasted with an isotropic antenna radiating a similar power. Gain is the relation between the radiation intensity from the antenna and from an isotropic antenna which accepts the same transmitted power contrasted with the antenna under examination.

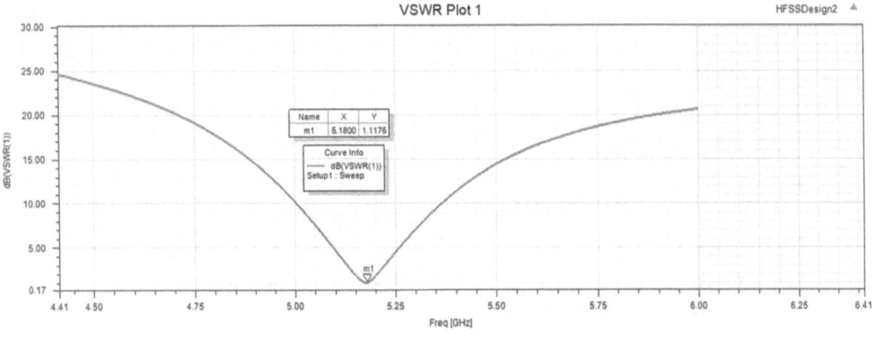

FIGURE 12.6 VSWR of the proposed antenna.

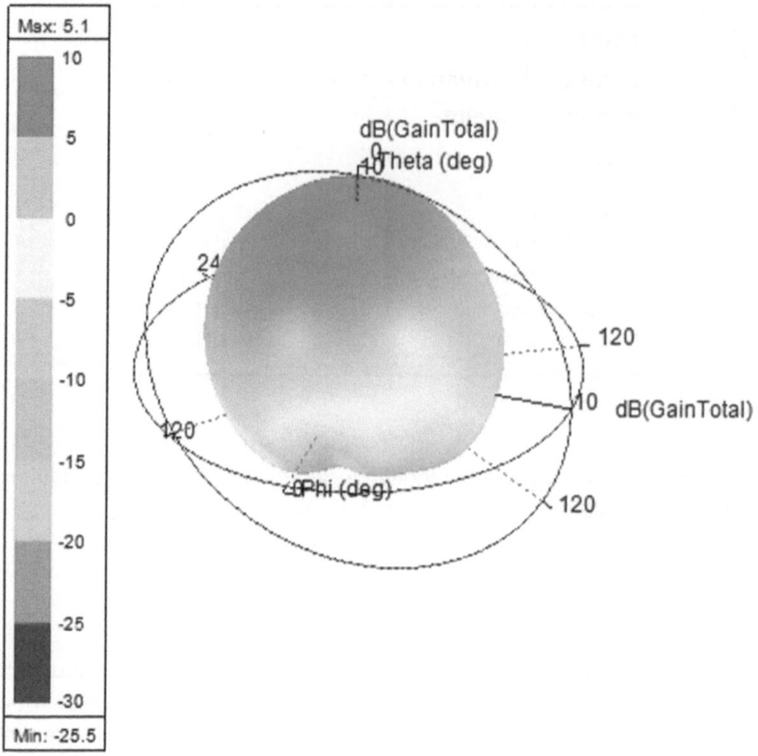

FIGURE 12.7 Simulated gain plot for the designed antenna.

12.3.4 Radiation Pattern

The radiation pattern is created as a function of the angular factors, $E(\theta, \varphi)$, from the outflow of the electric field of the antenna for the spherical coordinate system. Figure 12.8 demonstrates the radiation pattern of the designed antenna. The vertical axis represents a peak gain. The plane E is shaped by the direction of maximum radiation and the electric field towards that path. Likewise, the H plane is framed by the direction of maximum radiation and the magnetic field towards that path. The two planes are perpendicular to each other and their convergence characterizes the maximum radiation of the antenna.

Table 12.4 lists the comparison of the designed antenna with a few other similar WBAN antennas [14–16].

From this comparison, it is proved that the proposed antenna size is much smaller in size, better in gain, and offers higher bandwidth than other antennas.

12.4 PERFORMANCE INVESTIGATION FOR WBAN MODULATION SCHEMES:

To investigate the performance of the designed antenna in WBANs, parameters like transmitted power is compared with distance (d). From the designed antenna the

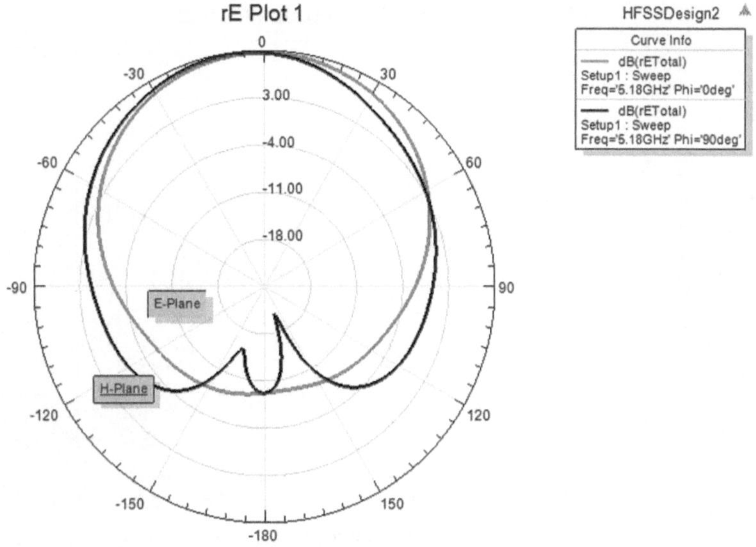

FIGURE 12.8 Simulated radiation pattern of designed antenna.

TABLE 12.4
Comparison of WBAN Antennas

Ref. No	Size (mm)	Resonant Frequency	Bandwidth (MHz)	Max Gain (dB)
[14]	75 × 42	5.8 GHz	230	3.12
[15]	40.15 × 40.15	5.8 GHz	370	0
[16]	41 × 48	5.8 GHz	280	1.12
Proposed Antenna	60 × 35	5.18 GHz	288.3	5.1

gain of 5dB is obtained, which provides the value for G_t and G_r as 5 dB for equation (12.1). G_t and G_r are same because it is a homogenous network. With the help of MATLAB®, equation (12.1) is implemented with the gain obtained from the designed antenna and the results are obtained for power transmission in the M-ary Modulation Schemes.

Figure 12.9 illustrates the effect of separation distance d between the sender and receiver node on the transmitted power for M-ary PSK. Similarly, Figure 12.10 illustrates the influence of distance (*d*) on the transmitted power for M-ary FSK. In case of M-ary modulation scheme, PSK, more the value of M, the more will be the power transmitted. In case of M-ary modulation scheme, FSK, the lesser the value of M, the more will be the power transmitted. The more the power transmitted, the sooner the node will die and hence network lifetime will be reduced. So, to increase the network lifetime keeping the network operational for a longer period, the value

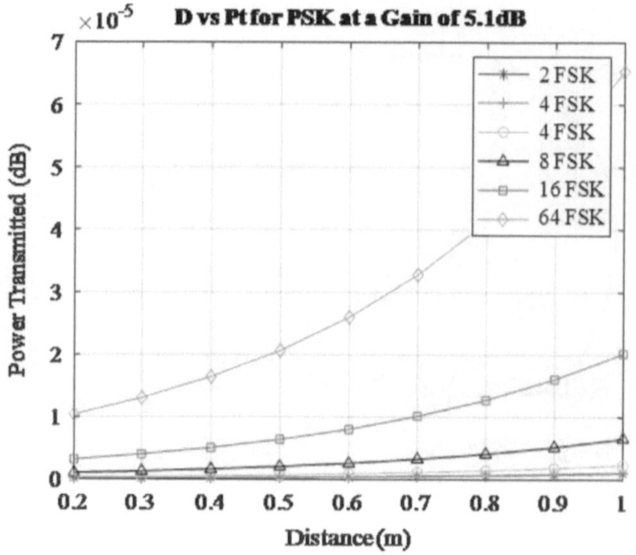

FIGURE 12.9 Power transmitted vs distance for M-ary PSK.

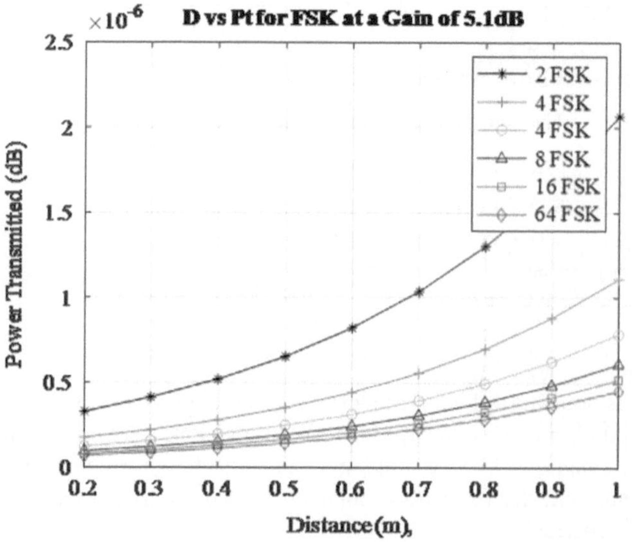

FIGURE 12.10 Power transmitted vs distance for M-ary FSK.

of M should be low in case of M-ary PSK, while in the case of M-ary FSK its value should be more. It is also observed that as a distance (*d*) increases the power required for transmission rises but comparably for M-ary FSK less power is required for transmission of data than M-ary PSK.

Table 12.5 illustrates the variation in the power transmitted corresponding to each M-ary modulation scheme obtained from equation (12.1) with the help of MATLAB.

From Table 12.5 also, it can be observed that with increasing value of M the transmitted power increases with respect to distance (*d*) in the case of M-ary PSK while it decreases in the case of M-ary FSK. But overall, less power is required for the transmission of data in M-ary FSK as compared to M-ary PSK. Hence it has resulted that the M-ary FSK modulation scheme is suited best for the designed antenna.

12.4.1 Specific Absorption Rate

As the proposed antenna is specifically designed for WBAN, therefore the calculation of SAR becomes a priority for the safety of human health. The maximum value of SAR that an antenna can have, while intended to apply to the human body, is 1.6 W/kg [17], as recommended by IEEE/ANSI/FCC.

Figure 12.11 shows the SAR distribution for the designed antenna and is equal to 3.11 W/kg at an input power of 50 mW. This SAR value is greater than the acceptable level. Therefore, a technique has been used to reduce the SAR value which for this designed antenna is more than the acceptable value. In this technique, a layer of Polyamide with a thickness of 1 mm has been installed below the ground plane [18]. Relative Permittivity (ε_r) of polyamide is 4.3. The length and width of this layer are the same as that of the ground plane. The thickness of this layer is very

TABLE 12.5
Comparison of WBAN Antennas

Modulation Scheme	*M*	Transmitted Power (mW)
M-ary FSK	2	11.3×10^{-10}
	4	6.06×10^{-10}
	8	4.29×10^{-10}
	16	3.33×10^{-10}
	32	2.83×10^{-10}
	64	2.47×10^{-10}
M-ary PSK	2	5.66×10^{-10}
	4	5.66×10^{-10}
	8	1.26×10^{-9}
	16	3.57×10^{-9}
	32	1.1×10^{-8}
	64	3.57×10^{-8}

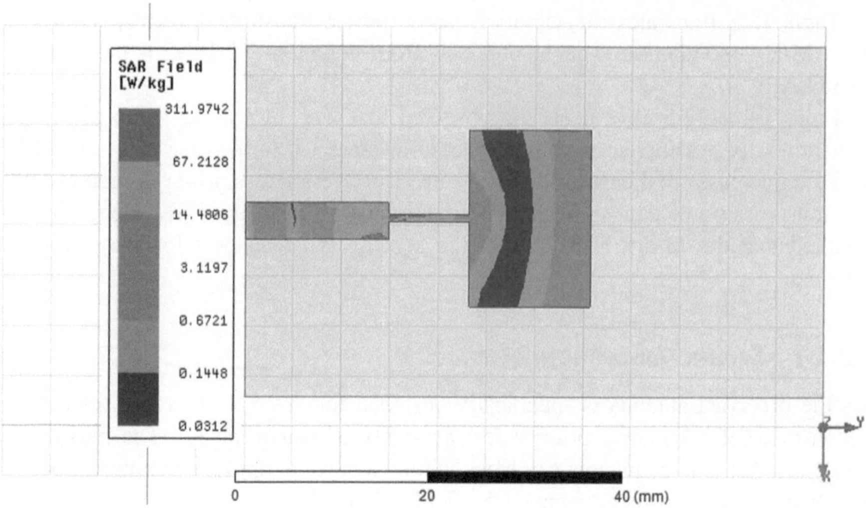

FIGURE 12.11 Illustration of SAR offered by the proposed antenna.

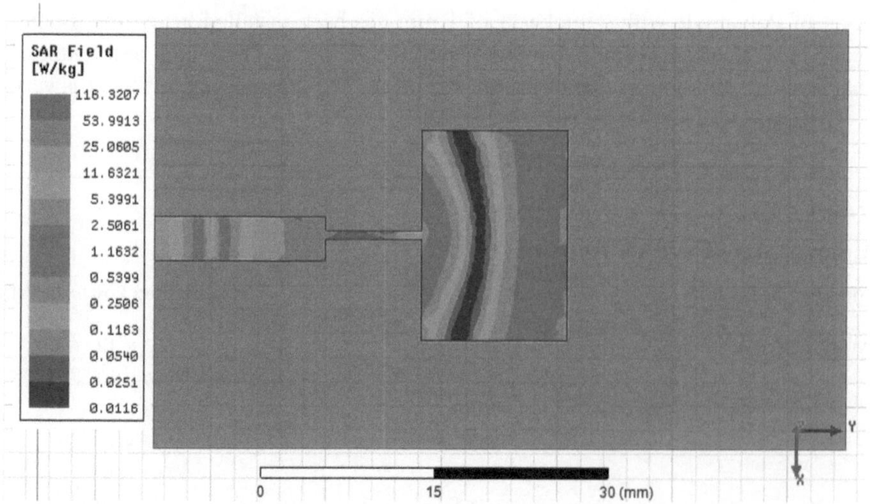

FIGURE 12.12 Illustration of SAR offered by the proposed antenna with the polyamide layer.

small, therefore its effect on antenna parameters is very small. Using this technique, SAR value is reduced up to 1.16 W/kg, which is within the acceptable levels. Figure 12.12 shows the reduced value of SAR for the proposed antenna. Therefore, the proposed antenna is harmless to human health and a better choice for WBAN applications.

REFERENCES

[1]. Suryadevara NK, Mukhopadhyay SC. Wireless sensor network based home monitoring system for wellness determination of elderly. *IEEE Sensors Journal.* 2012 Jan 3;2(6):1965–1972.

[2]. Singh T. Design and Analysis ofAdhoc on Demand Distance Vector Routing (AODV) inwirelesssensornetworkusingneutralnetwork. *Global Journa of Computers & Technology.* 2017 Jul 15;6(1):344–353.

[3]. Movassaghi S, Abolhasan M, Lipman J, Smith D, Jamalipour A. Wireless body area networks: A survey. *IEEE Communications Surveys & Tutorials.* 2014 Jan 14;16(3):1658–1686.

[4]. Cypher D, Chevrollier N, Montavont N, Golmie N. Prevailing over wires in healthcare environments: benefits and challenges. *IEEE Communications Magazine.* 2006 May 15;44(4):56–63.

[5]. Alomainy A, Sani A, Rahman A, Santas JG, Hao Y. Transient characteristics of wearable antennas and radio propagation channels for ultrawideband body-centric wireless communications. *IEEE Transactions on Antennas and Propagation.* 2009 Apr 7;57(4):875–884.

[6]. Thakuria K, Vivekananda AG. Analysis of bit error rate of different M-ary PSK modulation schemes in AWGN channel. *American Journal of Networks and Communications.* 2016;5(5): 82–90.

[7]. Zhao H, Gong Y, Guan YL, Tang Y. Performance analysis of $ M $-PSK/$ M $-QAM modulated orthogonal space–time block codes in keyhole channels. *IEEE Transactions on Vehicular Technology.* 2008 Jul 16;58(2):1036–1043.

[8]. Ntouni GD, Lioumpas AS, Nikita KS. Reliable and energy-efficient communications for wireless biomedical implant systems. *IEEE Journal of Biomedical and Health Informatics.* 2014 Jan 14;18(6):1848–1856.

[9]. Wu TY, Lin CH. Low-SAR path discovery by particle swarm optimization algorithm in wireless body area networks. *IEEE Sensors Journal.* 2014 Oct 2;15(2):928–936.

[10]. Tak J, Kwon K, Kim S, Choi J. Dual-band on-body repeater antenna for in-on-on WBAN applications. *International Journal of Antennas and Propagation.* 2013;2013: Article ID 10725. https://doi.org/10.1155/2013/107251

[11]. Sheikh SA, Gupta SH. Implementation and analysis of energy efficiency of M-ary modulation schemes for wireless sensor network. In *Data and Communication Networks* (pp. 57–68). Singapore: Springer, 2019.

[12]. Ahmad S, Islam MR, Haque MA, Mazed KA, Hasan RR, Fahim-Uz-Zaman M. Body implantable patch antenna for biotelemetry system. In *2018 International Electrical Engineering Congress (iEECON)* (pp. 1–4). New York: IEEE, 2018.

[13]. Yan S, Vandenbosch GA. Radiation pattern-reconfigurable wearable antenna based on metamaterial structure. *IEEE Antennas and Wireless Propagation Letters.* 2016 Feb 11;15:1715–1718.

[14]. Hong Y, Tak J, Choi J. An all-textile SIW cavity-backed circular ring-slot antenna for WBAN applications. *IEEE Antennas and Wireless Propagation Letters.* 2016 Apr 1;15:1995–1999.

[15]. Björninen T, Yang F. Low-profile head-worn antenna with a monopole-like radiation pattern. *IEEE Antennas and Wireless Propagation Letters.* 2015 Aug 31;15:794–797.

[16]. Kang DG, Tak J, Choi J. Low-profile dipole antenna with parasitic elements for WBAN applications. *Microwave and Optical Technology Letters.* 2016 May;58(5): 1093–1097.

[17]. Li G, Zhai H, Ma Z, Liang C, Yu R, Liu S. Isolation-improved dual-band MIMO antenna array for LTE/WiMAX mobile terminals. *IEEE Antennas and Wireless Propagation Letters*. 2014;13:1128–1131.

[18]. Sudha MN, Benitta SJ. Design of antenna in Wireless Body Area Network (WBAN) for biotelemetry applications. *Intelligent Decision Technologies*. 2016 Jan 1;10(4):365–371.

13 UWB Planar Microstrip Fed Antennas for Various Wireless Communication and Imaging Applications with Mitigation of Interference

Manish Sharma
Chitkara University Institute of Engineering and Technology, Chitkara University, Punjab, India

13.1 INTRODUCTION

Ultrawideband bandwidth (UWB) was released by the Federal Communication Commission (FCC) by the USA in 2002 for unlicensed use for a frequency band of 3.10 GHz–10.60 GHz. Microstrip patch antenna has given an edge which fulfills the requirement of applications for UWB band which includes imaging systems, ground penetrating Radar systems. Ultra-wideband was formerly known as pulse radio, but the International Telecommunication Union Radiocommunication Sector currently defines UWB as an antenna transmission for which emitted signal bandwidth exceeds the lesser of around 20% of the arithmetic center frequency: wall imaging systems and through wall Imaging system, surveillance system, medical application, etc. To design the above said application antenna, several designs have been reported in the literature [1–17]. Truncated ground plane with a dome-topped radiating patch provides wider impedance bandwidth of 2.65 GHz–13.0 GHz [1]. A U-type square patch that is combined with two parasitic tuning-stubs printed on one plane of the substrate with dimension 24×28×0.787 mm3 also provides measured bandwidth of 2.81 GHz–12.58 GHz [2]. Defected ground with elliptical slot and two types of the radiating patch, M-shaped radiator, beak-type radiator are the other reported structures used for UWB applications [3–5]. On the other hand, triangular wheel shape fractal antenna, fractal geometry obtained from the square patch is the

DOI: 10.1201/9781003187325-13

other fractal designs used for UWB applications [6–8]. Spanner-type feed line used to feed patch, using rectangular parasitic elements behind the patch, decagonal shaped with the truncated ground are the other geometries used for obtaining UWB bandwidth [9–13]. Quarter-circular truncations and rectangular slots offer lower resonating frequency in a wideband antenna [14]. Also, modification in a circular patch with parasitic elements and the hexagonal slotted antenna is also capable of providing wider useable bandwidth. There are different existing wireless technologies such as WiMAX, WLAN, etc. operating in the same UWB band which causes interference in the working band. Hence, a need arises to mitigate the interference caused which is achieved by introducing bandstop filters. These bandstop filters may be in the form of slot/stub/slit on the radiating patch, electromagnetic coupling (ECT) by using parasitic elements, Electromagnetic Band Gap structure (EBG), Meta-material, etc. A conformal antenna with a circular patch and etched dual concentric circles not only improves bandwidth but also notches the WLAN band [15–18]. Also, a unique notching methodology is discussed [19] where mushroom-type EBG is placed on the feed line. Attaching the strip to the radiating patch results in extra resonance at the Bluetooth band and a C-type etched slot on the radiating patch notches the WLAN band [20]. Similarly, by using an inverted U-shape slot on the feedline, placing a fractal structure near the feed line, etching a rectangular slot on the grounded CPW, etching a quarter-wavelength slot on the patch are the other reported methods to achieve single notched band characteristics microstrip antenna [21–25]. These single band antennas are converted to dual notched band functional antennas by using band two bandstop filters [26–46]. Etching a pair of E-shaped slots on patch and ground, using an L-shaped stub, and C-type parasitic element has the capability of rejecting dual bands [26–30]. Also, by using Dielectric Resonator Antennas (DRAs) on a radiating antenna, band notch characteristics are obtained [31–32]. Using EBG structure [33], also results in filtering action. In CPW fed antenna, a pair of etched inverted U-slot and etched rectangular strip in-ground is another method of removing interfering bands [34–46]. Similarly, notching of several interfering bands can be further improved by increasing the number of slots/stubs and other notching techniques which were discussed to achieve three notched band characteristics [47–71]. As reported [47], three notched bands are achieved by using externally connected quarter-wavelength stubs and etched C-type slot. Multiple notches are also achieved by embedding L-shaped stubs within the slotted polygonal patch, and interference is mitigated in patch antenna [48,49]. Etched elliptical slots with a single complementary split-ring resonator are also capable of dealing with interfering bands [50]. Modification in capacitively loaded loops near the feed line provides a notching of WiMAX/WLAN and X-band interference [51]. In [52], square slots etched within the microstrip line not only notches the interfering band but they are also used to tune these filters. Three interfering bands are notched by using stepped impedance resonator and forked shaped stubs [53]. Etched L-type slot in the CPW-feed antenna is also capable of eliminating interfering bands [54,55]. Two fractal antennas [56,57], one utilizing slots and other Sierpinski fractal, also ensure the filtering of interfering bands. Electromagnetic Band Gap structure is the other technique reported for the removal of interfering bands [58,59]. Externally added stubs, etching U-C type slots on radiating patch [61], spiral loop resonator attached with patch [62–69] are the other reported techniques to achieve band-notched characteristics. Two antennas with bandwidth ratio 10:1 reported as super-wideband finds elimination of interfering bands by using etched C-type slots and inverted T-type stub with the radiating patch. Elliptical slots, T-type stubs, Γ-shaped stubs, and π-type parasitic elements backed plane all lead to notching of interfering bands [70–71]. Positioning in general refers to the position of the people, equipment, and other objects. In the UWB positioning system, the time difference of the RF signal reaching the target is used and hence the distance between source and object is determined [72].

13.2 UWB TECHNOLOGY

UWB technology offers different advantages such as higher data rate (100 Mb/s), lower consumption of power limited to EIRP (Effective Isotropic Radiated Power) of –41.3 dBm/MHz, ease of fabrication of UWB devices used for UWB applications, and good immune response to multiple path fading. Figure 13.1(a) signifies power utilized for both indoor and outdoor mask applications with flat power consumption (EIRP) for both countries (USA, Japan). For applications in an indoor communication system, UWB bands are sub-divided into 14 different bands observed in Figure 13.1(b). Also, from Figure 13.1(c), it is clear that different existing wireless communication bands do interfere in the working of UWB bandwidth. These said bands are WiMAX (Wireless interoperability for microwave access), C-Band, Wireless Local Area Network, and X-band downlink/uplink bands. This interference is encountered by introducing bandstop filters. Also, UWB bandwidth is sometimes extended to cover applications in Ku, K, and Ka bands which are called super-wideband bandwidth having bandwidth ratio >10:1.

13.3 MICROSTRIP FED UWB ANTENNA

Figure 13.2 shows a microstrip fed UWB antenna for applications in the UWB band. The antenna is printed on a substrate with a radiating patch on one plane and ground on the opposite plane. The radiating patch is connected to 50 Ω transmission which is modeled by using the following equations:

$$Zo = \frac{120\pi}{\sqrt{\varepsilon reff\,[\frac{W}{h} + 1.393 + \frac{2}{3}\ln(\frac{W}{h} + 1.444)]}} \tag{13.1}$$

$$\varepsilon_{reff} = \frac{\varepsilon_r + 1}{2} + \frac{\varepsilon_r - 1}{2}\left[1 + 12\frac{h}{W}\right]^{\frac{-1}{2}} \tag{13.2}$$

Z_o is the impedance of the microstrip transmission line which is 50 Ω. W is the width of microstrip in mm, h is the height of the substrate in mm, ε_{reff} is the effective permittivity of the substrate and ε_r is the dielectric constant of the antenna. This equation holds good for the condition when $W/h < 1$. In the above shown

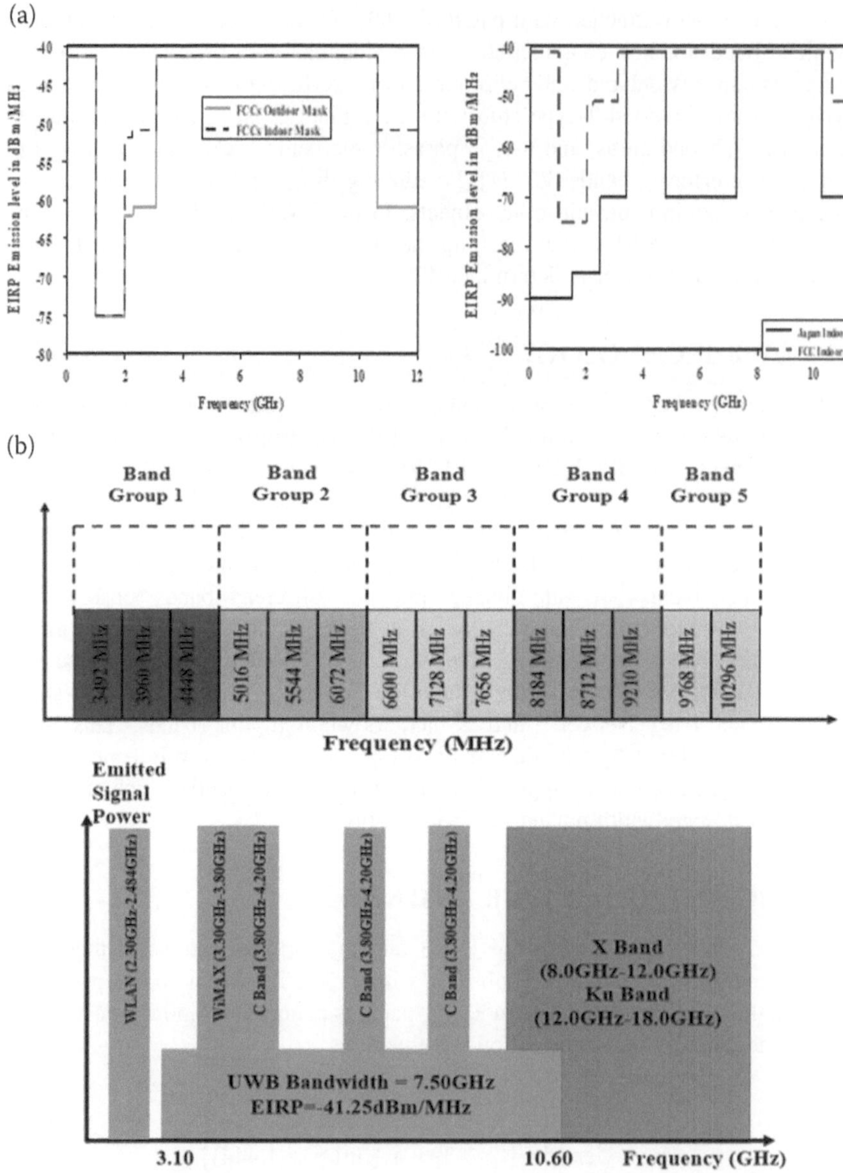

FIGURE 13.1 Frequency vs EIRP for Outdoor and Indoor UWB communication: USA and Japan (b) Band allocation for UWB and Interfering bands in UWB bandwidth.

microstrip antenna, the FR4 substrate is used [1] as dielectric material with dielectric constant 4.4 and height of 1.60 mm. W is fixed to 1.70 mm to match the impedance of the antenna. Further, the microstrip feed is connected to a 50 Ω SMA connector for signal input. The antenna shown above is capable of covering a bandwidth of 2.83 GHz–12.67 GHz as noted in Figure 13.3.

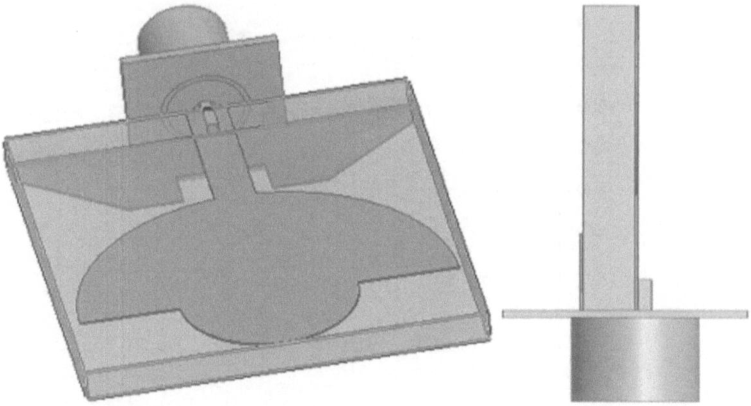

FIGURE 13.2 Microstrip fed UWB antenna: Design environment and side view.

Figure 13.4 represents radiation pattern of antenna offering dipole and omni-directional pattern in principal planes. This is the required bandwidth given by FCC standards and, hence, the antenna is useable for different imaging applications. To understand the impedance matching of the antenna, surface current density

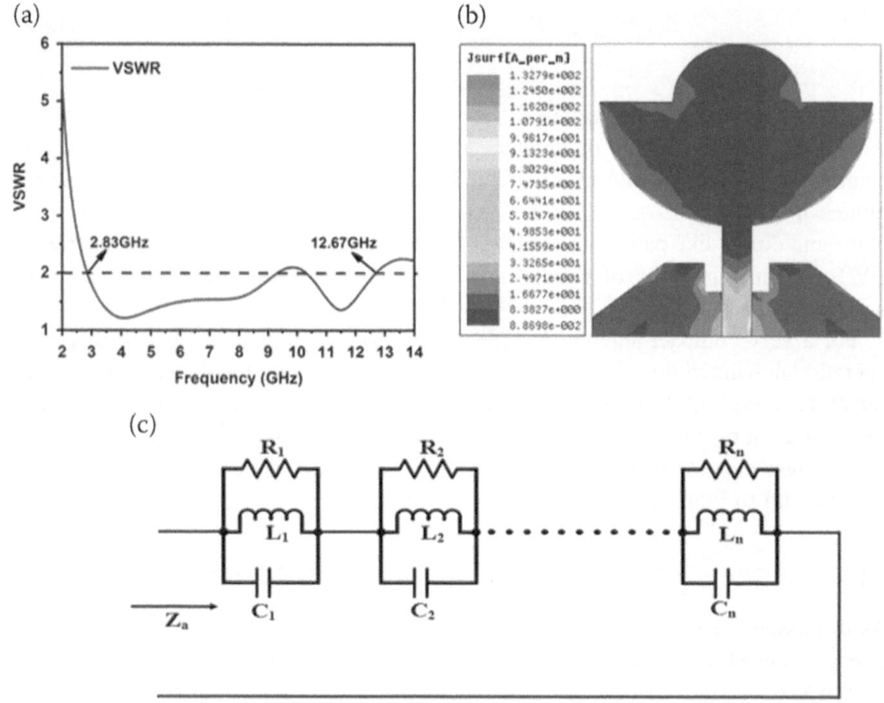

FIGURE 13.3 (a) VSWR characteristics of antenna, (b) Surface current density distribution at 6.85 GHz and (c) Equivalent circuit model.

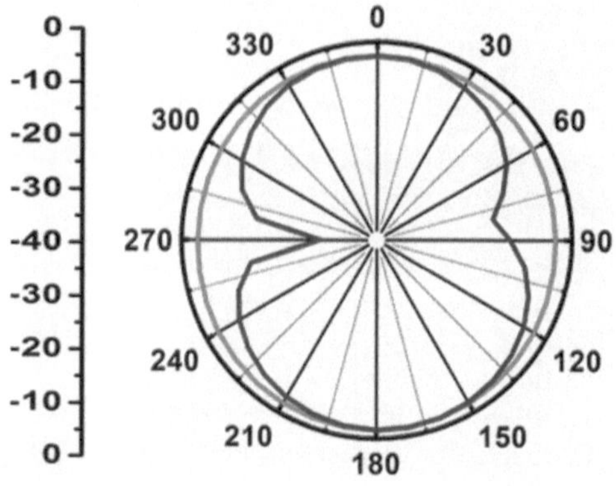

FIGURE 13.4 Radiation Pattern (principal planes).

distribution is studied at a designed center frequency of 6.85 GHz. It can be observed that surface current density is distributed evenly on the surface of the radiating patch which concludes that UWB antenna offers good matching of impedance. It can be also noted that the UWB antenna can be modeled as series-connected parallel RLC circuits with Z_a indicating input impedance.

$$Z_{La} = \sum_{j=1}^{n} [j\omega R_j L_j]/[R_j(1 - \omega^2 L_j C_j) + j\omega L_j] \tag{13.3}$$

(13.3) is used to calculate passive RLC components for each parallel circuit which is a frequency-dependent parameter. Microstrip fed patch antenna has to also characterize in a far-field environment. Figure 3.3 shows a 2-D radiation pattern plotted in principal E- and H-planes, exhibiting an Omni-directional pattern in H-plane and dipole like pattern in E-plane. Table 13.1 shows a comparison of different UWB antenna in terms of overall size, the substrate used, maximum gain, and operational bandwidth.

For a very compact antenna in [15], there was a records gain of 3.50 dBi with operational bandwidth of 2.50–12.20 GHz. On the other hand, when the same substrate is used in [7], impedance bandwidth is increased. However, in general, the FR4 epoxy substrate is used to design UWB antennas as reported in Table 13.1. Also, when compared to Rogers RT Duroid substrate, gain in FR4 substrate is achieved up to 9.50 dBi.

13.4 MICROSTRIP FED UWB SINGLE NOTCHED BAND ANTENNA

As discussed earlier, while UWB antenna works at the front-end, it suffers interference caused by several existing working wireless communications such as WiMAX or WLAN bands. Several techniques have been reported to eliminate interference, including by suing slots on either radiating or ground, using parasitic elements, embedding resonant structure in feedline, etc.

TABLE 13.1
Comparison of Different UWB Antenna

Ref	Size (mm^3)	Substrate	Maximum Gain (dBi)	Bandwidth (GHz)
[2]	24 × 28 × 0.787	Rogers RTDuroid5870ε_r = 2.33, tanδ = 0.0012	5 dBi	2.76–12.80
[3]	45 × 45 × 1.575	Rogers RTDuroid5880ε_r = 2.20, tanδ = 0.0009	7 dBi	3.00–14.0
[4]	36 × 36 × 1.60	FR4 ε_r = 4.40, tanδ = 0.024	—	2.38–12.40
[7]	60.1 × 60.1 × 1.53	FR4 ε_r = 4.30, tanδ = 0.024	—	2.355–15.0
[8]	24 × 36 × 1.52	FR4 ε_r = 4.3, tanδ = 0.024	4.97	3.14–11.86
[9]	24 × 17 × 1.60	FR4 ε_r = 4.40, tanδ = 0.020	5.18	2.94–22.2
[10]	20 × 20 × 1.60	FR4 ε_r = 4.40, tanδ = 0.020	3.34	2.70–11.0
[11]	26 × 10 × 0.80	FR4 ε_r = 4.40, tanδ = 0.020	5.00	3.00–12.0
[12]	35 × 35 × 1.60	FR4 ε_r = 4.40, tanδ = 0.020	5.00	2.30–12.80
[13]	20 × 25 × 1.50	FR4 ε_r = 4.30, tanδ = 0.0250	5.10	3.10–10.80
[14]	20 × 20 × 1.60	FR4 ε_r = 4.30, tanδ = 0.0250	3.40	3.13–14.07
[15]	15 × 19 × 1.60	FR4 ε_r = 4.30, tanδ = 0.0250	3.50	2.50–12.20
[16]	49 × 48.5 × 0.80	FR4 ε_r = 4.3, tanδ = 0.0250	9.50	2.77–10.40

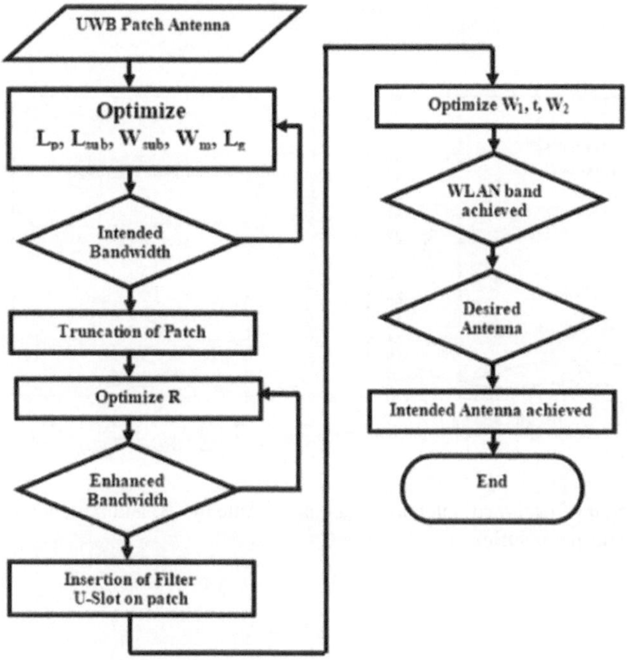

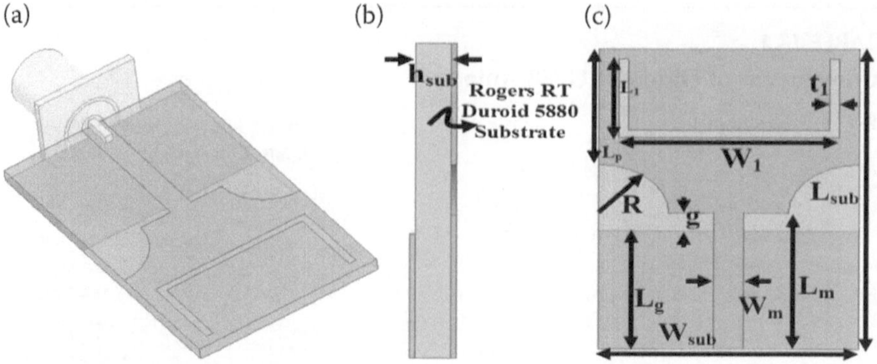

FIGURE 13.5 Configuration of Single Notch antenna with slant, side and front view (with notched band).

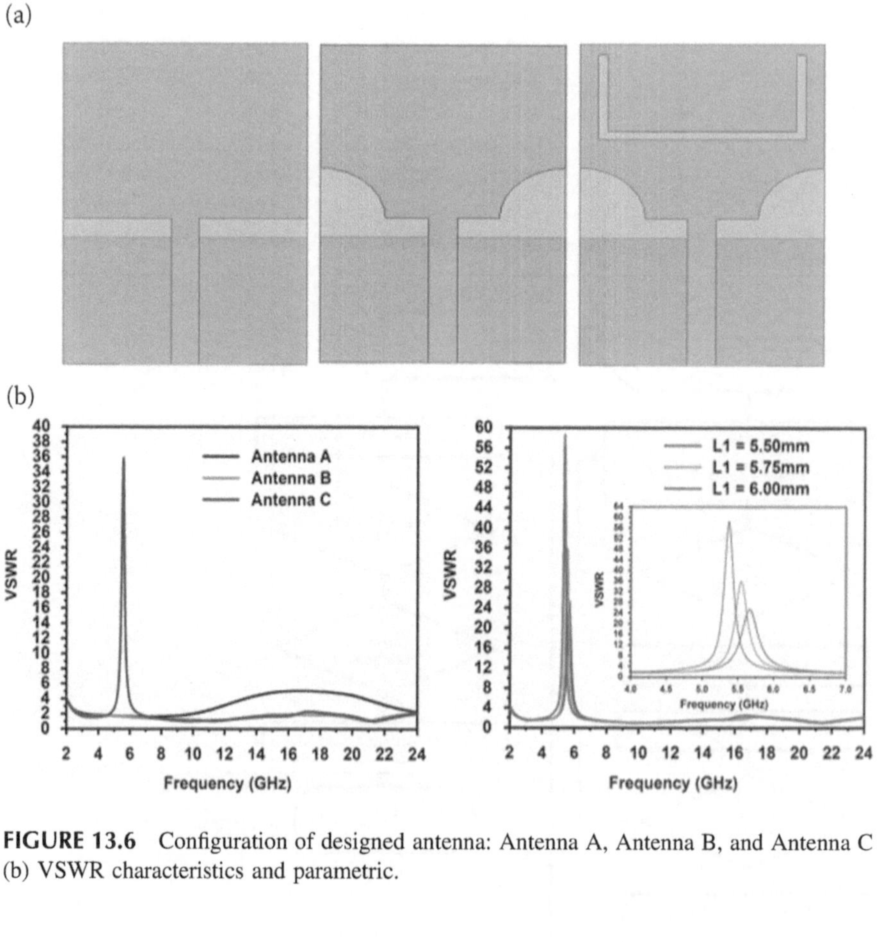

FIGURE 13.6 Configuration of designed antenna: Antenna A, Antenna B, and Antenna C (b) VSWR characteristics and parametric.

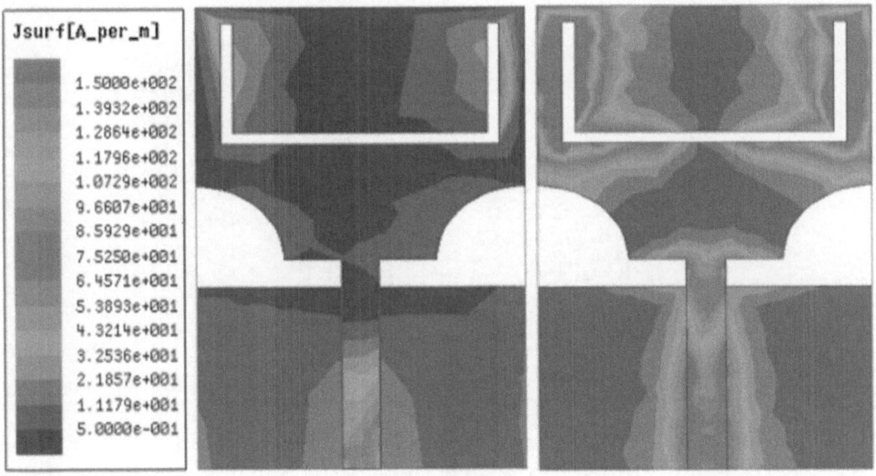

FIGURE 13.7 Surface current density distribution at 6.85 GHz and 5.50 GHz.

TABLE 13.2
Comparison of Different Single Notched Band UWB Antenna

Ref	Size (mm^3)	Substrate	Notched Band (GHz)	Bandwidth (GHz)
[18]	70 × 70 × 3	Thick PDMSε_r = 2.77, tanδ = 0.076	5.00–600	3.80–8.30
[19]	48 × 50 × 1	Taconicε_r = 2.65	5.150–5.825	2.74–11.00
[21]	34 × 34 × 1.58	FR4ε_r = 4.40, tanδ = 0.020	5.00–6.00	3.10–10.60
[22]	24.4 × 33.3 × 1.60	FR4ε_r = 4.40, tanδ = 0.020	5.15–5.825	3.10–10.60
[24]	26 × 27 × 1.60	FR4ε_r = 4.40, tanδ = 0.020	4.67–6.21	2.15–13.95
[25]	13 × 22 × 0.787	Taconic TLY Substrate FR4ε_r = 2.20, tanδ = 0.0009	4.90–6.10	2.50–23.5

Flow chart design

The working of the microstrip fed notched band antenna is studied [25] by designing the antenna in a simulation environment. In this design, a U-type slot is etched which is capable of rejecting WLAN interfering band as shown in Figure 13.5. The radiating patch is printed on one plane of the substrate and ground on the other side with 50 Ω microstrip feed shown in Figure 13.5 and the dimension

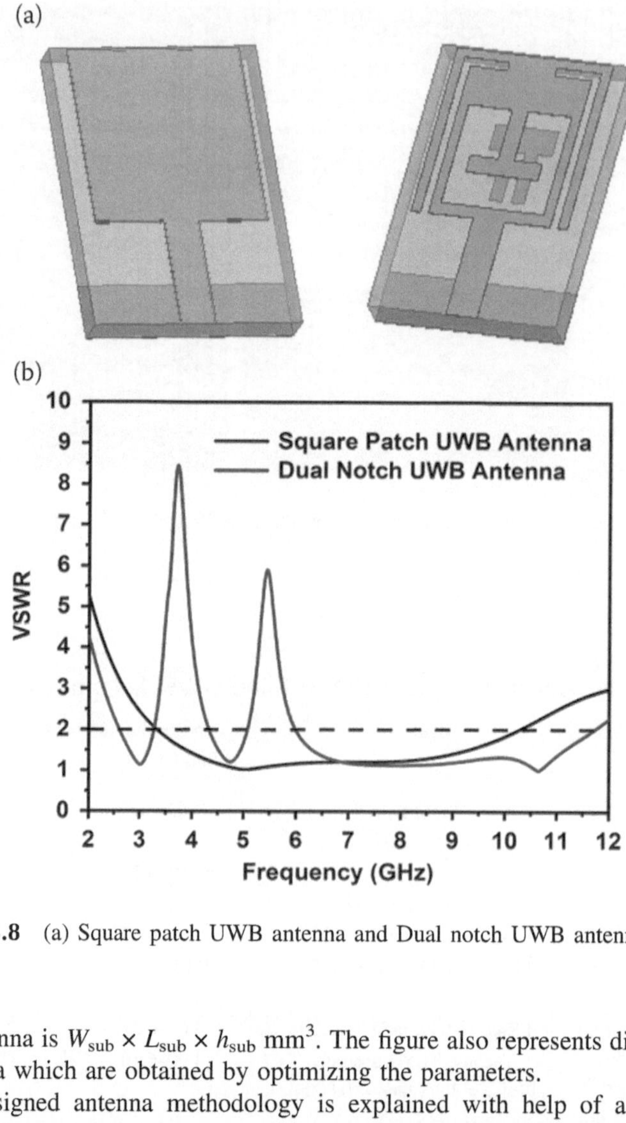

FIGURE 13.8 (a) Square patch UWB antenna and Dual notch UWB antenna (b) VSWR comparison.

of the antenna is $W_{sub} \times L_{sub} \times h_{sub}$ mm^3. The figure also represents dimensions of the antenna which are obtained by optimizing the parameters.

The designed antenna methodology is explained with help of a flow chart. Initially, all the required antenna parameters are optimized to achieve impedance matching by using a square patch and rectangular ground plane. The patch is modified by etching a quarter circle at both bottom edges, resulting in wide operational bandwidth. This antenna is modified by introducing a U-type band stop filter to eliminate the WLAN interfering band. Finally, the required antenna is obtained which is applicable for different wireless applications.

UWB antenna is designed in 3 steps by developing stepwise shown in Figure 13.6(a). Antenna A consists of a rectangular radiating patch and ground plane which provides partial bandwidth of 3.14 GHz–9.99 GHz which is noted from Figure 13.6(b). This is Antenna A. Antenna B is obtained by modifying Antenna A to achieve the intended operating bandwidth of 2.80 GHz–23.23. Further by

TABLE 13.3
Comparison Table for Dual-notched Band UWB Antennas

Ref.	Size (mm^3)	Substrate	Method used for Filtering Bands	Interfering Band (GHz)	Bandwidth (GHz)	Gain (dBi)
[26]	25 × 18 × 1	FR4	E-shaped slots on patch & ground	WiMAXWLAN	2.55–21.65	—
[28]	26 × 26 × 0.762	Taconic RF Substrate	Split Ring Resonators	WLANX-Band	2.90–11.60	6.00
[30]	16 × 28 × 0.80	FR4	Rectangular Slit	WiMAXWLAN	3.10–10.85	5.56
[33]	42 × 50 × 1.60	FR4	DG-ECBG	WiMAXWLAN	2.53–11.86	6.32
[34]	21 × 28 × 1.60	FR4	U-slotRectangular Slot	WLANX-Band	3.10–11.0	4.32
[36]	14 × 20 × 1.60	FR4	Rectangular Stubs	WiMAXWLAN	2.60–11.60	5.00
[37]	47 × 47 × 0.675	Silicon	U SlotI Slot	WiMAXWLAN	0.68–16.23	5.05
[38]	11 × 132 × 1.60	FR4	Inverted U-type slots	WiMAXWLAN	3.10–17.30	4.40
[39]	24 × 35 × 1.60	Rogers	L-shaped slots	WiMAXWLAN	3.10–11.0	4.85
[41]	20 × 23 × 1.60	FR4	Split Ring Circular Slots	WLANX Band	3.20–10.50	4.83
[42]	12 × 18 × 1.60	FR4	I-T shape Parasitic Structure	WLANX Band	3.00–15.7	4.85
[43]	12 × 18 × 1.60	FR4	Hook-type slitΓ-type strips	WiMAXWLAN	2.80–13.30	6.00
[44]	37 × 40 × 1.60	FR4	ω-type slotT-shape slot	WiMAXWLAN	2.50–11.50	5.00

introducing a band-stop filter on the radiating patch, the WLAN band (4.74 GHz–6.55 GHz) is rejected while operating bandwidth of 2.67 GHz–23.87 GHz is maintained. This notched band is calculated by using the following formula

$$F_{WLAN} = \frac{c}{2L_{WLAN}\sqrt{\varepsilon_e}} \tag{13.4}$$

It can be observed that when length L_1 is changed from 5.50 mm to 6.00 mm with a step size of 0.25 mm, shifting of bandwidth is also observed from higher to lower frequency side as notched center frequency F_{WLAN} is inversely proportional to the length of slot L_{WLAN}.

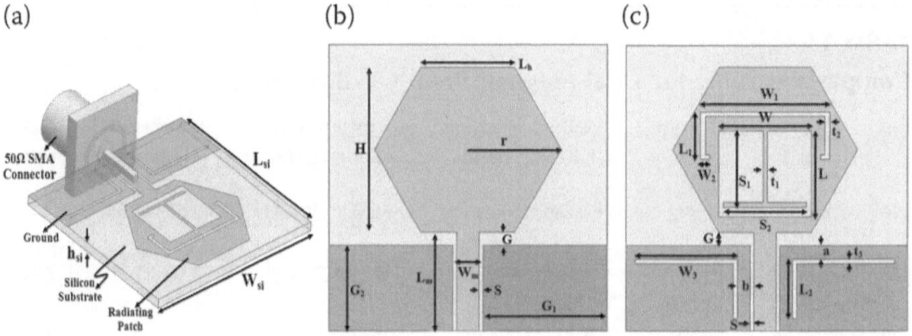

FIGURE 13.9 Triple notched UWB antenna with Inclined plane, Front view (without notched bands) and Front view (with notched bands).

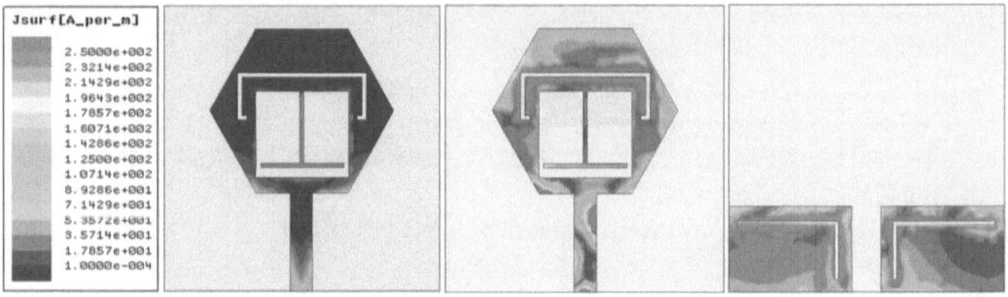

FIGURE 13.10 Simulated SCD distribution 3.44 GHz, 5.54 GHz, and 7.48 GHz.

Figure 13.7 shows the surface current density distribution of the designed antenna at 6.85 GHz and 5.50 GHz. Surface current density is evenly distributed over the surface of the radiating patch for 6.85 GHz as it is operating frequency of interest. For notched frequency 5.50 GHz, maximum surface current density distribution is observed around the U-type slot suggesting a high mismatch of impedance and there is no radiation taking place. This concludes that all the signals are reflected leading to notched band characteristics.

Table 13.2 shows a comparison of different single notched band antennas designed for UWB applications. A different substrate such as PDMS, FR4, and Taconic TLY are used to design these above said notched band antennas. It can be seen that [18] has a large dimension of antenna which covers comparatively less impedance bandwidth, while [25] which is very compact in size using very low permittivity in design provides larger impedance bandwidth which can be not only used for UWB application but are applicable for X, Ku, and K bands.

13.5 MICROSTRIP FED UWB DUAL AND TRIPLE NOTCHED BAND ANTENNAS

As discussed earlier, a single notched band is only capable of mitigating either WiMAX or WLAN interference. When there are more interfering bands, these microstrip fed UWB antennas are needed to be converted to dual or also triple interfering bands UWB antennas.

TABLE 13.4
Comparison Table for Triple-Notched Band UWB Antennas

Ref	Size (mm^2)	Operating Bandwidth (GHz)	Interference Band (GHz)	Bandwidth of Notched Bands	Substrate	Radiation Efficiency
[47]	12 × 19	2.95–12.0	WiMAX/WLAN/X Band	3.32–3.815.20–6.107.30–8.20	FR4	95
[48]	14 × 21	3.00–12.0	WiMAX/WLAN/X Band	3.21–3.655.00–5.627.85–8.45	RO4003	—
[50]	35 × 35	2.21–12.83	WiMAX/WLAN/X Band	2.96–3.735.12–6.078.04–8.65	FR4	—
[52]	27 × 30.5	3.10–10.60	WiMAX/WLAN/X Band	3.31–3.884.96–6.237.90–8.70	FR4	—
[54]	15 × 15	2.98–13.34	WiMAX/WLAN/X Band	3.40–3.954.50–6.347.110–7.802	Silicon	90
[68]	20 × 20	2.55–13.0	WiMAX/WLAN/X Band	2.99–3.825.00–6.007.26–7.95	FR4	90
[69]	18 × 22	2.74–10.57	WiMAX/WLAN/X Band	3.17–3.894.87–6.197.30–7.86	Rogers	91.8

Figure 13.8(a) shows the dual notched band UWB antenna. A square patch with the rectangular ground provides a simple UWB antenna with impedance bandwidth 3.31 GHz–10.30 GHz noted from Figure 13.8(b). It can be converted to the antenna with mitigation of interference by adding bandstop filters. Here, the π-type parasitic element is used backed plane which improves impedance bandwidth. An inverted T-type stub is added within the rectangular slot of the patch which eliminates the WIMAX/C interfering band, while the l-shaped slot removes WLAN interfering.

Table 13.3 shows a comparison table for dual notched band characteristics UWB antenna. Here, different methods are used to obtain band top filters such as using different type slots/stubs/stubs, Split ring resonators, EBG, and parasitic elements backed plane. Antennas are fabricated on FR4, Rogers substrate and a maximum gain of 6 dBi is obtained on FR4 substrate. Antenna size varies for different antennas as matching of impedance has to be achieved.

Figure 13.9 shows a triple notched band UWB antenna with a different configuration. Here, notched bands are achieved by using a C-type slot, inverted T-shaped stub on the patch, and by etching a pair of rotated L-type slit in the ground [54]. The antenna is capable of achieving a bandwidth of 2.98 GHz–13.34 GHz.

Figure 13.10 shows the surface current distribution at centered notched band frequencies. As observed in all the three cases, the maximum surface current density is distributed either within stub or around slots used for filter unwanted bands. This signifies that a high mismatch of impedance is observed leading to all the input signals reflected.

Table 13.3 makes the comparison of the triple notched band microstrip fed UWB antenna. Reported antennas are fabricated by using a different substrate such an FR4, Rogers, etc. As observed, notched band characteristics have been obtained by different filtering techniques such as stub/slot, ECT, Split Ring resonator, etc. Moreover, two superwideband are also reported [70–71] which are capable of covering superwideband bandwidth of bandwidth ratio>10:1 (Table 13.4).

13.6 CONCLUSIONS

UWB antennas play a major role in a short communication system that applies to different wireless system applications. This chapter has discussed different aspects of the UWB antennas design methodology where the different radiating patch has been utilized to obtain UWB bandwidth. Also, sometimes there arises a need to include higher bandwidth along with UWB bandwidth to cover higher microwave bands called super-wideband antennas. It has been observed that antenna size has been reduced by using the different methodologies of design thereby easily integrated with other circuitry. Mitigation of interference is well encountered by all the UWB antennas where different filters design has been used either on radiating patch or ground or also as parasitic elements.

REFERENCES

[1]. M. Koohestani, M. N. Moghadasi, and B. S. Virdee, "Miniature microstrip-fed ultra-wideband printed monopole antenna with a partial ground plane structure," *IET Microwaves, Antennas & Propagation*, vol. 5, no. 14, pp. 1683–1689, 2011.

[2]. M. Koohestani and M. Golpour, "U-shaped microstrip patch antenna with novel parasitic tuning stubs for ultra wideband applications," *IET Microwaves, Antennas & Propagation*, vol. 4, no. 7, pp. 938–946, 2010.
[3]. A. Elboushi, O.M. Ahmed, and A.R. Sebak, "Study of elliptical slot UWB antennas with a 5.0-6.0 GHz band-notch capability," *Progress in Electromagnetics Research C*, vol. 16, pp. 207–222, 2010.
[4]. M. K. Shrivastava, A. K. Gautam, and B. K. Kanaujia, "An M-shaped monopole-like slot UWB antenna," *Microwave and Optical Technology Letters*, vol. 56, no. 1, pp. 127–131, 2014.
[5]. R. Chandel, A. K. Gautam, and B. K. Kanaujia, "Microstrip-line FED beak-shaped monopole-like slot UWB antenna with enhanced band width," *Microwave and Optical Technology Letters*, vol. 56, no. 11, pp. 2624–2628, 2014.
[6]. R. Kumar and P. Malathi, "On the design of CPW-fed Ultra Wideband Triangular heel shape fractal antenna," *International Journal of Microave and Optical Technology*, vol. 5, no. 2, pp. 89–93, 2010.
[7]. R. Kumar and A. G. Kokate, "On the design of square shape fractal antenna with matching strip for UWB applications," *International Journal of Electronics*, vol. 100, no. 7, pp. 881–889, 2013.
[8]. M. O. Dwairi, M. S. Soliman, A. A. Alahmadi, S. H. A. Almalki, and I. I. M. Abu Sulayman, "Design and performance analysis of fractal regular slotted-patch antennas for ultra-wideband communication systems," *Wireless Personal Communications*, vol. 105, no. 3, pp. 819–833, 2019.
[9]. R. N. Tiwari, P. Singh, and B. K. Kanaujia, "A modified microstrip line fed compact UWB antenna for WiMAX/ISM/WLAN and wireless communications," *AEU - International Journal of Electronics and Communications*, vol. 104, pp. 58–65, 2019.
[10]. S. Hota, S. Baudha, B. B. Mangaraj, and M. Varun Yadav, "A novel compact planar antenna for ultra-wideband application," *Journal of Electromagnetic Waves and Applications*, vol. 34, no. 1, pp. 116–128, 2019.
[11]. C.-X. Mao and Q.-X. Chu, "Miniaturization of UWB antenna by asymmetrically extending stub from ground," *Journal of Electromagnetic Waves and Applications*, vol. 28, no. 5, pp. 531–541, 2014.
[12]. J. Lim, S. Oh, S. Lee, W. Yoon, and J. Lee, "Enhanced broadband common-mode filter based on periodic electromagnetic bandgap structures," *Microwave and Optical Technology Letters*, vol. 60, no. 12, pp. 2932–2937, 2018.
[13]. S. Baudha and M. V. Yadav, "A novel design of a planar antenna with modified patch and defective ground plane for ultra-wideband applications," *Microwave and Optical Technology Letters*, vol. 61, no. 5, pp. 1320–1327, 2019.
[14]. S. Baudha, A. Basak, M. Manocha, and M. V. Yadav, "A compact planar antenna with extended patch and truncated ground plane for ultra wide band application," *Microwave and Optical Technology Letters*, vol. 62, no. 1, pp. 200–209, 2019.
[15]. S. Hota, S. Baudha, B. B. Mangaraj, and M. V. Yadav, "A compact, ultrawide band planar antenna with modified circular patch and a defective ground plane for multiple applications," *Microwave and Optical Technology Letters*, vol. 61, no. 9, pp. 2088–2097, 2019.
[16]. S. Guruswamy, R. Chinniah, and K. Thangavelu, "A printed compact UWB Vivaldi antenna with hemi cylindrical slots and directors for microwave imaging applications," *AEU - International Journal of Electronics and Communications*, vol. 110, pp. 152870, 2019. https://doi.org/10.1016/j.aeue.2019.152870
[17]. B. Roy, S. K. Chowdhury, and A. K. Bhattacharjee, "Symmetrical hexagonal monopole antenna with bandwidth enhancement under UWB operations," *Wireless Personal Communications*, vol. 108, no. 2, pp. 853–863, 2019.

[18]. B. Mohanadzade, R.B.V.B. Simorangkir, R.M. Hashmi, Y.C. Oger, M. Zhadabov, R. Sauleau, "A conformal band-notched Ultrawideband antenna with monopole-like radiation characteristics," *IEEE Antennas Wireless and Propagation Letters*, vol. 19, no. 1, pp. 203–207, 2020.

[19]. L. Peng, B-J. Wen, X-F. Li, and S-M. Li, "CPW fed UWB antenna by EBGs with rectangular notched-band," *IEEE Access*, vol. 4, pp. 9542–9552, 2016.

[20]. S. Yadav, A.K. Gautam, B.K. Kanaujia, and K. Rambabu, "Design of band-rejected UWB planar antenna with integrated Bluetooth band," *IET Microwaves, Antennas and Propagation*, vol. 10, no. 14, pp. 1528–1533, 2016.

[21]. Y. K. Choukiker and S. K. Behera, "Modified Sierpinski square fractal antenna covering ultra-wide band application with band notch characteristics," *IET Microwaves, Antennas & Propagation*, vol. 8, no. 7, pp. 506–512, 2014.

[22]. B. Biswas, R. Ghatak, and D. R. Poddar, "UWB monopole antenna with multiple fractal slots for band-notch characteristic and integrated Bluetooth functionality," *Journal of Electromagnetic Waves and Applications*, vol. 29, no. 12, pp. 1593–1609, 2015.

[23]. M. Karthikeyan, R. Sitharthan, T. Ali, and B. Roy, "Compact multiband CPW fed monopole antenna with square ring and T-shaped strips," *Microwave and Optical Technology Letters*, vol. 62, no. 2, pp. 926–932, 2019.

[24]. R. K. Garg, M. V. D. Nair, S. Singhal, and R. Tomar, "A new type of compact ultra-wideband planar fractal antenna with WLAN band rejection," *Microwave and Optical Technology Letters*, vol. 62, no. 7, pp. 2537–2545, 2020.

[25]. N. Hussain, M. Jeong, J. Park, S. Rhee, P. Kim, and N. Kim, "A compact size 2.9-23.5 GHz microstrip patch antenna with WLAN band-rejection," *Microwave and Optical Technology Letters*, vol. 61, no. 5, pp. 1307–1313, 2019.

[26]. M. Akbari, N. Rojhani, M. Saberi, and R. Movahedinia, "Dual band-notched monopole antenna with enhanced bandwidth for ultra-wideband wireless communications," *The Journal of Engineering*, vol. 2014, no. 8, pp. 415–419, 2014.

[27]. K. Srivastava *et al.*, "Integrated GSM-UWB Fibonacci-type antennas with single, dual, and triple notched bands," *IET Microwaves, Antennas & Propagation*, vol. 12, no. 6, pp. 1004–1012, 2018.

[28]. Z. Li, C. Yin, and X. Zhu, "Compact UWB MIMO vivaldi antenna with dual band-notched characteristics," *IEEE Access*, vol. 7, pp. 38696–38701, 2019.

[29]. H. Oraizi and N. Valizade Shahmirzadi, "Frequency- and time-domain analysis of a novel UWB reconfigurable microstrip slot antenna with switchable notched bands," *IET Microwaves, Antennas & Propagation*, vol. 11, no. 8, pp. 1127–1132, 2017.

[30]. Y.-P. Zhang and C.-M. Li, "Design of small dual band-notched UWB slot antenna," *Electronics Letters*, vol. 51, no. 22, pp. 1727–1728, 2015.

[31]. M. Abedian, S. K. A. Rahim, S. Danesh, S. Hakimi, L. Y. Cheong, and M. H. Jamaluddin, "Novel design of compact UWB dielectric resonator antenna with dual-band-rejection characteristics for WiMAX/WLAN bands," *IEEE Antennas and Wireless Propagation Letters*, vol. 14, pp. 245–248, 2015.

[32]. I. B. Vendik, A. Rusakov, K. Kanjanasit, J. Hong, and D. Filonov, "Ultrawideband (UWB) Planar antenna with single-, dual-, and triple-band notched characteristic based on electric ring resonator," *IEEE Antennas and Wireless Propagation Letters*, vol. 16, pp. 1597–1600, 2017.

[33]. N. Jaglan, B. K. Kanaujia, S. D. Gupta, and S. Srivastava, "Design of band-notched antenna with DG-CEBG," *International Journal of Electronics*, vol. 105, no. 1, pp. 58–72, 2017.

[34]. C. Liu, T. Jiang, Y. Li, and J. Zhang, "A compact wide slot antenna with dual band-notch characteristic for ultra-wideband applications," *International Journal of Electronics*, vol. 100, no. 8, pp. 1134–1146, 2013.

[35]. M. A. Salamin, W. A. E. Ali, S. Das, and A. Zugari, "Design and investigation of a multi-functional antenna with variable wideband/notched UWB behavior for WLAN/X-band/UWB and Ku-band applications," *AEU - International Journal of Electronics and Communications*, vol. 111, 2019.

[36]. R. Kumar and Y. Kamatham, "Fork shaped with inverted L-stub resonator UWB antenna for WiMAX/WLAN rejection band," *AEU - International Journal of Electronics and Communications*, vol. 110, pp. 152881, 2019.

[37]. K. Sharma *et al.*, "Reconfigurable dual notch band antenna on Si-substrate integrated with RF MEMS SP4T switch for GPS, 3G, 4G, bluetooth, UWB and close range radar applications," *AEU - International Journal of Electronics and Communications*, vol. 110, 2019.

[38]. J. Acharjee, K. Mandal, S. K. Mandal, and P. P. Sarkar, "A compact printed monopole antenna with enhanced bandwidth and variable dual band notch for UWB applications," *Journal of Electromagnetic Waves and Applications*, vol. 30, no. 15, pp. 1980–1992, 2016.

[39]. G. Gao, B. Hu, C. Yang, S. Wang, and R. Zhang, "Design of a dual band-notched UWB antenna and improvement of the 5.5 GHz WLAN notched characteristic," *Journal of Electromagnetic Waves and Applications*, vol. 33, no. 14, pp. 1834–1845, 2019.

[40]. H. U. Bong, M. Jeong, N. Hussain, S. Y. Rhee, S. K. Gil, and N. Kim, "Design of an UWB antenna with two slits for 5G/WLAN-notched bands," *Microwave and Optical Technology Letters*, vol. 61, no. 5, pp. 1295–1300, 2019.

[41]. W. A. Awan, A. Zaidi, N. Hussain, A. Iqbal, and A. Baghdad, "Stub loaded, low profile UWB antenna with independently controllable notch-bands," *Microwave and Optical Technology Letters*, vol. 61, no. 11, pp. 2447–2454, 2019.

[42]. Y. Ojaroudi, S. Ojaroudi, and N. Ojaroudi, "A novel 5.5/7.5 GHz dual band-stop antenna with modified ground plane for UWB communications," *Wireless Personal Communications*, vol. 81, no. 1, pp. 319–332, 2014.

[43]. N. Ojaroudi, N. Ghadimi, and Y. Ojaroudi, "Compact multi-resonance monopole antenna with dual band-stop property for UWB wireless communications," *Wireless Personal Communications*, vol. 81, no. 2, pp. 563–579, 2014.

[44]. G. K. Pandey, H. S. Singh, P. K. Bharti, and M. K. Meshram, "Design and analysis of Ψ-shaped UWB antenna with dual band notched characteristics," *Wireless Personal Communications*, vol. 89, no. 1, pp. 79–92, 2016.

[45]. N. Jaglan, B. K. Kanaujia, S. D. Gupta, and S. Srivastava, "Design and development of an efficient EBG structures based band notched UWB circular monopole antenna," *Wireless Personal Communications*, vol. 96, no. 4, pp. 5757–5783, 2017.

[46]. M. Moosazadeh, A. M. Abbosh, and Z. Esmati, "Design of compact planar ultra-wideband antenna with dual-notched bands using slotted square patch and pi-shaped conductor-backed plane," *IET Microwaves, Antennas & Propagation*, vol. 6, no. 3, pp. 290–294, 2012.

[47]. S. Doddipalli and A. Kothari, "Compact UWB antenna with integrated triple notch bands for WBAN applications," *IEEE Access*, vol. 7, pp. 183–190, 2019.

[48]. H. Hosseini, H. R. Hassani, and M. H. Amini, "Miniaturised multiple notched omnidirectional UWB monopole antenna," *Electronics Letters*, vol. 54, no. 8, pp. 472–474, 2018.

[49]. S. u. Rehman and M. A. S. Alkanhal, "Design and system characterization of ultra-wideband antennas with multiple band-rejection," *IEEE Access*, vol. 5, pp. 17988–17996, 2017.

[50]. D. Sarkar, K. V. Srivastava, and K. Saurav, "A compact microstrip-fed triple band-notched UWB monopole antenna," *IEEE Antennas and Wireless Propagation Letters*, vol. 13, pp. 396–399, 2014.

[51]. J. Wang, Y. Yin, and X. Liu, "Triple band-notched ultra-wideband antenna using a pair of novel symmetrical resonators," *IET Microwaves, Antennas & Propagation*, vol. 8, no. 14, pp. 1154–1160, 2014.

[52]. C. You-Zhi, Y. Hong-Chun, and C. Ling-Yun, "Wideband monopole antenna with three band-notched characteristics," *IEEE Antennas and Wireless Propagation Letters*, vol. 13, pp. 607–610, 2014.

[53]. C. Zhang, J. Zhang, and L. Li, "Triple band-notched UWB antenna based on SIR-DGS and fork-shaped stubs," *Electronics Letters*, vol. 50, no. 2, pp. 67–69, 2014.

[54]. M. Sharma, Y. K. Awasthi, and H. Singh, "CPW-fed triple high rejection notched UWB and X-band antenna on silicon for imaging and wireless applications," *International Journal of Electronics*, vol. 106, no. 7, pp. 945–959, 2019.

[55]. K. Srivastava, A. Kumar, B. K. Kanaujia, and S. Dwari, "Integrated amateur band and ultra-wide band monopole antenna with multiple band-notched," *International Journal of Electronics*, vol. 105, no. 5, pp. 741–755, 2017.

[56]. A. H. Nazeri, A. Falahati, and R. M. Edwards, "A novel compact fractal UWB antenna with triple reconfigurable notch reject bands applications," *AEU - International Journal of Electronics and Communications*, vol. 101, pp. 1–8, 2019.

[57]. A. Gorai, A. Karmakar, M. Pal, and R. Ghatak, "Multiple fractal-shaped slots-based UWB antenna with triple-band notch functionality," *Journal of Electromagnetic Waves and Applications*, vol. 27, no. 18, pp. 2407–2415, 2013.

[58]. E. Thakur, N. Jaglan, and S. D. Gupta, "Design of compact triple band-notched UWB MIMO antenna with TVC-EBG structure," *Journal of Electromagnetic Waves and Applications*, vol. 31, no. 11, pp. 1601–1615, 2020.

[59]. A. Ghosh, T. Mandal, and S. Das, "Design and analysis of triple notch ultrawide-band antenna using single slotted electromagnetic bandgap inspired structure," *Journal of Electromagnetic Waves and Applications*, vol. 33, no. 11, pp. 1391–1405, 2019.

[60]. T. Li, H. Zhai, L. Li, and C. Liang, "Monopole antenna for GSM, Bluetooth, and UWB applications with dual band-notched characteristics," *Journal of Electromagnetic Waves and Applications*, vol. 28, no. 14, pp. 1777–1785, 2014.

[61]. N. A. Murugan, R. Balasubramanian, and H. R. Patnam, "Printed ultra-wideband monopole U-slotted antenna for triple band-rejection," *Journal of Electromagnetic Waves and Applications*, vol. 30, no. 12, pp. 1532–1544, 2016.

[62]. S. Natarajamani, S. K. Behera, and S. K. Patra, "A triple band-notched planar antenna for UWB applications," *Journal of Electromagnetic Waves and Applications*, vol. 27, no. 9, pp. 1178–1186, 2013.

[63]. M. Alibakhshikenari, B. S. Virdee, and E. Limiti, "Compact single-layer traveling-wave antenna designusing metamaterial transmission lines," *Radio Science*, vol. 52, no. 12, pp. 1510–1521, 2017.

[64]. B. Hammache, A. Messai, I. Messaoudene, and T. A. Denidni, "A compact ultra-wideband antenna with three C-shaped slots for notched band characteristics," *Microwave and Optical Technology Letters*, vol. 61, no. 1, pp. 275–279, 2018.

[65]. A. Yadav, R. P. Yadav, and A. Alphones, "CPW fed triple band notched UWB antenna: slot width tuning," *Wireless Personal Communications*, vol. 111, no. 4, pp. 2231–2245, 2019.

[66]. P. S. Bakariya, S. Dwari, and M. Sarkar, "A triple band notch compact UWB printed monopole antenna," *Wireless Personal Communications*, vol. 82, no. 2, pp. 1095–1106, 2015.

[67]. S. Haroon, K. S. Alimgeer, N. Khalid, B. T. Malik, M. F. Shafique, and S. A. Khan, "A low profile UWB antenna with triple band suppression characteristics," *Wireless Personal Communications*, vol. 82, no. 1, pp. 495–507, 2014.

[68]. M. Sharma, Y. K. Awasthi, and H. Singh, "Design of compact planar triple band-notch monopole antenna for ultra-wideband applications," *Wireless Personal Communications*, vol. 97, no. 3, pp. 3531–3545, 2017.
[69]. M. Sharma, Y. K. Awasthi, H. Singh, R. Kumar, and S. Kumari, "Compact UWB antenna with high rejection triple band-notch characteristics for wireless applications," *Wireless Personal Communications*, vol. 97, no. 3, pp. 4129–4143, 2017.
[70]. M. Sharma, "Superwideband triple notch monopole antenna for multiple wireless applications," *Wireless Personal Communications*, vol. 104, no. 1, pp. 459–470, 2018.
[71]. S. Singh, R. Varma, M. Sharma, and S. Hussain, "Superwideband monopole reconfigurable antenna with triple notched band characteristics for numerous applications in wireless system," *Wireless Personal Communications*, vol. 106, no. 3, pp. 987–999, 2019.
[72]. A. Alarifi, A. Al-Salman, M. Alsaleh, A. Alnafessah, S. Al-Hadhrami, M.A. Al-Ammar, and H.S. Al-Khalifa, "Ultra widwband indooor positioning technologies: analysis and recent advances," *Sensors*, vol. 16, no. 5, pp. 1–36, 2016.

14 Spline Based Ultra Wideband Antenna and Design

A. J. Sharath Kumar[1] *and Dr. T. P. Surekha*[2]
[1]Assistant Professor, Department of Electronics and Communication Engineering, Vidyavardhaka College of Engineering, Mysuru, Karnataka, India
[2]Professor, Department of Electronics and Communication Engineering, Vidyavardhaka College of Engineering, Mysuru, Karnataka, India

14.1 SPLINE BASED PRINTED MONOPOLE UWB ANTENNA AND DESIGN

To define antenna geometric parameters, the spline-based design exploits the best features of flexibility and enables quick synthesis. Figure 14.1 displays the diagram of the proposed UWB antenna constructed using ANSYS HFSS. Spline based radiator with six control points is printed on Rogers R04350B substrate with permittivity 3.58. The designed antenna dimension is 35 × 32 × 0.508 mm^3. The coordinates of the control points (P_1, P_2, P_3, P_4, P_5, P_6) are (5.8, −0.52, 0.508), (4, −6, 0.508), (−4, −10, 0.508), (4.10, 0.508), (5.8, 3, 0.508), and (5.8, −0.52, 0.508) respectively. Microstrip line feeding is used to excite the antenna. At the bottom of the substrate, partial ground plane with 11 mm length is printed. This is done to add extra reactance resulting in broadband response. Simulated return loss and VSWR plots are shown in Figures 14.2 and 14.3 respectively. From Figure 14.2 it is noticed that the designed antenna has return loss less than −10 dB between 3 to 16 GHz and thus is said to work in the UWB frequency band. From Figure 14.3 it is seen that within the frequency range of 3 to 16 GHz VSWR is kept less than two. The 3D gain plots are depicted in Figure 14.4. It is observed that the antenna has omni-directional gain pattern at 5 GHz and 7 GHz.

As seen in Table 14.1, the optimization techniques such as particle swarm optimization (PSO) and genetic algorithm (GA) are taken into account in [17] and [18] to determine the antenna shape. The implementation of these methods of optimization technique results in high computational cost, limited capabilities for

DOI: 10.1201/9781003187325-14

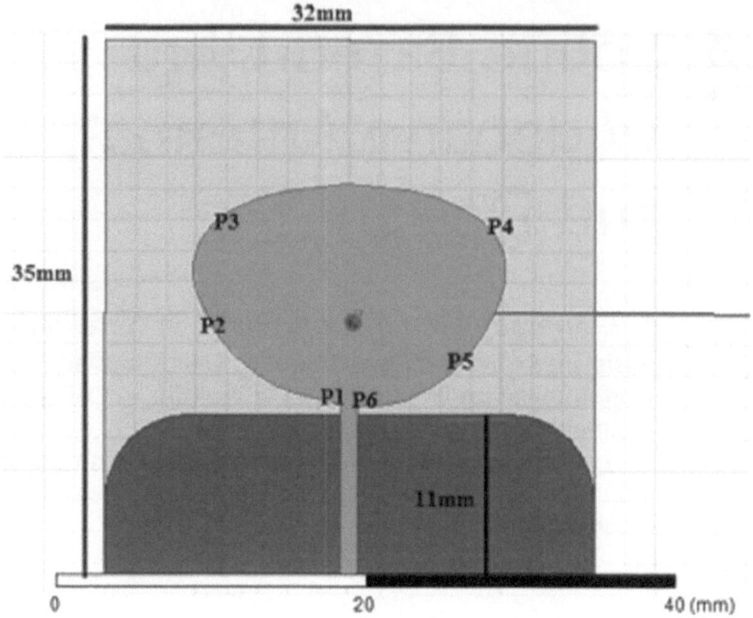

FIGURE 14.1 Proposed Spline based ultra wideband antenna.

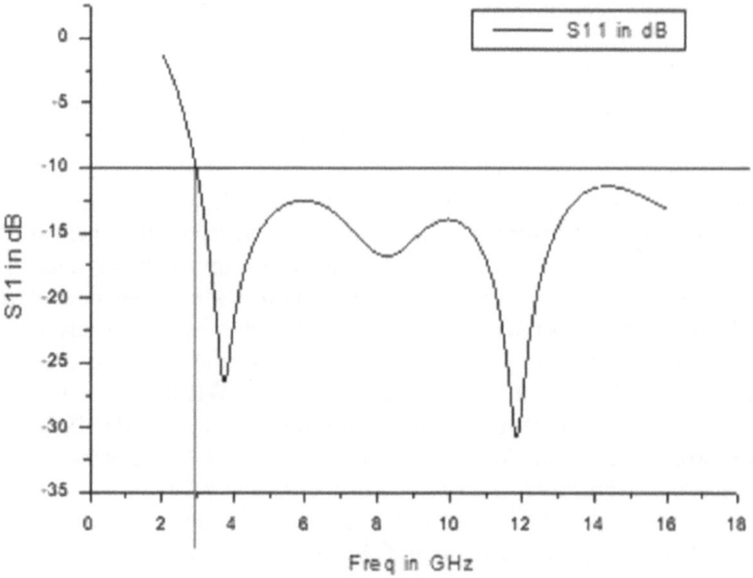

FIGURE 14.2 Reflection coefficient graph of proposed spline based ultra wideband antenna.

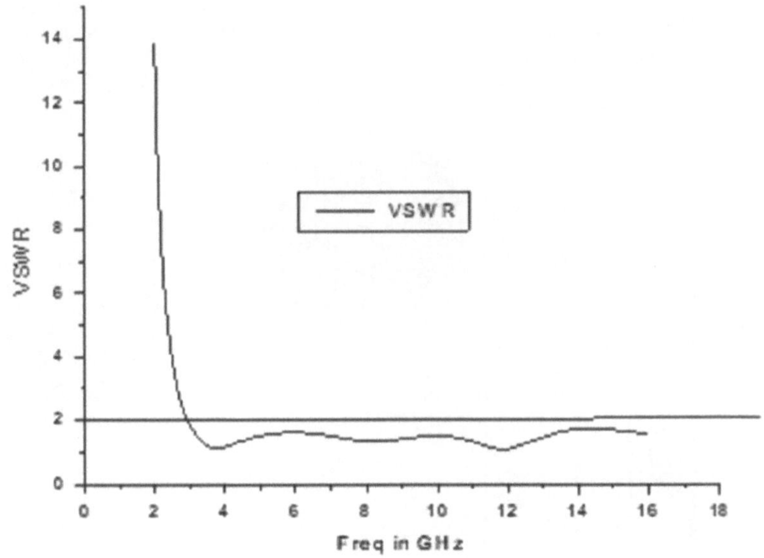

FIGURE 14.3 VSWR plot of proposed spline based UWB antenna.

addressing constraints, and parameter tuning. For example, the fitness function considers only certain frequencies that result in $S_{11} < -10$ dB in both [17] and [18], while no consideration is paid to other parameters, such as bulkiness of the antenna. In the proposed antenna, it can be shown that relative to previous works, broad bandwidth is accomplished with a compact design.

14.2 MINIATURIZED SPLINE BASED UWB ANTENNA AND DESIGN

The major task of this research is to reduce the volume and obtain excellent efficiency of the antenna over the UWB spectrum. Using HFSS simulator, the antennas are simulated. Tests are carried out using TG124A Spectrum analyzer. This segment discusses the study of the proposed compact antenna.

14.3 H-EMBEDDED SPLINE BASED ANTENNA ULTRA WIDEBAND ANTENNA

The configuration of the proposed antenna is shown in Figure 14.5. The geometry of the spline antenna is embedded with H-shaped structure as seen in Figure 14.6. The size is diminished due to this injection. Details of antenna dimension are presented in Table 14.2. The fabricated antenna is depicted in Figure 14.7. The simulated and measured UWB Spline antenna return loss plot with a double H-structure is shown

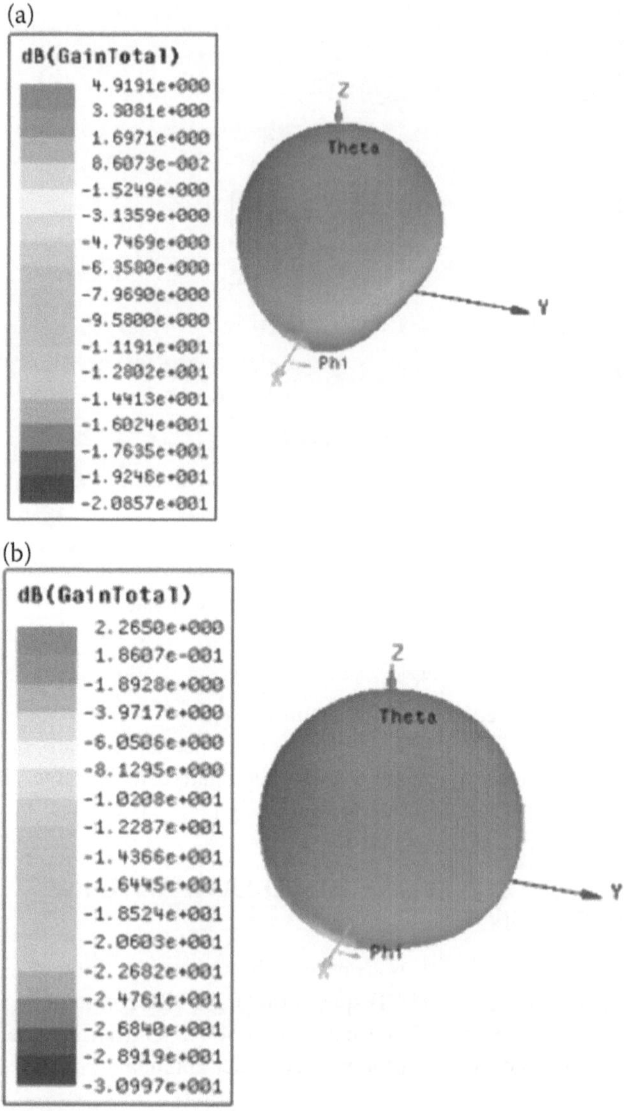

FIGURE 14.4 A Gain at 5 GHz, b Gain at 7 GHz.

in Figure 14.8. The designed antenna resonates between 4–13.1 GHz in simulation and 3.7–12.9 GHz in measurements. VSWR in the resonant bandwidth is less than two (Figure 14.9).

14.4 RESHAPED SPLINE-BASED ULTRA WIDEBAND ANTENNA

The goal of reducing the size is assumed to be satisfied by the structural changes of the original spline-based ultra wideband antenna. The amended configuration of the

TABLE 14.1
Comparison with Previous Works

Paper	Size in mm^3	Operating Frequency	Comments
[17]	35 × 32 × 0.508	3.7 to 9.2 GHz	PSO optimization, distortions not considered
[18]	40 × 31 × 0.762	2.92 GHz to 14.9 GHz	GA optimization, distortion at high frequency
This antenna	35 × 32 × 0.508	3.1 to 16 GHz	Less distortion at high frequency

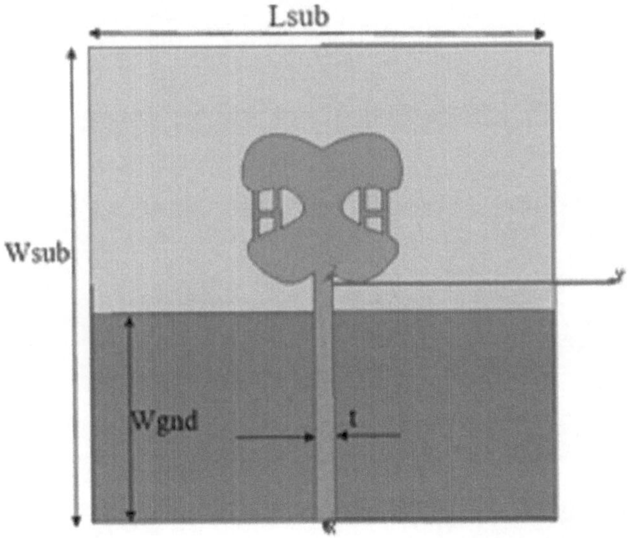

FIGURE 14.5 Proposed H-embed Spline based ultra wideband antenna.

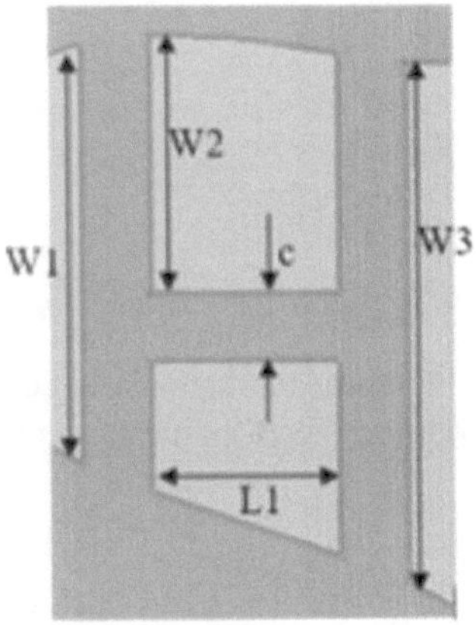

FIGURE 14.6 H-embedded structure.

TABLE 14.2
Antenna Dimension

Parameter	Dimension in mm
W_{sub}	25
L_{sub}	25
h	1
W_{gnd}	11
L_{gnd}	L_{sub}
t	1
W_1	2
W_2	1.24
W_3	2.4
L_1	0.9
c	0.3

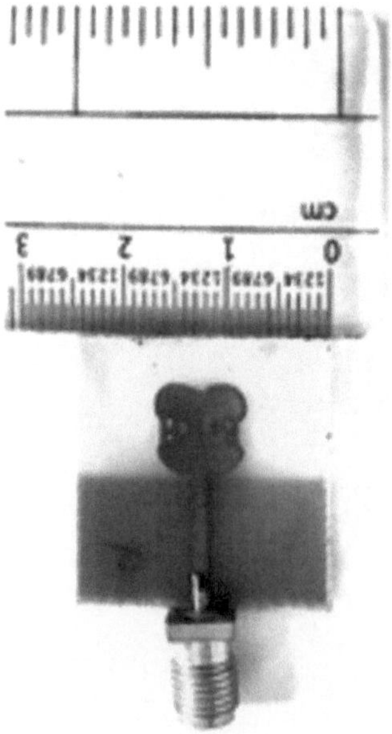

FIGURE 14.7 Fabricated spline antenna.

antenna is presented in Figure 14.10. By changing the position of control points 2, 4, 8, and 9, the suggested configuration is obtained. The fabricated antenna is depicted in Figure 14.11. The simulated and measured modified UWB Spline antenna

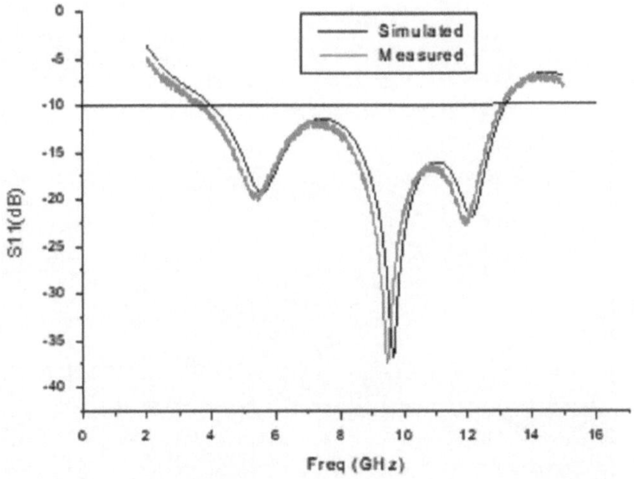

FIGURE 14.8 Reflection coefficient graph of H-embed ultra wideband spline antenna Figure 14.9 VSWR plot of proposed H-embed spline based UWB antenna.

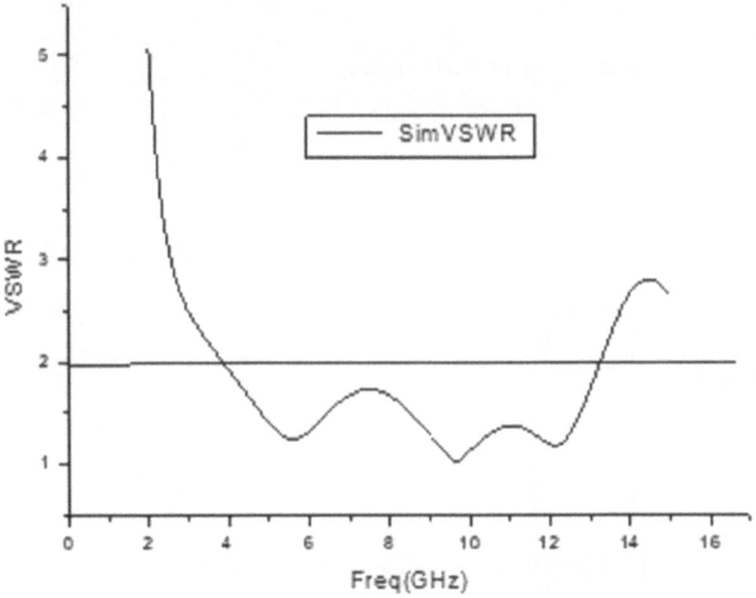

FIGURE 14.9 VSWR plot of UWB spline antenna with double H-shaped structure.

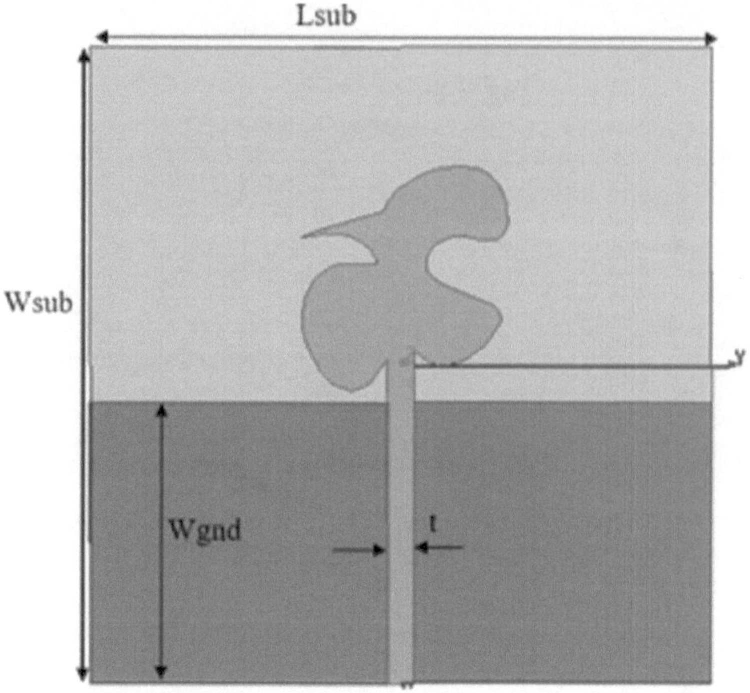

FIGURE 14.10 Modified Spline UWB antenna.

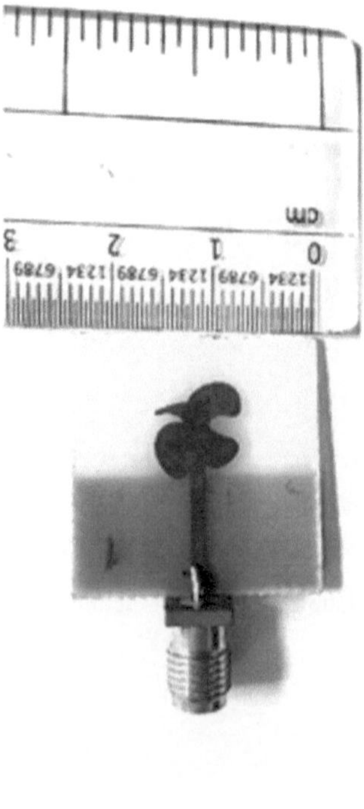

FIGURE 14.11 Fabricated antenna.

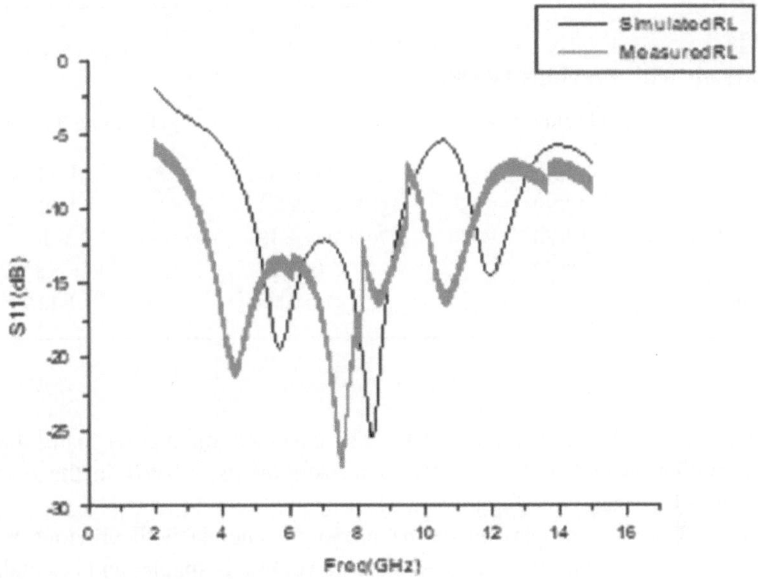

FIGURE 14.12 Reflection coefficient graph of altered spline antenna.

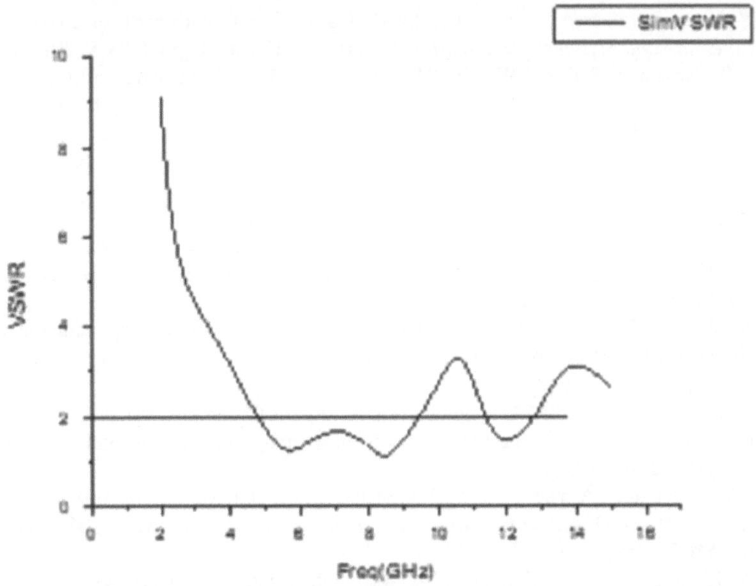

FIGURE 14.13 VSWR graph of altered spline antenna.

TABLE 14.3
Comparison with Previous Works

Reference	Publication Year	Size in mm^3	Operating Band in GHz
[19]	2019	36 × 24 × 1.6	3.1–14.6
[20]	2019	41.5 × 41.5 × 1	2–12
[21]	2020	70 × 70 × 3.18	3.4–3.7
[22]	2020	70 × 45 × 3	1–3.5
This antenna		25 × 25 × 0.508	3.7–12.9

return loss plot is shown in Figure 14.12. The designed antenna resonates 4.8–9.4 GHz in simulation and 3.35–9.4 GHz in measurements. VSWR in the resonant bandwidth is less than two (Figure 14.13).

Table 14.3 shows comparison of the proposed antenna with previous works. Among the other antennas the proposed antenna volume is smaller and is suitable to operate in the ultra wideband region.

REFERENCES

1. Y. Xu et al., "Design of a notched-band vivaldi antenna with high selectivity," *IEEE Antennas and Wireless Propagation Letters* vol. 17, no. 1, pp. 62–65, 2018.
2. I. T. Nassar and T. M. Weller, "A novel method for improving antipodal Vivaldi antenna performance," *IEEE Transactions on Antennas and Propagation* vol. 63, no. 7, pp. 3321–3324, 2015.
3. Bang, Jihoon, Juneseok Lee, and Jaehoon Choi. "Design of a wideband antipodal vivaldi antenna with an asymmetric parasitic patch." *Journal of Electromagnetic Engineering and Science* vol. 18, no. 1, pp. 29–34, 2018.
4. Z. Yang, L. Zhang, and T. Yang, "A microstrip magnetic dipole yagi-uda antenna employing vertical i-shaped resonators as parasitic elements," *IEEE Transactions on Antennas and Propagation* 2018.
5. X. Wei, J. Liu, and Y. Long, "Printed log-periodic monopole array antenna with a simple feeding structure," *IEEE Antennas and Wireless Propagation Letters* vol. 17, no. 1, pp. 58–61, 2018.
6. L. Ge and K-M. Luk, "A magneto-electric dipole for unidirectional UWB communications," *IEEE Transactions on Antennas and Propagation* vol. 61, no. 11, pp. 5762–5765, 2013.
7. J-Y. Li et al., "A wideband high-gain cavity-backed low-profile dipole antenna," *IEEE Transactions on Antennas and Propagation* vol. 64, no. 12, pp. 5465–5469, 2016.
8. A. Elsherbini and K. Sarabandi, "UWB high-isolation directive coupled-sectorial-loops antenna pair," *IEEE Antennas and Wireless Propagation Letters* vol. 10, pp. 215–218, 2011.
9. M. Li, and K-M. Luk, "A differential-fed UWB antenna element with unidirectional radiation," *IEEE Transactions on Antennas And Propagation* vol. 64, no. 8, pp. 3651–3656, 2016.

10. R. A. Moody and S. K. Sharma, "Ultrawide bandwidth (UWB) planar monopole antenna backed by novel pyramidal-shaped cavity providing directional radiation patterns," *IEEE Antennas and Wireless Propagation Letters* vol. 10, pp. 1469–1472, 2011.
11. P. Kumar, G. C. Ghivela and J. Sengupta, "Optimized N-sided polygon shaped microstrip patch antenna for UWB application," *2019 10th International Conference on Computing, Communication and Networking Technologies (ICCCNT)*, Kanpur, India, 2019, pp. 1–7, doi: 10.1109/ICCCNT45670.2019.8944343.
12. Z. Zhang, S. Yang, M. Liu, S. Deng and L. Li, "Design of an UWB microstrip antenna with DGS based on genetic algorithm," *2019 21st International Conference on Advanced Communication Technology (ICACT), PyeongChangKwangwoon_Do*, Korea (South), 2019, pp. 228–232, doi: 10.23919/ICACT.2019.8701944.
13. F. N. Mohd Isa, N. H. Kamaludin, N. F. Abdul Malek and S. Y. Mohamad, "Compact UWB Slotted Pentagonal Patch Antenna for Radar and Communication," *2019 IEEE International Symposium on Antennas and Propagation and USNC-URSI Radio Science Meeting*, Atlanta, GA, USA, 2019, pp. 1393–1394, doi: 10.1109/APUSNCURSINRSM.2019.8888353.
14. R. Verma and R. P. S. Gangwar, "Design & Simulation of a Compact UWB Microstrip Filtenna," *2019 Women Institute of Technology Conference on Electrical and Computer Engineering (WITCON ECE)*, Dehradun Uttarakhand, India, 2019, pp. 235–240, doi: 10.1109/WITCONECE48374.2019.9092924.
15. Z. Ul Islam, I. Nadeem, Z. Ahmed, F. K. Lodhi, F. B. Zarrabi and M. Haneef, "UWB Microstrip Antenna with U-Shape Slot for WLAN Band-Suppression Application," *2019 International Conference on Communication Technologies (ComTech)*, Rawalpindi, Pakistan, 2019, pp. 26–29, doi: 10.1109/COMTECH.2019.8737791.
16. M. N. Hasan, P. Singh and M. D. Nadeem, "Omnidirectional UWB Antenna Loaded with Rectangular Loop for Band Notch Characteristics," *2019 6th International Conference on Signal Processing and Integrated Networks (SPIN)*, Noida, India, 2019, pp. 425–429, doi: 10.1109/SPIN.2019.8711662.
17. L. Lizzi, F. Viani, R. Azaro, P. Rocca and A. Massa, "An innovative spline-based shaping approach for ultra-wideband antenna synthesis," *2007 IEEE Antennas and Propagation Society International Symposium*, Honolulu, HI, 2007, pp. 757–760, doi: 10.1109/APS.2007.4395604.
18. M. John and M. J. Ammann, "Spline based geometry for printed UWB antenna design," *2007 IEEE Antennas and Propagation Society International Symposium*, Honolulu, HI, 2007, pp. 761–764, doi: 10.1109/APS.2007.4395605.
19. M. O. Al-Dwairi, A. Y. Hendi, M. S. Soliman and M. A. Nisirat, "Design of A Compact Ultra-Wideband Antenna for Super-Wideband Technology," *2019 13th European Conference on Antennas and Propagation (EuCAP)*, Krakow, Poland, 2019, pp. 1–4.
20. M. Min, L. Guo, W. Che and W. Yang, "A Miniaturized Ultra-Wideband Dipole Antenna Utilizing A Concaved Arm," *2019 IEEE MTT-S International Wireless Symposium (IWS)*, Guangzhou, China, 2019, pp. 1–3, doi: 10.1109/IEEE-IWS.2019.8804093.
21. Z. Liu, L. Zhu and N. Liu, "A Compact Omnidirectional Patch Antenna with Ultra-Wideband Harmonic Suppression," *IEEE Transactions on Antennas and Propagation*, doi: 10.1109/TAP.2020.2995422.
22. S. Kim and S. Nam, "Compact Ultra-Wideband Antenna on Folded Ground Plane," *IEEE Transactions on Antennas and Propagation*, doi: 10.1109/TAP.2020.2977818.

15 Design Strategy of Wearable Textile Antenna

P. Potey and K. Tuckley
[1]LTCE, Navi Mumbai, India
[2]IIT Bombay, Mumbai, India

15.1 INTRODUCTION

One thought provoking development in the field of wireless wearable technology are body worn textile antennas. Fabric with uniform structure and bendable construction materials are two primary requirements for wearable textile antennas. This antenna has huge demand in wireless communication purpose, which includes tracing, direction finding, mobile computing, community safety, and health monitoring. Also in the realm of textiles antennas, many developments have been achieved. These textile antennas can be worn by an individual and used in connection with sensing system in body area network (BAN). This BAN application can be used for a number of reasons. Body worn textile antennas made from fabric offers a flexible and cost effective solution when compared to conventional antennas. This wearable technology is also attractive for medical applications due to the inherent low radiated power in its system design and the low power operation. The textile antennas can be easily incorporated into special uniforms and fashion garments using a simple stitching process.

Generally, conformal antennas are made up of flexible materials or any fabric as compare to traditional inflexible antennas. Basically fabric antenna consists of three parts: ground plane, radiating patch, and substrate material sandwiched between these two. For ground plane and radiating part conductive fabric material are used, which is also called electro textiles. For antenna substrate, any common fabric material can be used. It can be logically anticipated reason for the use of fabric or cloth in antenna design concealed in the reason for the purpose they are projected. This intellectual wearable system connotes an innovative idea of clothes apart from guarding the human skin from the climatic conditions; it also provides additional function like tracking, navigation, and communication. Wearable textile antennas are acknowledged as body worn gazette those are assimilated in the "smart garment". Textiles are considered as a reasonable resource of design as they are generally used in daily life and also easily available. In body worn application, planar design structure is the most reasonable solution because antenna ground

DOI: 10.1201/9781003187325-15

plane proficiently protects the skin from radiation and also radiates perpendicular to the antenna plane; thus, they are frequently used in body worn uses. Some significant antenna characteristic such as efficiency and bandwidth are mostly governed by permittivity and thickness of textile. Likewise radiating patch conductivity of them must be good enough because this is a vital design factor.

This chapter is organized as follows: designing of square patch antenna using various textile substrate materials such as cordura, polyester, cotton, and lycra having different dielectric properties. Then the optimization and simulation of antenna by using simulation software IE3D is done for the investigation of antenna parameters by keeping biomedical application in cognizance. Apart from this, the design strategies of wearable antennas are explained meticulously and some practical issues are examined and tested.

15.2 LITERATURE SURVEY

Being a smart garment system is an instinctively implicit system, the motive for the usage of fabric as material for antennas design lies in its application itself. This smart antenna system represents a new concept of communication in addition to usual purposes like protecting the body skin from dust, water, sunlight, heat, cold, etc. It also offers supplementary functions like detection, actuation, and communication. Various structures of textile antennas like most popular patches, slots, monopoles, patches, arrays are choice of maximum researchers for wireless body BAN applications and could easily be incorporated in wearable fashion garments. Fabric antennas directly assimilated into the clothes are connected with sensors to send information wirelessly via textile antennas.

Along with this, in the last few years fabric antennas have gained popularity in the research area of [1] wireless communication, with vital significance for many applications starting from patient monitoring [2] to tracking of soldier, likewise entertainment area [3] and in athletics too [4]. The designing of WTA requires a light weight, low loss, and compact structure. Particularly, some features are suitable for placement on the human body and complete incorporation into apparel. Due to such reason most of the available textile antenna designs are patch, often referred to as microstrip (MS). MS designs are very popular as they have received researchers' remarkable attention during the last few years. The brainchild reported [5] in the 1950s. Various types of design strategies of wearable patch are applied and proposed by different designers. Salonen et al. in 1999 pictured the first wearable antenna [6] as wearable patch represents a natural choice although it's not a fabric antenna. It is design for GSM application suitable for wearable application with planer inverted-F structure. Secondly, the very first body worn antenna [7] was reported in laboratories of Philips by Massey et al. Oodles of wearable textile antennas beginning from the first prototypes were investigated and proposed [8–11] by designers worldwide.

In wet performance investigation of fabric antenna, many researchers have used normal drinking water, rain water [12,13], and water from the sea to demonstrate the effect of moisture and also salty water to sign post effect of sweating on a wearable antenna. For depicting textile climatic situations [14] the effect of rain and sweat water, usually tap water is showered. Corresponding parameters like resonance frequencies and return loss of antenna are measured after the drying process. For the further investigation on worst climatic conditions textile antennas tested under the freezing state. For the analysis of dry to wet and wet to dry fluctuating effect in [15], the antenna is soaked in water until it is completely drenched. Reflection coefficient in drenched condition and later drying is calculated. For the examination in [16], wearable textile antenna for the application of iridium/gps is fabricated and salt absorption assessment is done to observe salt soaking influence. In this case return loss is calculated with various dissimilar situations. Applying 5% of salt on antenna working at 1565–1625.5 MHz, depicts variation in resonance frequency. The testified investigates in [17], shown laundry influence on performance of antenna by dampening it with the help of moist paper, showing discrepancy in the working of antenna by fluctuating the substrate permittivity. In the reported research in [18], various five wearable antennas are designed with different substrate and conditioned at weather check for the time period of around 12 hours at fixed temperature 25°C. For examining sweating influence on body worn textile antenna working, normal drinking water or salt solution is employed. Though the sweat of humans is not a mixture of salty water or salt solution and it contains a number of additional elements [19], those are absolutely ignored for minimalism in moisture or sweat investigation survey. Dielectric constant of human sweat is different; however, dielectric constant of salty water and normal drinking water are approximately the same.

When a wearable antenna is worn by the user, body sweat present on human skin will not keep this antenna in a dry state. Hence it is essential to inspect precisely the influence on wearable antenna with human sweat splash.

15.3 DESIGN PROCEDURE OF TEXTILE ANTENNA

For the designing and fabrication purpose of body-worn textile antenna, three important parameters are operating: frequency, dielectric constant, and height of substrate material. For operating frequency, according to application, for which antenna is to be designed, selection of frequency is done. Generally biomedical application uses the frequency in ISM band. Therefore this antenna must operate at the frequency range of 2.45 GHz.

Four different textile materials having dissimilar values of permittivity at single frequency, that is 2.45 GHz, are selected. The reason behind selection of low value substrate dielectric constant is that surface wave losses get reduced and bandwidth increases.

Since thickness textile materials are very low to be used as a substrate in antenna designing, hence to achieve desired height, layering or stacking method is utilized. By this process required thickness of the substrate material can be easily attained. But it introduces heterogeneousness in the substrate material due to existence of extra layers of air between two fabrics material. Parameters values used for proposed design are f_0 = 2.45 GHZ, ε_r = 1.90 for cordura and polyester, ε_r = 1.60 for cotton, ε_r = 1.50 for Lycra, as mentioned in Table 15.1 and height h = 1.5 mm is

TABLE 15.1
Fabric Dielectric Properties

Sr. No	Substrate	Permittivity	Loss Tangent
1	Cotton	1.60	0.0400
2	Polyester	1.90	0.0045
3	Cordura	1.90	0.0098
4	Lycra	1.50	0.0093

common for all substrate. Figure 15.1 depicts the geometry structure of rectangular patch textile antenna.

To trace the feeding point position, center of patch is considered as origin and is specified by two coordinates (*X*, *Y*). Theoretically the feeding position must be found at a point where the input impedance value lies at 50 ohms, and generally it is consider as length divide by six. Since this will not result, return loss is mostly negative. Therefore, a trial and error method is applied for various location of the feeding. For this work IE3D software is used to simulate and model the proposed design. This simulation software IE3D is built on method of electromagnetic computational, i.e., moment of method usually known as MOM.

15.4 PERFORMANCE ANALYSIS OF PROPOSED ANTENNA

Four antennas with dissimilar textile materials such as cotton, cordura, Lycra, and polyester at the same frequency but unequal values of permittivity and loss tangent performance are analyzed. Results of the projected idea are briefly listed in Table 15.2, which gives relative performance analysis of the design of four antennas.

From the results analysis it is summarized that the antenna constructed with polyester textile material shows good performance when equated with the remaining three. It is due to the less value of material tan delta as compare to the other three fabrics. Directive antenna is a major requirement in some particular

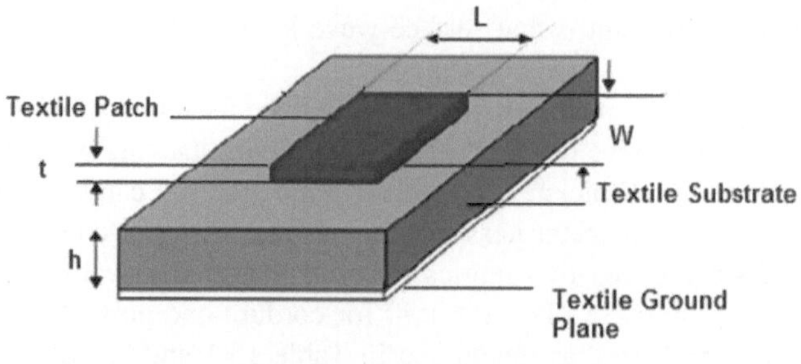

FIGURE 15.1 Schematic of textile rectangular patch antenna.

TABLE 15.2
Result analysis of all four design textile antennas.

Fabric Material	*L**W*	S11	Gain (dBi)	BW (dB)	Directivity (dBi)	Efficiency (η)	3 dB Beam Width	Impedance
Cotton	46*53	−32	3.8	0.097	8.44	35%	73°, 78°	48-j0.789
Polyester	43*50	−35	6.8	0.040	7.90	76%	75°, 87°	48-j1.849
Cordura	43*50	−29	5.9	0.050	7.93	64%	75°, 87°	50-j2.450
Lycra	48*54	−31	6.8	0.048	8.59	67%	73°, 75°	47-j0.740

applications like wireless body area network (WBAN) antenna and antenna design with cordura, and Lycra textile can be used for such application.

In this case, directivity of Lycra is comparable with cordura, and thus both designs can be used. When two body worn antennas need to communicate at very small distance, for this scenario low gain antenna is a significant prerequisite. Thus cotton textile design is perfect for such cases where communication with low data rates with relatively close range is needed. Figure 15.2 indicates antenna pictures using cotton and polyester fabric.

From this investigation it is concluded that by using normal textile material wearable textile antenna is constructed with very minimal budget and it can be easily assimilated into fashion garments. Low value of substrate material permittivity is beneficial in this designing as it diminishes the effect of losses due to surface wave and likewise it accommodates wave propagation inside substrate. A basic requirement to increase efficiency is selection of low loss substrate material, and this is fulfilled with the use of fabric. Deviation in substrate dielectric constant too disturbs the antenna bandwidth.

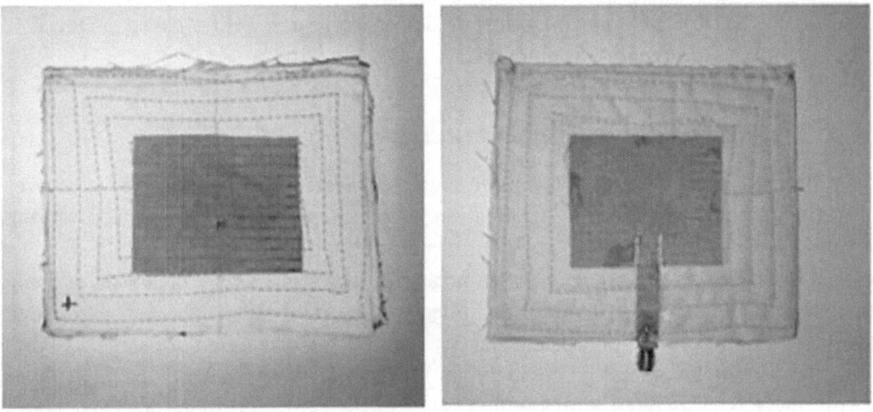

FIGURE 15.2 Antenna prototype employing cotton and polyester.

15.5 DESIGN ASPECT OF WTA

This section provides design requirement criteria for wearable textile antenna. Rectangular patch design with denim fabric and its specifications are proposed and studied thoroughly.

15.5.1 Material Selection Criteria

Selection of proper material for the substrate is an essential design factor in antenna because it determines the electrical performance. The primary technical prerequisite of the substrate for the body worn antenna are high viewing density, robustness, durability against frequent bending, temperature, and moisture tolerance. From the wearer's point of view this antenna should be flexible to conform to the human body, comfortable to the operator, and easily available. For real-world application it needs to be cost-effective, easily incorporate into fashion garments, and provide consistent results in erratic climatic conditions.

Denim cloth with style name 2440Y royal blue color through rope dying yarn die process is used. The preshrink finish process is chosen to avoid shrinking. This shrinking method is utilized to avoid future shrinking of fabric after assembly of the patch. Entire design dimensions can change due to shrinking which results in a shift of the frequency of resonance. Generally, conducting material thin copper foil or conductive fabrics are used as a metallic patch in antenna fabrication. Substrate material which is non-conducting provides the necessary spacing between the ground plane and radiating patch and it also gives good mechanical backing to the radiating patch. The thickness of the substrate for the basic design ranges from 0.01 to 0.04 free-space wavelength. The value of the dielectric constant lies in the range of 1 to 10 and it is divided into three categories. Fabric material has a value of dielectric constant between 1 and 2. The possible fabric material is fabrics, foam, polystyrene, honeycomb, etc. Materials with relative dielectric constant values in the range of 2 to 4, those materials are glass Fiber, Teflon. The final and third category is materials dielectric constant available in the range of 4 to 10. Specimens of such materials are alumina, ceramic, quartz. The most frequently used material in case of the flexible antenna is cotton or denim, whereas Teflon-based material is most popular in conventional antennas.

15.5.2 Association of Conducting Plane

To achieve desired thickness, substrate materials need to be layered thus adhesives are utilized for this purpose. Application of adhesive secures connection among several layers. Those layers may be substrate or conductive. Applying adhesive is a simple straightforward process. After arranging the first textile layers of suitable dimensions, the adhesive layers are applied on the material which need associated together. To continue, appropriate heat is applied which causes adhesive to bond with the textile layer and removal of undesired layers. Afterward the next layer textile layer is associated with the first layer in the preferred material. Heat is applied to this next layer to unite these two sheets to one another. Heat application ensures no air gap in any two layers where adhesive is applied. It is not easy to apply a uniform layer of adhesive; application of glue introduces heterogeneity on the areas where it accumulates. Sometimes glue application may act as a layer of insulator between patch and conductive fabric. Besides, adhesive will become usually stiff once it get dried and may affect the flexibility of material.

During union of antenna different layers, retaining the original dimension is a necessary factor. Through this association, layer of bonding agent deposition may enter into the superficial layer of substrate. These superficial layers sometimes change sheet resistance and the dielectric constant of substrate material. Appropriate care is required during assembly of substrate, radiating upper patch, and below ground plane. Slight pressure or stress imposes on the assembled antenna and heat excursions to union region may give rise to problems like electrical opening in joined area over the period of time. For the reduction of the environmental pressure and to increase electrical consistency in union a well-designed mechanical locking is required.

Flexible copper tape is used for conductive parts for ground plane and radiating patch. This bendable adhesive copper tape has thickness of 0.02 mm. To confirm a perfect bond, jeans cloth material is washed and dry, then pressure with the help of finger at the rate of 3 to 10 psi (0.02 to 0.10 Mpa) is used at 20°C (68 F) to 25°C for the absorption of conductive particles with the substrates materials. Various methods are utilized by different investigator for thermal effect of conducting material and dielectric substrate. However, this technique of connection confirms association firmness at the time of washing activity. The adhesive solution which is used is water resistive. For the substitute of copper, conductive textile can be employed.

15.5.3 Interconnection of SMA Connector

In case of body worn fabric antennas, connection with a measuring device is a major challenge. This connection functions as the power-driven linking of surface fixed assemblage to the antenna ground plane on the lower side of the patch and it also ensures electrical contact of feed line on the reverse side. Bonding of SMA or N-type connector in textile antenna with help of normal soldering process is a difficult task; thus, superior precaution is required. Because solder temperature is 330°C to 355°C, it may brutally destruct the textile material while soldering. Consequently, sterling silver conducting soldering glue is used. It is obtainable in nozzle, and it also improves connection. The nozzle is tublike rim and flat on both sides which evades rolling and has good fingertip control.

15.6 DESIGN PROCEDURE AND SPECIFICATIONS

Several methods are available for fabrication of wearable textile antennas. Most popular methods are inkjet printing, screen printing, use of conductive fabric, and use of flexible adhesive metal sheet. Screen printing and inkjet printing are user friendly, fast, and inexpensive methods of fabrication. This method is also reasonable when mass production is required. If required, various malleable patterns and high resolution designs can be achieved on textiles. The resolution factor is also to be governed by the textile structure, surface roughness, weaving of fabric and material thickness. Direct use of conductive textile is the most popular and reasonable method. Different forms of conductive textiles are available such as mesh of copper, conductive thread embroidery, conductive ribbons, conductive adhesive tape, and conductive material spray. In this work normal machine sewing technique is used. Specification use in designing of wearable textile patch is summarized in Table 15.3.

The procedure mentioned in the above section is used for designing semi textile antennas. Denim fabric use for fabrication has permittivity of 1.6 and tan delta 0.02 at the frequency of 2.45 GHz. Figure 15.3 shows the top view and bottom view of the fabricated denim patch. For simulation and modeling of proposed semi-textile antenna, IE3D simulation software is used.

15.7 TEST AND MEASUREMENTS

A brief description of several measurements and tests required for designing and investigation and of body worn textile antenna are shared.

TABLE 15.3
Design Dimension of Semi-Textile Antenna

Parameter	Dimension (mm)
Length (L) × Width (W)	46 × 51
height of substrate (h)	2
Ground plane size	96 × 101
Feed point	37
Feed line width	7.49
Inset feed	0.1

FIGURE 15.3 Fabricated semi textile denim antenna.

15.7.1 Preparation of Sweat Solution

For making of artificial sweat, composition in [19] is employed and this configuration is the nearest parodist to true human body perspiration. Key components such as sodium chloride with amount of 2.923%, lactic acid with 1.44%, Squalene 3%, fatty acids 16.4%, Urea 1.2%, and Cholesterol 0.8% with P.H level of 6.5 are used in this preparation. However other components are ignored as its influence is insignificant. Readymade sweat solutions are also obtainable in the pharmaceutical market, which contains five most plentiful crystals, seventeen amino acids, and five metabolites with the pH value of 5.5 is used straight. This non-stabilized solution can be retained in a cool and waterless place and also it has one year expiry.

15.7.2 Measurement of Textile Dielectric Properties

Normally, dielectric properties of any textile material depend on surface irregularity of operation and temperature. Different researchers have used several techniques like Transmission Line Method, Resonance Method, Non resonance method, Method of moment (MOM), and Free Space Method.

For the measurement of dielectric properties of used cloth or fabric material dielectric device name broadband spectrometer is used. Fabric material is cut into a small circular piece of 3 mm diameter. This sample is kept in a sample holder of spectrometer which is shown in Figure 15.4.

Then procedure mentioned below is followed for further measurement.

FIGURES 15.4 Measurement of fabric properties on dielectric broadband spectrometer.

The steps of operation of broadband spectrum analyzer for frequency of 2.45 GHz are as follows:

1. Perform Line calibration of instrument
2. Calibrate the sample holder under short and open conditions using air and Teflon as dielectric
3. Load the sample in between the holder
4. Choose appropriate values of frequency and run the measurement.

The following graph shows the result of a completed experiment, variation of dielectric constant of used denim fabric with frequency range from 1 GHz to 3 GHz.

In this measurement, real and imaginary part of permittivity is calculated at room temperature shown in Figure 15.5. Then fabric is wet and again the same procedure is followed to observe the change in values of permittivity with moisture. However the experiment can also be done with a different temperature range.

15.7.3 Sweat Regain Test for Fabric

Sweats retain capability of the cloth or textile is stated as the amount of sensitivity of textile material to the perspiration. The sweat retention of any fabric is described by the physical construction of fabric. Thicker structure fabric may have higher regain capacity. Sweat captivation changes the physical and mechanical characteristics of fabric. When sweat enters in the fabric it can change its size, dimension, toughness, and dielectric constant. Apart from this it can create bulginess, which alters individual thread thickness used in viewing, which results in variations in overall patch thickness. Drying after sweat absorption may cause the presence of white salty layered patches on the surface of the fabric. While describing electrical properties of the textile material this sweat regain situation is an important aspect. Equilibriums the stage of any textile or cloth when it is kept on biological tissues at the temperature of 25 degree Celsius, it loses or gains sweat at the slowly

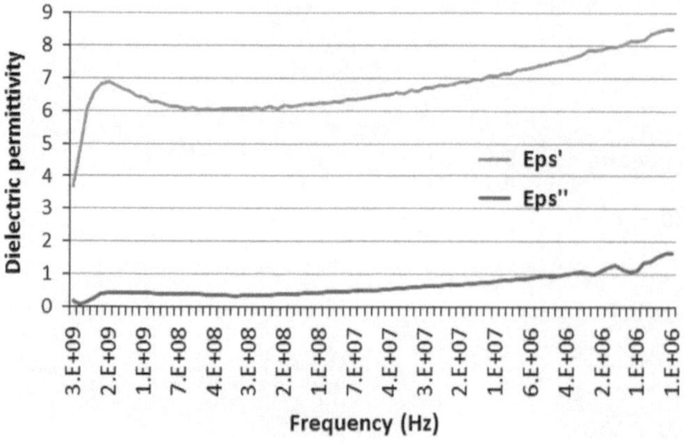

FIGURE 15.5 Permittivity graph of denim material.

diminishing rate until it goes to stable state, and afterwards no additional variation. It happens because of vaporization of fluid particles from the textile in a quantified time which is equal to molecules number involved. Content of sweat in any textile is a fraction of weight of sweat absorbed textile at equilibrium with not soaked or unsoaked fabric. In the beginning for the measurement of denim fabric sweat retain test the normal 20 cm × 20 cm denim cloth is weight is calculated. After that sweat solution is sprinkled on denim fabric, and the fabric remains as it is to settle the sweat. When it comes at equilibrium position, its weight is calculated and then it is equated with the weight of not soaked or unsoaked fabric. From this experiment it is determined that denim cloth possesses 6% of sweat retain ability.

$$\text{Sweat Retain} = [(\text{weight of sweat absorbed textile} - \text{weight of unsoaked textile})] \\ /(\text{Weight of unsoaked textile}) \times 100\%$$

Weight of unsoaked textile = 114 gms
Weight of sweat soaked fabric at equilibrium = 120 gms

$$\text{Sweat Retain} = [(120 - 114)/121] \times 100\% = 6\% \tag{15.1}$$

From this above test it is concluded that used textile shows 6% sweat retain ability.

15.7.4 Substrate Material Thickness Measurement

Thickness of textile materials is very important for materials to be used as a substrate, thus the desired thickness is achieved by layering or stacking together individual pieces of fabric material. By this way of layering of the fabric material, essential thickness required for substrate can be achieved. To get required thickness a half meter of processed denim cloth having thick around 0.3 mm is cut out into small pieces of chosen dimension. These cutout sections are arranged together into sheets to attain the necessary thickness. These arranged sections of thickness are calculated with the help of instrument Vernier caliper to confirm its essential thickness which is shown in Figure 15.6. Finally, fabric to fabric joining is established by stitching with the help of a sewing machine. For creating association between fabric layers, several stitching styles can be used. In high electromagnetic field regions, special precaution is taken during sewing because such regions are very delicate. Textile material relative permittivity can be altered to maintain firmness during sewing. There is a possibility of shift in resonance frequency up to 100 MHz due to low dense stitch.

15.7.5 Preshrink Process

Most of the textile or cloth materials shrink after washing. If any new fabric is used without washing for the fabrication process due to its shrinking tendency after washing. Shrinkage magnitude is determined by the quality of threads used in the fabric weaving. For maintaining antenna dimensional constancy, shrinking is an important factor in antenna performance. Overlooking the shrinking process may

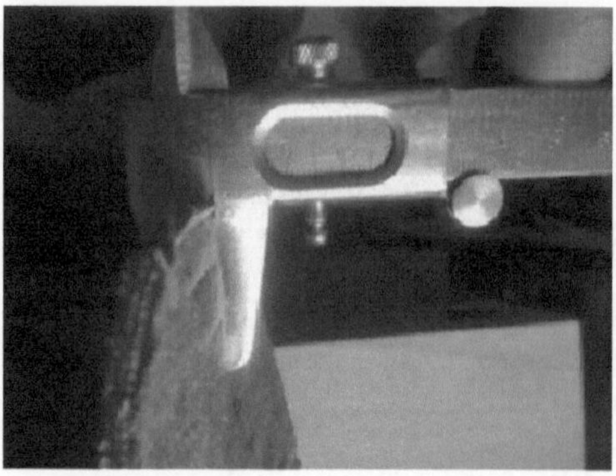

FIGURE 15.6 Thickness measurement on Vernier caliper.

bring down antenna dimensions after washing. To make antenna shrinking free and to avoid a reduction in antenna dimension due to shrinking after washing, the fabric is completely immersed in hot water for 24 hours meticulously. This immersed textile is removed, rinse, spun, and dried in natural light around 6 to 8 hours. Finally, hot air dryers are blown on it so that it can dry totally. The entire procedure is repeated to ensure that the cloth material is not shrinking further.

For this process, two equal-size cloth sections with dimensions 18×18 cm are considered. One sample among these two is kept as it is and another is allowed to go through a shrinking process. After the entire shrinking process, length and width of both the cloth are measured. Using the below formula shrinkage in length and width is measured as,

Sample 2 dimensions after washing:

Length of sample 2 = 17 cm

Width of sample 2 = 15 cm

$$\text{Total shrinkage} = [(\text{size of sample 1}) - (\text{size of sample 2})]/\text{size of sample 1}] \times 100 \tag{15.2}$$

$$\text{Shrinkage in Length} = [(18 - 17)/18] \times 100 = 5.55\% \tag{15.3}$$

$$\text{Shrinkage in Width} = [(18 - 15)/18] \times 100 = 16.66\% \tag{15.4}$$

By this method shrinking test of fabric is carried out and thus it is concluded that denim material undergoes 5–6% shrinkage in length and 16–17% shrinkage in width. Width-wise shrinkage is more as compared to lengthwise shrinkage this is due fabric weaving method.

15.8 RESULT AND PERFORMANCE ANALYSIS

Comparison plot among simulated and measured return loss verses frequency is depicted in Figure 15.7 and it shows good matching between both the

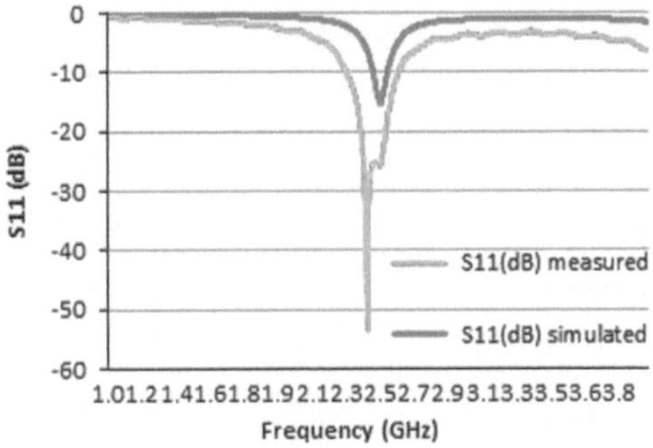

FIGURE 15.7 Measured and simulated result of textile antenna.

measurements. The simulated resonance frequency of the antenna is 2.5 GHz and measured frequency is at 2.44 GHz. This slight discrepancy is due to few thread appearing at the border of cloth while designing the antenna prototype.

Some undesired thread coming out at the borders of fabric while cutting creates variation in patch dimension. Magnetic and electric fields of antennas are aligned with width and length respectively. The flow of current electric fields is affected marginally. Thus changes in changing patch dimensions leads to shifting in resonance frequency of an antenna. Also, normally fabric materials are not uniform in nature because of existence of openings on it, where air may enter easily. Essentially this porous texture of textile is not taken into account during simulation, and in fact there is no provision to consider this fact. To demonstrate, synthesis method is assumed by straight considering experimental measured values. Therefore this assembly tends to increase losses and thus low efficiency. The solution to this inadequacy is more keen examination is carried out.

In order to increase efficiency, in depth analysis of antenna losses has to be completed in the future. This design model is located at human arm having 65 mm bending radius, for the examination of antenna with bending situation. Figure 15.8 depicts variation in reflection coefficient with respect to bending radius. Owing to bend, change in reflection coefficient towards lower side is observed, but still it is an inacceptable range. To calculate the effect of human sweat, artificial sweat solution is sprayed at design prototype with the help of a sprayer to realize a relative perspiration. According to the arrangement of human skin pores prayer nozzle is approximately set. Biological tissues on human skin consist of approximately 400 to 1100 sweat pores on the skin. But these glands of sweating vary according to male/female and skin type. However position of the pores on the skin does not vary with respect to time. Over a fingerprint one pore to another pore dimension is approximately 0.5 mm. With respect to available database around 200 to 300c.c sweat is dispersed from human skin in 12 hours. Adult male arm location is sagaciously selected to mount design prototype which nearly vanishes 2.5cc sweat in 24 hours.

To continue this process, the prototype was progressively sprayed with artificial sweat solution 2.5 cc, 2 cc, and 1.5 cc to signify Indian climatic conditions. Effect of perspiration on the working on antenna with respect to reflection coefficient for

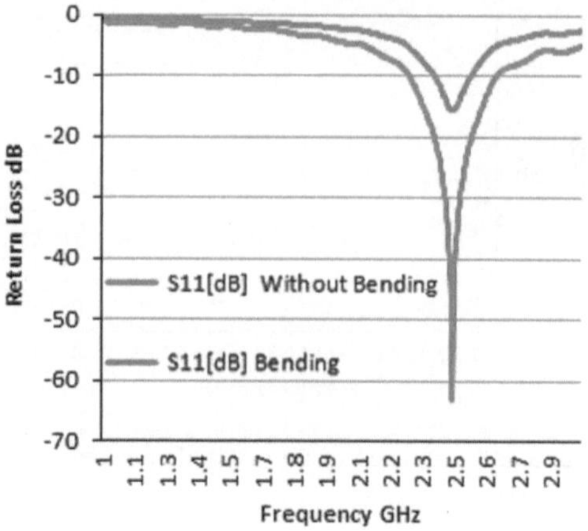

FIGURE 15.8 Comparison of S11 with and without bending.

various Indian climatic is presented in Figure 15.9. From all climatic situations it is concluded that about 10% ominously decrease in antenna performance with slight shift in resonant frequency.

From three different sweat conditions, slight reduction in return loss is observed. This reduction in S11 is due to change in permittivity of fabric material due to presence of moisture. Slight shift in resonance frequency towards lower range with increasing sweat is also observed. Also resonance peak expands and becomes low profound with growing sweat; implicate more losses happen in the fabric material. The presence of sweat entering into the structure increases the permittivity, as could be predicted. It is also clear that fabric having high

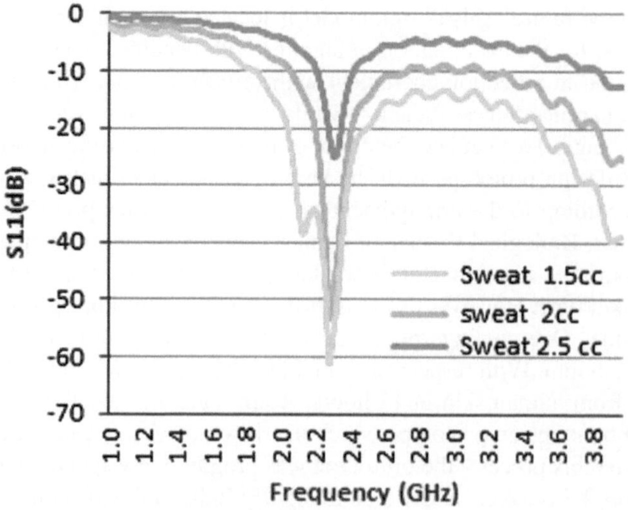

FIGURE 15.9 Sweat analysis result of textile antenna.

moisture regain are more liable to humidity variations. The antenna design prototype is placed on the male arm, to examine antenna behavior in specific state, and its reflection coefficient was measured. Figure 15.10 presented in the comparison between the measured on body return loss of antenna with free space return loss.

In a real environment, the antenna worn on-body is prone to mechanical deformations such as bending. For bending analysis, the prototype is placed on adult human arm having 65 mm bending radius. This examination shows the variation in S11 with respect to bending radius. The bending situation was examined for the radius of 65 mm, which are values that can be found in reality. The setup for measurement of bending conditions is shown in Figure 15.11.

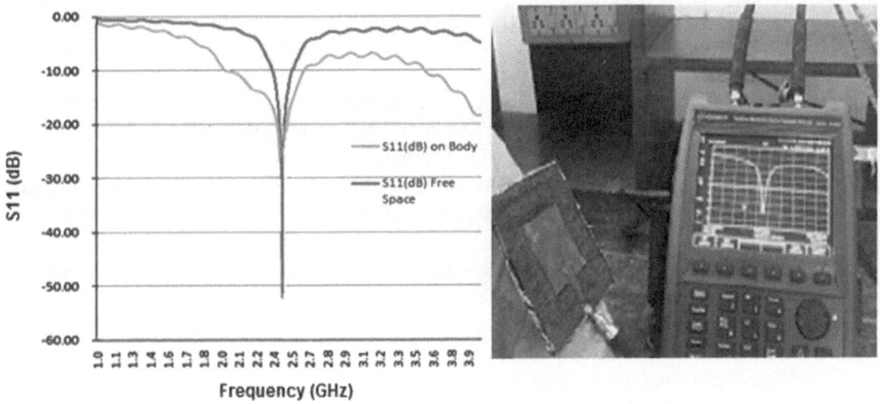

FIGURE 15.10 Performance of antenna in presence of human body.

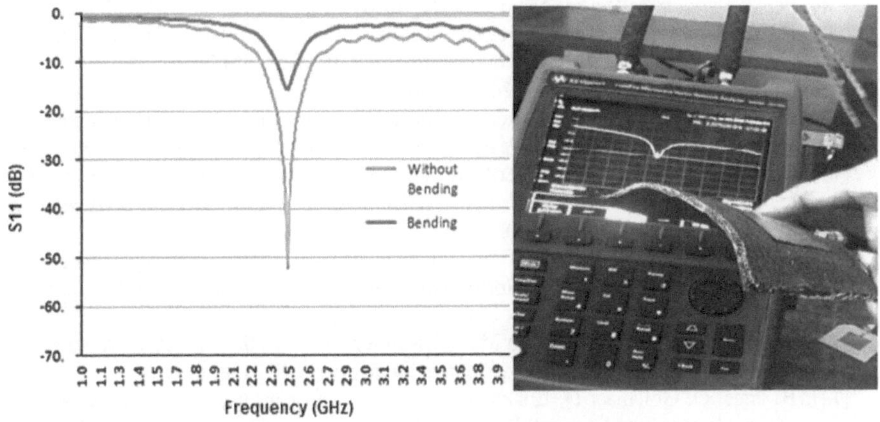

FIGURE 15.11 Antenna performances under bending condition.

15.9 CONCLUSION AND FUTURE ASPECTS

This chapter addresses the optimization and simulation of four antennas with different fabric materials for investigation on various parameters of antennas. Also the examination of dielectric properties of fabric material and its effect on antenna design is carried out. However, in the second part some design methodology of wearable textile antenna optimization, fabrication, and testing of proposed (Semi-textile) prototype is done. To give a feel of practical challenges when antenna placed on human body is executes along with this the practical circumstances like bending and sweating.

REFERENCES

[1]. P. Hall, Y. Hao, *Antennas and Propagation for Body-centric Wireless Communications*, second ed., Artech House, Norwood, MA, 2012, p. 400.
[2]. D. Curone, G. Dudnik, G. Loriga, G. Magenes, E. Secco, A. Tognetti, A. Bonfiglio, Smart garments for emergency operators: results of laboratory and field tests, in: *2008 30thAnnual International Conference of the IEEE Engineering in Medicine and Biology Society*, 2008.
[3]. M. Orth, J. Smith, E. Post, J. Strickon, E. Cooper, Musical jacket, SIGGRAPH '98: ACM SIGGRAPH 98 Electronic Art and Animation Catalog (1998) 38.
[4]. S. Coyle, D. Morris, K. Lau, D. Diamond, N. Moyna, Textile-based wearable sensors for assisting sports performance, in: *BSN '09: Proceedings of the 2009 Sixth InternationalWorkshop on Wearable and Implantable Body Sensor Networks*, Washington, DC, USA, 2009.
[5]. G.A. Deschamps, Microstrip microwave antennas, in: *Third USAF Symposium on Antennas*, 1953.
[6]. P. Salonen, M. Syd€anheimo, M. Keskilammi, M. Kivikoski, A small planar inverted-Fantenna for wearable applications, in: *Third International Symposium on Wearable Computers*, October 19, 1999.
[7]. P. Massey, Mobile phone fabric antennas integrated within clothing, in: *Antennas and Propagation, 2001. Eleventh International Conference on*, 2001.
[8]. T. Kellom€aki, W. Whittow, J. Heikkinen, L. Kettunen, 2.4 GHz plaster antennas for health monitoring, in: *EuCAP 2009, the 3rd European Conference on Antennas and Propagation*, Berlin, March 23, 27, 2009.
[9]. A. Tronquo, H. Rogier, C. Hertleer, L. Van "Langenhove, Robust planar textile antenna for wireless body LANs operating in 2.45 GHz ISM band," *IEEE Electron Device Letters*, **3** (42), pp. 142e143, 2006.
[10]. M. Klemm, I. Locher, Tr€oster, A novel circularly polarized textile antenna for wearable applications, in: *34th European Microwave Conference 2004*, 2004.
[11]. L. Vallozzi, H. Rogier, C. Hertleer, "Dual polarized textile patch antenna for integration into protective garments," *IEEE Antennas and Wireless Propagation Letters*, **7**, pp. 440e443 (2008).
[12]. W. Hoand W. F. Hall, "Measurements of the dielectric properties of seawater and NaCl solutions at 2.65 *GHz," Journal of Geophysics ResearchWiley Online Library*, **78**, pp. 6301–6315 (1973).
[13]. W. Ho, A. W. Love, M. J. Van Melle, "Measurements of the dielectric properties of *sea* water at 1.43 GHz," *NASA Contractor Report*, CR-2458, (1974).
[14]. B. Ivsic, G. Golemac, D. Bonefacic, "Performance of wearable antenna exposed to adverse environmental conditions," *Applied Electromagnetics and Communications*, 2013, pp. 163–178 (2013). 10.1109/ICECom.2013.6684727.

[15]. H. S. Zhang, S. L. Chai, K. Xiao, L. F. Ye, "Numerical and experimental analysis of wideband e-shape patch textile antenna," *Progress In Electromagnetics Research*, **45**, pp. 163–178 (2013).
[16]. J. Lilja, P. Salonen, T. Kaija, and P. D. Maagt, "Design and manufacturing of robust textile antennas for harsh environments," *IEEE Transaction Antennas Propagation*, **60**, pp. 4130–4140 (2012).
[17]. R. Yahya, M. R. Kamarudin, and N. Seman, "Investigation on CPW koch antenna durability for microwave imaging," in: *PIERS Proc., Taiwan*, pp. 498–501, Marcha, 2013.
[18]. C. Hertleer, A.V. Laere, H. Rogier, L.V. Langenhove, "Influence of relative humidity on textile antenna performance," *Textile Research Journal*, **80**, pp. 177–183 (2009).
[19]. C. Callewaert, B. Buysschaert, "Artificial sweat composition to grow and sustain a mixed human axillary microbiome", *Journal of Microbiological Methods*, **103**, pp. 6–8 (2014).

Part VII

Case Studies

16 UWB Deterministic Channel Modeling

B. Bansal
Jaypee Institute of Information Technology, Noida, Uttar Pradesh, India,

16.1 UWB CONCEPTS AND SIGNALS

UWB communication employs narrow pulses with very small value of duty cycle (smaller than 0.5% [2]) where the duty cycle is related to the pulse on and off time and it is defined as

$$\text{Duty cycle} = \frac{T_{\text{on}}}{T_{\text{on}} + T_{\text{off}}} \tag{16.1}$$

This property results in very low transmitted power required for UWB devices.

There exist different types of narrow pulses that satisfy the bandwidth and transmit power limitations given by the Federal Communications Commission (FCC) [11, 12] and thus can be used for UWB systems. Gaussian monocycle is one of the possible UWB pulses that is expressed as [12]

$$e_{inc}(t) = \sqrt{\frac{\tau}{\sqrt{\frac{\pi}{2}}}}\frac{-2t}{\tau^2}e^{-\left(\frac{t}{\tau}\right)^2} \tag{16.2}$$

with τ representing the scaling factor. The FD counterpart for (16.2) is written as

$$E_{inc}(f) = \sqrt{\frac{\tau}{\sqrt{\frac{\pi}{2}}}}\tau\sqrt{\pi}\,(j2\pi f)\,e^{-(\pi f\tau)^2} \tag{16.3}$$

Another important UWB pulse is the Gaussian doublet pulse that is expressed as [12]

$$e_{inc}(t) = \frac{1}{\tau}\sqrt{\frac{\tau}{3\sqrt{\frac{\pi}{2}}}}\left(1 - \frac{-2t^2}{\tau^2}\right)e^{-\left(\frac{t}{\tau}\right)^2} \tag{16.4}$$

with its FD form given as

DOI: 10.1201/9781003187325-16

$$E_{inc}(f) = 2\sqrt{\pi}\sqrt{\frac{\tau}{3\sqrt{\frac{\pi}{2}}}}(\pi f\tau)^2 e^{-(\pi f\tau)^2} \tag{16.5}$$

Satisfying the FCC mask, the fourth-order Gaussian pulse is also one of the candidates for UWB communication and is defined as [12]

$$e_{inc}(t) = \frac{-2}{\tau^2}\sqrt{\frac{\tau^7}{2^{\frac{7}{2}}\frac{(9\times7\times5\times3\sqrt{\pi})}{32}}}\left(\frac{24t^2}{\tau^4} - \frac{8t^4}{\tau^6} - \frac{6}{\tau^2}\right)e^{-\left(\frac{t}{\tau}\right)^2} \tag{16.6}$$

with its FD representation given as

$$E_{inc}(f) = \sqrt{\frac{\tau^7}{2^{\frac{7}{2}}\frac{(9\times7\times5\times3\sqrt{\pi})}{32}}}\tau\sqrt{\pi}(2\pi f)^4 e^{-(\pi f\tau)^2} \tag{16.7}$$

Figure 16.1 shows above discussed UWB pulses with $\tau = 0.1$ ns.

16.2 UWB REGULATIONS

All radio communication is subject to different laws and regulations about power output in certain frequency bands. This is to prevent the interference to the other users in nearby or the same frequency bands. Figure 16.2 shows FCC regulated spectrum characteristics for different indoor and outdoor applications of UWB communication. This defines a wide range of 7.5 GHz from 3.1 GHz to 10.6 GHz for communication purposes. For indoor communication, the UWB devices are allowed with a power spectral density (PSD) of −41.3 dBm/MHz [13]. Because of this power emission limit, the UWB communication can coexist with other wireless services while providing the minimum interference to them. In 0.96–1.61 GHz band, the power level for UWB systems is kept very low to avoid any harmful effects for other services like global positioning system (GPS), mobile telephony, etc. [12].

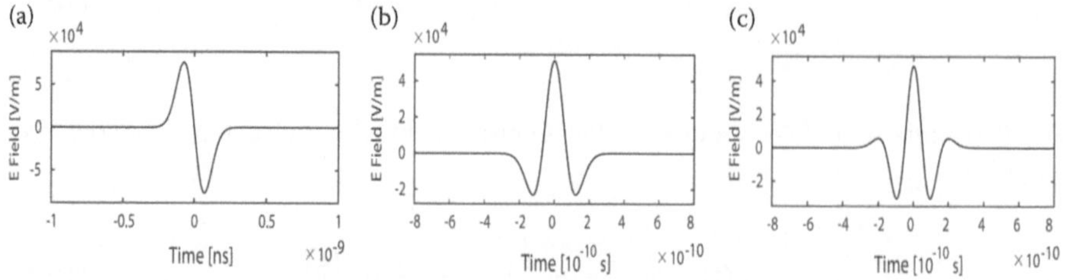

FIGURE 16.1 Excitation Gaussian pulses for UWB communication, (a) Gaussian monocycle, (b) Gaussian doublet, (c) Gaussian fourth-order.

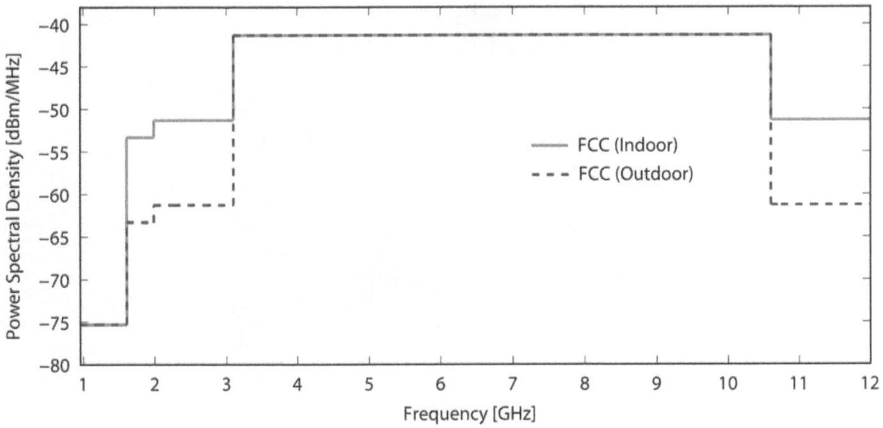

FIGURE 16.2 FCC mask for UWB communications [14].

16.3 UWB ADVANTAGES

The UWB communication offers several advantages over a narrowband communication system.

High Data Rate: The high data rate feature of UWB is evident from the well-known Shannon-Hartley's channel capacity formulation which can be defined as

$$C = BW\ log_2(1 + SNR) \quad \text{bits per second} \tag{16.8}$$

with SNR and C respectively representing the signal-to-noise power ratio at Rx and the channel capacity of an information transmission channel (having bandwidth BW Hz) where the capacity signifies the number of information bits that can be transmitted over the channel (in one second). Thus, the UWB technology that uses very large bandwidth signals can provide data transmission at a very high speed.

Multipath Immunity: Figure 16.3 shows the multipath phenomenon that is caused by reflection, diffraction, or scattering of the electromagnetic (EM) energy by objects in-between the Tx and Rx. In the case of UWB communication, as the UWB pulses are of very short duration, there is less probability that different reflected pulses will arrive within one pulse width and will cause interference. In other words, because of the extremely short pulse width, the different pulses can be resolved in the TD and thus inter symbol interference (ISI) can be mitigated. This multipath immunity results in fine resolution for UWB communication.

Ability to Share the Spectrum: A valuable aspect of UWB communication is the spectrum sharing capability with other users. Figure 16.4 shows the general idea of UWB's co-existence with other radio services like GPS, PCS, and WLAN, where GPS stands for global positioning systems, PCS for personal communication service, and WLAN for wireless local area network. Because of low PSD values for UWB systems (as approved by FCC), they provide minimal or no interference to other wireless services and thus can coexist with them. In addition, they reside below the narrowband Rx noise floor and thus can be treated as unintentional radiators.

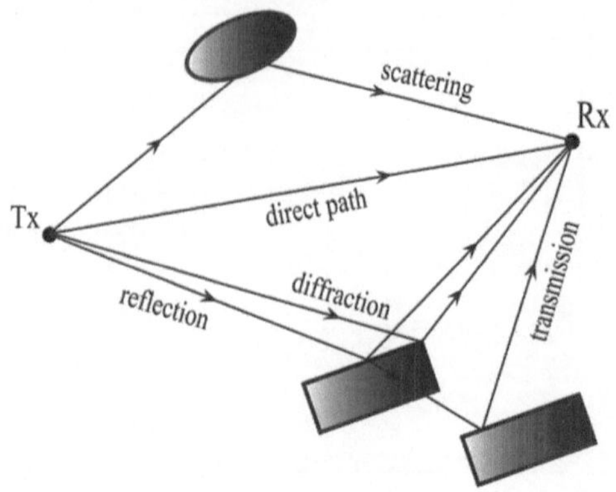

FIGURE 16.3 Multipath propagation in wireless communication.

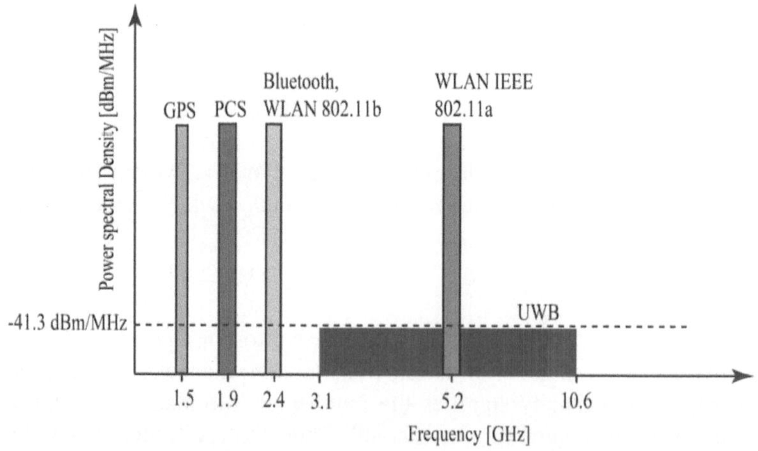

FIGURE 16.4 UWB spectrum with some existing radio services [15].

Low Probability of Detection: The low emission limit of UWB systems leads to enhanced security in communication. Because of their very low PSD value, UWB systems are immune to detection and interception. This feature of UWB technology is of particular interest for military applications, such as covert communication and radar. This is also useful for wireless consumer applications, where security of the data remains the main concern.

16.4 UWB PROPAGATION CHANNEL AND ADVANTAGES OF TD ANALYSIS

The wireless channel plays an important role in deciding the performance of any wireless communication system. To build a system with all UWB capabilities, it is important to gain insight into the UWB propagation channel.

16.4.1 UWB Propagation Channel

In comparison to the narrowband channel, the UWB propagation channel behaves in a different way. For narrowband systems having small bandwidth, the Rx response is generally flat and so the frequency dispersion effects are not accounted significantly [16,17]. But in UWB systems with large bandwidth of UWB signals, different frequency components of the signal experience different losses and dispersions [18].

The primary difference between the UWB and the narrowband channel is related to the frequency dependent transfer characteristics or impulse response (IR) of UWB channel. While considering multipath propagation and interaction with environmental objects, each multipath component (MPC) in conventional narrowband systems gets delayed, attenuated, and shifted in phase [19]. Thus in conventional narrowband propagation, the IR of a generalized multipath channel is defined as

$$h(t) = \sum_{i=1}^{N} \alpha_i \delta(t - t_i) \tag{16.9}$$

with N denoting the different MPCs and α_i and t_i as the complex amplitude and the delay of the ith MPC respectively. Here, it can be seen that the interaction of a MPC with objects is frequency independent and therefore the IR of single component is simply an impulse function.

On the contrary side, the UWB communication is characterized by frequency dependent transfer characteristics or IR. Because of huge bandwidth of UWB signals, the UWB propagation depends on the frequency properties of the different electromagnetic materials [20]. Due to this frequency dependent channel behavior, the spectrum of the transmitted signal changes during propagation and consequently the received UWB pulse is generally distorted in shape [10]. In other words, different attenuations are encountered by different components of the UWB signal. Thus for UWB channels, the IR of a single component reduces to a distorted pulse $\chi_n(t)$ rather than some ideal impulse signal. Considering different propagation mechanisms like in Figure 16.3, the generalized IR of a multipath UWB system is given by [21]

$$h(t) = \begin{pmatrix} \sum_{p=1}^{K} A_p \delta(t - t_p) + \sum_{q=1}^{L} A_q d_q(t) * \delta(t - t_q) + \\ \sum_{r=1}^{M} A_r r_r(t) * \delta(t - t_r) + \sum_{s=1}^{N} A_s \Gamma_s(t) * \delta(t - t_s) \end{pmatrix} \tag{16.10}$$

where A_p, A_q, A_r, and A_s are the real values. The symbols K, L, M, and N show the number of MPCs for different propagation mechanisms. The terms $d_q(t)$, $r_r(t)$, and $\Gamma_s(t)$ denote the IRs due to diffraction, reflection, and transmission respectively and these are responsible for distortion in the UWB signal.

Channel modeling is fundamental in analyzing the system performance and channel parameter estimation [21]. In the context of UWB communication, the channel can be modeled by using either the empirical method [22] or the physics-based approach. In the context of physics-based method, the ray tracing tool that is based on physics principles proves more useful in understanding the UWB propagation. According to [10,21,23,24], physics-based signal processing has been used widely for UWB systems. The frequency-dependent dispersive effects result in UWB signal distortion and so impact the Rx design [23]. According to studies [23], the performance of correlation Rx strongly depends on the pulse distortion where the distortion can lead to performance degradation at the output of the correlation Rx.

16.5 ADVANTAGES OF TD ANALYSIS

There are two possible ways to study the distortion in UWB pulse shape. The first method includes the FD approach and applying the inverse fast Fourier transform (IFFT) for obtaining the TD results. According to the second method, direct TD approach is used and so the effect of all frequencies is counted simultaneously. For UWB signals having large bandwidth, the TD approach is generally preferred in comparison to the IFFT-FD one which calculates the propagation effects for different frequencies individually and thus becomes time consuming. Using TD solution, it also becomes possible to know the channel impulse response behavior which can provide more insight into the environment characteristics. It should be noted that the impulse response can be convolved with the source pulse to know the response at the Rx. This Rx response implicitly includes the shape distortion and thus can improve the overall performance of the system [10].

In multipath propagation, by considering different rays as different paths, the shape knowledge of the pulse transmitting through the path can be easily determined [10,21,23,24]. The TD analysis can give more insight into the UWB propagation as the nature of impulse response (contribution of different paths) for various paths are resolvable in TD. Due to very short pulse duration, say 1 ns, two different pulses have the path difference of usually greater than 1 feet and thus the two pulses remain resolvable in nature. This results in the possibility of separate impulse behavior for different paths which consequently provides valuable insight into the system performance. On the contrary side, it should be noted that it is quite challenging to derive the TD solutions because one should be able to get the closed form of inverse Fourier transform of the FD counterparts.

In [21], for studying the UWB performance, new generalized TD propagation model has been used that uses the diffraction theory and can analyze the distortion in the transmitted pulse. The TD expressions have been derived while considering the different geometric configurations like half plane and dielectric slab. TD modeling is presented for different UWB propagation mechanisms like geometrical optics (GO) rays and diffraction through different objects like half plane or some dielectric slab.

16.6 TD SOLUTIONS FOR UWB DIFFRACTION

The diffraction theory plays an important role in wireless propagation as it is able to predict the signal strength even into the deep shadow environments of objects. Further it can account for considerable signal prediction in worst scenarios of line-of-sight (LOS) or non-line-of-sight (NLOS) propagation. The millimeter wave (mm-wave) signals, characterized by wavelengths in the order of millimeters and can prove significant for future cellular communication, experience diffraction phenomenon more strongly than the reflection mechanism [25–27].

UTD [28] is the most important diffraction theory and plays a vital role in analyzing the UWB diffraction [29]. Figure 16.5 shows the diagram for UWB diffraction where single diffracted (path P-Q-R) and double diffracted (path P-S-T-R) rays are received at the Rx. As shown here, the shape of received pulses is very different from the transmitted pulse shape and that depends on the diffraction effects while propagating through the wireless channel. It should be noted that the GO field discontinuity is generally compensated by using diffraction which makes the overall field (through GO shadow boundaries) continuous and smooth as well. But due to edge termination, the edge diffracted field becomes discontinuous and that can be made continuous by using the concept of corner diffraction [30].

Considering the significance of TD analysis, the TD solutions are presented here for UWB diffraction (edge and corner diffraction both) in environments made up of lossy materials.

16.7 EDGE DIFFRACTION OF UWB SIGNALS

Figures 16.6 and 16.7 show single diffraction (or amplitude diffraction) by a wedge and double diffraction (or slope diffraction) by two wedges. The heights of Tx, Rx, and wedge are represented by h_t, h_r, and h_w respectively while s_j represents the different travel distances. The incident and diffracted angles are respectively shown by ϕ' and ϕ and the wedge angle is defined as a_{int}.

Considering Figure 16.6, the FD expression for the diffracted field at Rx is given by [29]

$$E^d_{Rx,s,h}(\omega) = \frac{E_{inc}(\omega)}{s_1} D_{1s,h} A(s_2) \exp(-jk(s_1 + s_2)) \tag{16.11}$$

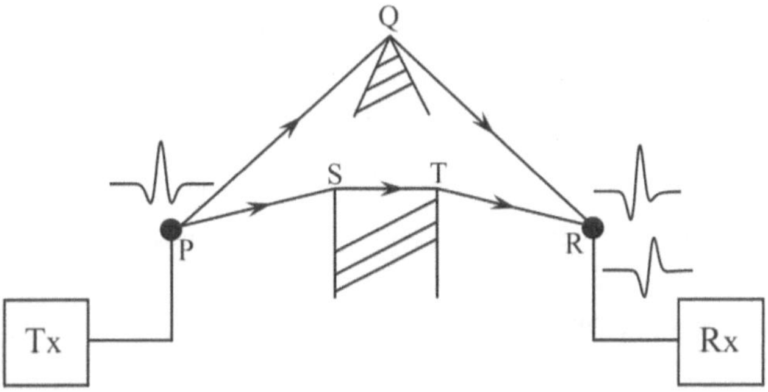

FIGURE 16.5 Diagram for single diffraction and double diffraction of UWB transmitted pulse.

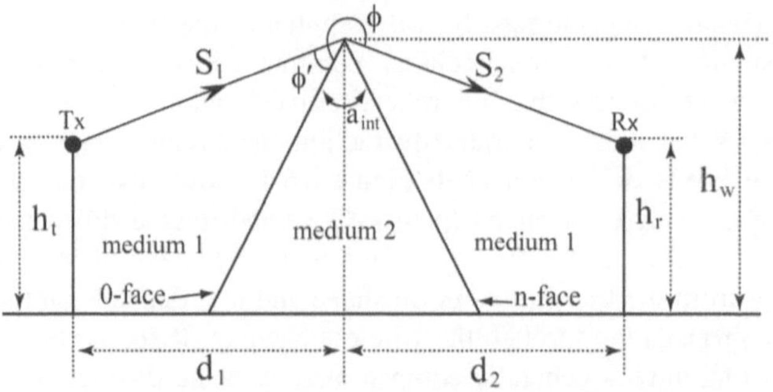

FIGURE 16.6 Wireless scenario representing the single wedge diffraction.

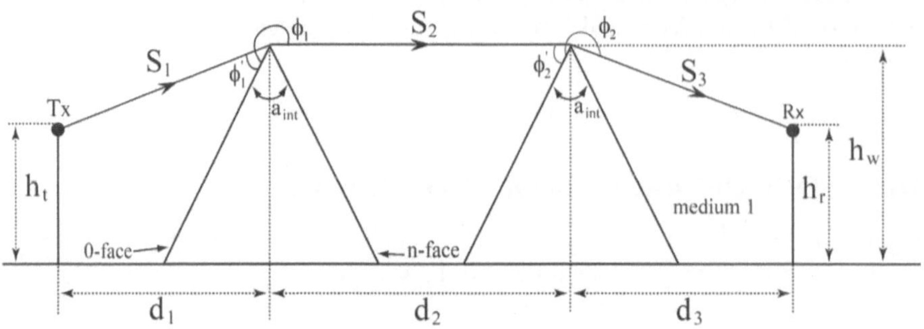

FIGURE 16.7 Wireless scenario representing the double wedge diffraction.

where $E_{inc}(\omega)$ is the electric field incident at the Tx. The subscripts s and h represent the soft and hard polarizations respectively. The parameter $A(s_2) = \sqrt{\frac{s_1}{(s_2(s_1+s_2))}}$ is the spreading factor and k is the wavenumber given by $k = \frac{\omega}{c}$, where ω is the angular frequency and c is the speed of light. The parameter D_1 is the UTD diffraction coefficient given as [31]

$$D_{1s,h} = G\,[R_{0s,h}R_{ns,h}D^1 + D^2 + R_{0s,h}D^3 + R_{ns,h}D^4] \tag{16.12}$$

with parameter $G = \frac{1}{2}$ when $\phi' = 0^\circ$ and $G = 1$ for other cases. The reflection coefficient associated with the 0- and n-face of wedge is defined by $R_{0,n}$. The diffraction coefficient term $D^i\,(i = 1, 2, \ldots, 4)$ is given as

$$D^i = \frac{-e^{\frac{-j\pi}{4}}}{2n\sqrt{2\pi k}}\,cot\,(a_i)\,F\,[2kLn^2sin^2a_i] \tag{16.13}$$

where $n = \left(2 - \frac{a_{int}}{\pi}()\right)$ and $L = \frac{(s_1 s_2)}{(s_1 + s_2)}$ are the wedge index parameter and distance parameter respectively with $F[x]$ as the transition function. The different terms a_i depend upon the incident and diffracted angles.

For Figure 16.7, the total field strength at Rx consists of amplitude and slope diffraction and is given by [29]

$$E^d_{Rx,s,h}(\omega) = \left(E_2 D_{2s,h} + \frac{1}{2}\frac{\partial E_2}{\partial n} ds_2\right) A(s_3) \exp(-jks_3) \tag{16.14}$$

where E_2 represents the diffracted field that is received at the second wedge and can be found out in the same way as (16.11). The diffraction term D_2 depends on ϕ_2 and ϕ'_2 with $A(s_3) = \sqrt{\frac{(s_1 + s_2)}{(s_3(s_1 + s_2 + s_3))}}$ denoting the spreading factor and $ds_2 = \left(\frac{1}{jk}\right)\left(\frac{\partial D_2}{\partial \phi'_2}\right)$ as the slope component of diffraction. The expression for $\frac{\partial E_2}{\partial n}$, that is the slope term of the received field, is written as

$$\frac{\partial E_2}{\partial n} = \left(\begin{aligned} & -\frac{1}{s_2}\frac{\partial E_2}{\partial \phi_1} \\ & = -\frac{A(s_2)}{s_2 s_1} E_{inc}(\omega) \frac{\partial D_1}{\partial \phi_1} \exp(-jk(s_1 + s_2)) \end{aligned} \right. \tag{16.15}$$

Using (16.11) and (16.15), (16.14) is now given as

$$E^d_{Rx,s,h}(\omega) = \left[\begin{aligned} & \frac{E_{inc}(\omega)}{s_1} A(s_2) A(s_3) D_1 D_2 \exp(-jk(s_1 + s_2 + s_3)) \\ & - \frac{1}{2}\frac{A(s_2)A(s_3)}{s_1 s_2} E_{inc}(\omega) \left(\frac{1}{\sqrt{jk}}\frac{\partial D_1}{\partial \phi_1}\right)\left(\frac{1}{\sqrt{jk}}\frac{\partial D_2}{\partial \phi'_2}\right) \exp(-jk(s_1 + s_2 + s_3)) \end{aligned} \right] \tag{16.16}$$

This expression depends on the partial derivative terms which can be solved by using [32].

Now the TD counterpart for (16.11) is given by

$$e^d_{Rx,s,h}(t) = \frac{A(s_2)}{s_1} e_{inc}(t) * d_{1s,h}(t) * \delta\left(t - \frac{s_1 + s_2}{c}\right) \tag{16.17}$$

where $e_{inc}(t)$ defines the source pulse at the Tx with $\delta(x)$ as the impulse function and symbol * standing for the convolution. The coefficient $d_{1s,h}(t)$ is the TD counterpart for (16.12) and is expressed as [31]

$$d_{1s,h}(t) = G \left[\begin{aligned} & r_{0s,h}(t) * r_{ns,h}(t) * d^1(t) + d^2(t) + \\ & r_{0s,h}(t) * d^3(t) + r_{ns,h}(t) * d^4(t) \end{aligned} \right] \tag{16.18}$$

with $r_{0/n,s,h}(t)$ as the TD version of $R_{s,h}$ [33]. The parameter $d^i(t)$ is related to D^i and is further defined as [29]

$$d^i(t) = -\frac{Ln}{2\pi\sqrt{2c}}\frac{\sin(2a_i)}{\sqrt{t}\,(t+\gamma_i)}u(t) \tag{16.19}$$

with $u(t)$ defining the step function and $\gamma_i = \frac{2Ln^2 sin^2(a_i)}{c}$.

The TD expression for (16.16) is written as

$$e^d_{Rx,s,h}(t) = \left[\begin{array}{l}\left\{\frac{A(s_2)A(s_3)}{s_1}\left(e_{inc}(t) * d_1(t) * d_2(t) * \delta\left(t - \frac{s_1+s_2+s_3}{c}\right)\right)\right\} \\ -\left\{\frac{1}{2}\frac{A(s_2)A(s_3)}{s_1 s_2}\left(e_{inc}(t) * d_1^{der}(t;\phi_1) * d_2^{der}(t;\phi'_2) * \delta\left(t - \frac{s_1+s_2+s_3}{c}\right)\right)\right\}\end{array}\right] \tag{16.20}$$

where $d_1(t)$ and $d_2(t)$ can be derived similar to (16.18). The TD coefficients $d_1^{der}\left(t;\phi_1\right)$ and $d_2^{der}(t;\phi'_2)$ are given as [34]

$$\begin{aligned} d_1^{der}(t;\phi_1) &= L^{-1}\left(\frac{1}{\sqrt{jk}}\frac{\partial D_1}{\partial \phi_1}\right) \\ &= -\frac{L_1}{\sqrt{2\pi}}[\{-r_0(t) * r_n(t) * F_s^1(t)\} + F_s^2(t) + \{r_0(t) * F_s^3(t)\} \\ &\quad - \{r_n(t) * F_s^4(t)\}] \\ &\quad + \frac{c}{2n\sqrt{2\pi}}\left[\cot a_1\left\{r_0(t) * r_n^{der}(t;\phi_1) * \frac{\gamma_1}{\sqrt{t+\gamma_1}}\right\}\right. \\ &\quad \left. + \cot a_4\left\{r_n^{der}(t;\phi_1) * \frac{\gamma_4}{\sqrt{t+\gamma_4}}\right\}\right] \end{aligned} \tag{16.21}$$

$$\begin{aligned} d_2^{der}(t;\phi'_2) &= L^{-1}\left(\frac{1}{\sqrt{jk}}\frac{\partial D_2}{\partial \phi'_2}\right) \\ &= -\frac{L_2}{\sqrt{2\pi}}[\{r_0(t) * r_n(t) * F_s^1(t)\} - F_s^2(t) + \{r_0(t) * F_s^3(t)\} \\ &\quad - \{r_n(t) * F_s^4(t)\}] \\ &\quad - \frac{c}{2n\sqrt{2\pi}}\left[\cot a_1\left\{r_n(t) * r_0^{der}(t;\phi'_2) * \frac{\gamma_1}{\sqrt{t+\gamma_1}}\right\}\right. \\ &\quad \left. + \cot a_3\left\{r_0^{der}(t;\phi'_2) * \frac{\gamma_3}{\sqrt{t+\gamma_3}}\right\}\right] \end{aligned} \tag{16.22}$$

where

$$F_s^i(t) = \frac{\sqrt{\gamma_i}}{2(t+\gamma_i)^{\frac{3}{2}}}u(t),\ \gamma_i = \frac{2Ln^2 sin^2(a_i)}{c} \tag{16.23}$$

The TD terms $r_n^{der}(t;\phi_1)$ and $r_0^{der}(t;\phi'_2)$ are the reflection coefficients derivatives [29].

Now the single diffracted field is shown in Figure 16.8 where Gaussian doublet has been used at the Tx with a special case of soft polarization. Further it is stated that dry concrete has been used here as the dielectric material [31]. The perfect match between the two results validates the TD approach. The high attenuation in the received signal is because of the dielectric parameters of the construction material (that is dry concrete here), whereas the distortion is due to the frequency dependent diffraction or simply frequency dependent propagation channel.

Considering the same excitation pulse and dielectric material, Figure 16.9 shows a double diffracted field for different polarizations. Here also, both the results are in perfect match, and that implies TD accuracy. It should be noted that this multiple diffraction case experiences more attenuation and signal distortion as compared to the single diffraction scenario of Figure 16.8.

Having soft polarization case, the impulse behavior for above simulated scenarios is shown in Figure 16.10. Because of a diffraction mechanism that strongly depends on frequency, the pulse width here is very different from the ideal impulse signal. Thus, after convolution with the source pulse, the received signal at the Rx gets distorted shape in general.

16.8 CORNER DIFFRACTION OF UWB SIGNALS

Figure 16.11 shows corner diffraction by a three-dimensional (3-D) wedge structure where Q_E is the point where edge diffraction takes place. As it has been discussed earlier also, due to edge termination, the edge diffracted field becomes discontinuous, and that can be made continuous by using the concept of corner diffraction.

Considering this wedge as a dielectric structure, the corner diffracted field is given as [30]

$$\begin{bmatrix} E_{\beta_{0c}}^{cd} \\ E_{\phi}^{cd} \end{bmatrix} \sim \left\{ \begin{matrix} \begin{bmatrix} C_a(Q_E) & C_b(Q_E) \\ C_c(Q_E) & C_d(Q_E) \end{bmatrix} \begin{bmatrix} E_{\beta_c}^{inc}(Q_C) \\ E_{\phi'}^{inc}(Q_C) \end{bmatrix} \\ \frac{e^{-j\pi/4}}{\sqrt{2\pi k}} \frac{\sqrt{\sin\beta_c \sin\beta_{0c}}}{\left(\cos\beta_{0c} - \cos\beta_c\right)} F\left[kL_c a\left(\pi + \beta_{0c} - \beta_c\right)\right] \frac{e^{-jkr_4}}{r_4} \end{matrix} \right\} \tag{16.24}$$

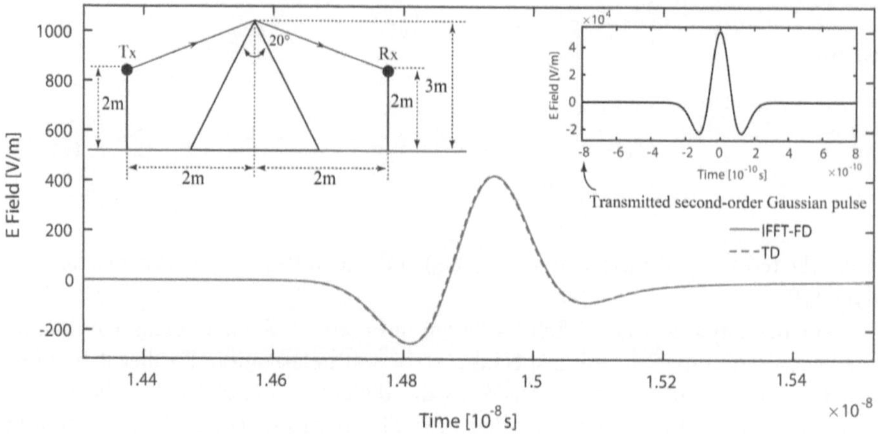

FIGURE 16.8 Single diffracted field at Rx with Gaussian doublet pulse used at the Tx.

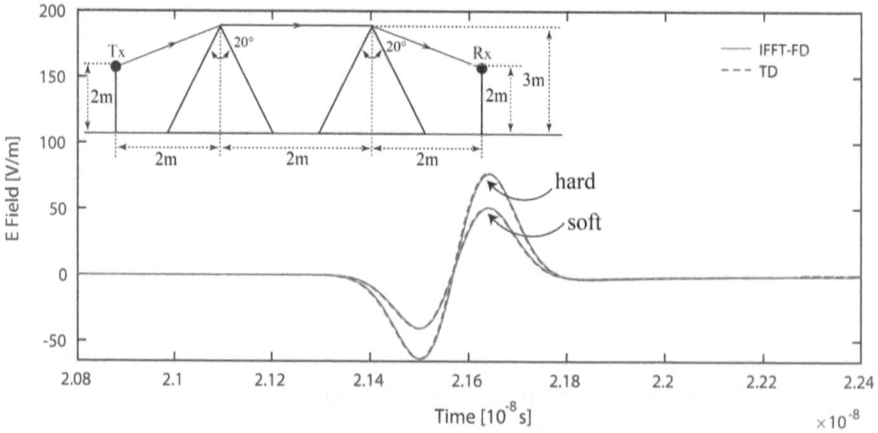

FIGURE 16.9 Double diffracted field at Rx for different polarizations.

where $E^{cd}_{\beta_{0c}}$ and E^{cd}_{ϕ} respectively represent the parallel and perpendicular polarized fields of the corner diffraction. The different terms $E^{inc}_{\beta'_c}(Q_C)$ and $E^{inc}_{\phi'}(Q_C)$ correspond to the excited field at the corner and these are ray optical in nature. The angles ϕ' and ϕ are the incident and diffracted angles respectively with $L_c = \frac{(r_3 r_4)}{(r_3 + r_4)}$ and $a(\gamma) = 2cos^2\left(\frac{\gamma}{2}\right)$. The corner diffraction coefficient $C(\theta)$ that is computed at Q_E, is expressed as [30]

$$C_a(Q_E) = (R_{\|}cos^2\alpha - R_{\perp}sin^2\alpha)(C_n(\phi + \phi') + C_0(\phi + \phi')) \quad (16.25)$$

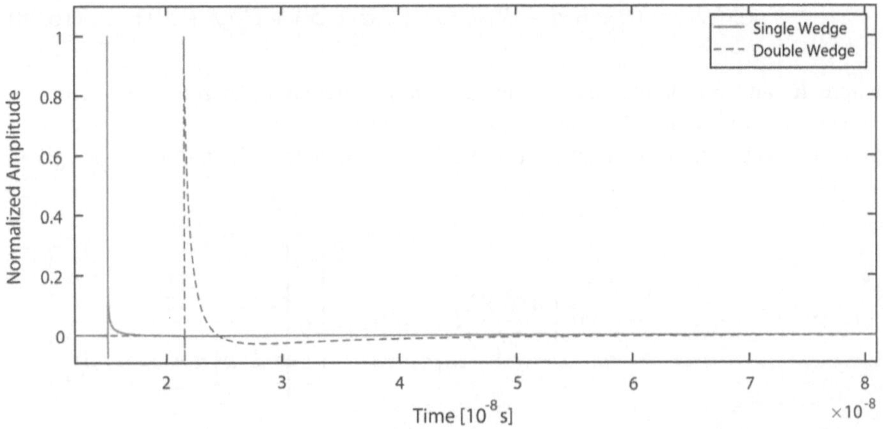

FIGURE 16.10 Impulse behavior for different simulated scenarios of diffraction.

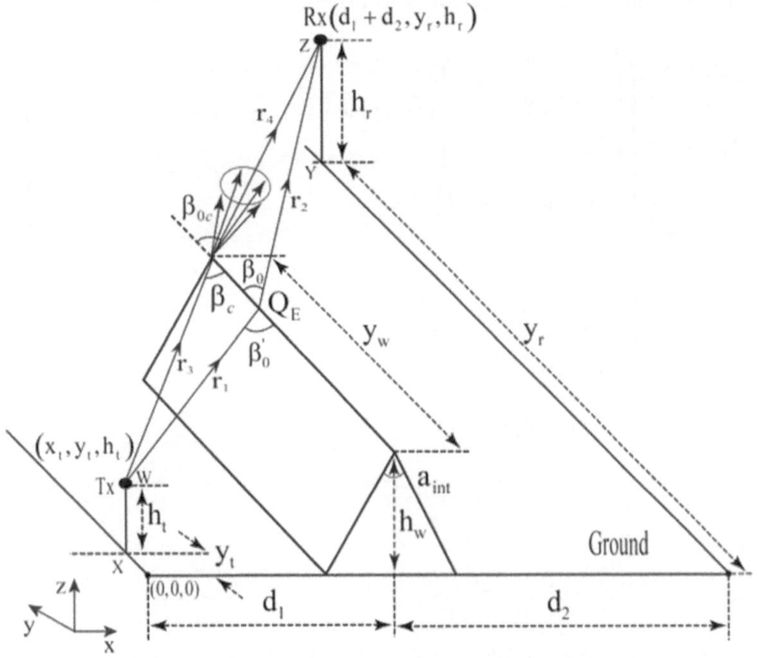

FIGURE 16.11 Corner diffraction by a 3-D wedge shaped obstacle.

$$C_b(Q_E) = (R_{\|} + R_{\perp})\sin\alpha\cos\alpha(C_n(\phi + \phi') + C_0(\phi + \phi')) \tag{16.26}$$

$$C_c(Q_E) = -(R_{\|} + R_{\perp})\sin\alpha\cos\alpha(C_n(\phi + \phi') + C_n(\phi + \phi')) \tag{16.27}$$

$$C_d(Q_E) = (R_{\perp} cos^2\alpha - R_{\|} sin^2\alpha)(C_n(\phi + \phi') + C_n(\phi + \phi')) \qquad (16.28)$$

where $R_{\|}$ and $R_{\perp}$ define the parallel and perpendicular polarized components of reflection coefficient. The angle parameter α is related to the different planes of incidence and can be referred from [35]. The parameters $C_0(\gamma)$ and $C_n(\gamma)$ are given as [30]

$$C_n^0(\gamma) = \frac{-e^{-j\pi/4}}{2n\sqrt{2\pi k}\ \sin\beta_0}\ cot\left(\frac{\pi \mp \gamma}{2n}\right) F\left[kLa^{\mp}(\gamma)\right] \left| F\left[\frac{\frac{La^{\mp}(\gamma)}{\lambda}}{kL_c a\left(\pi + \beta_{0c} - \beta_c\right)}\right]\right| \qquad (16.29)$$

where L is the distance parameter as $L = \left(\frac{(r_1 r_2)}{(r_1 + r_2)}\right) sin^2\beta_0$.

The TD formulations corresponding to (16.24) can be written as:

$$e_{\beta_{0c}}^{cd}(t) \sim \left(\begin{array}{l} \left[\frac{\sqrt{\sin\beta_c \sin\beta_{0c}}}{r_4\left(\cos\beta_{0c} - \cos\beta_c\right)}\right]\left[\left(c_a(t) * e_{\beta_c}^{inc}(t)\right) + \left(c_b(t) * e_{\phi'}^{inc}(t)\right)\right] \\ *f(t) * \delta\left(t - \frac{r_4}{c}\right) \end{array} \right) \qquad (16.30)$$

$$e_{\phi}^{cd}(t) \sim \left(\begin{array}{l} \left[\frac{\sqrt{\sin\beta_c \sin\beta_{0c}}}{r_4\left(\cos\beta_{0c} - \cos\beta_c\right)}\right]\left[\left(c_c(t) * e_{\beta_c}^{inc}(t)\right) + \left(c_d(t) * e_{\phi'}^{inc}(t)\right)\right] \\ *f(t) * \delta\left(t - \frac{r_4}{c}\right) \end{array} \right) \qquad (16.31)$$

with

$$c_a(t) = \left[r_{\|}(t) cos^2\alpha - r_{\perp}(t) sin^2\alpha\right] * \left(c_n(t;\ \phi + \phi') + c_0(t;\ \phi + \phi')\right) \qquad (16.32)$$

$$c_b(t) = \sin\alpha \cos\alpha\left[r_{\|}(t) + r_{\perp}(t)\right] * \left(c_n(t;\ \phi + \phi') + c_0(t;\ \phi + \phi')\right) \qquad (16.33)$$

$$c_c(t) = -\sin\alpha \cos\alpha\left[r_{\|}(t) + r_{\perp}(t)\right] * \left(c_n(t;\ \phi + \phi') + c_0(t;\ \phi + \phi')\right) \qquad (16.34)$$

$$c_d(t) = \left[r_{\perp}(t) cos^2\alpha - r_{\|}(t) sin^2\alpha\right] * \left(c_n(t;\ \phi + \phi') + c_0(t;\ \phi + \phi')\right) \qquad (16.35)$$

where $r_{\|}(t)$ and $r_{\perp}(t)$ are the TD reflection coefficients for parallel and perpendicular

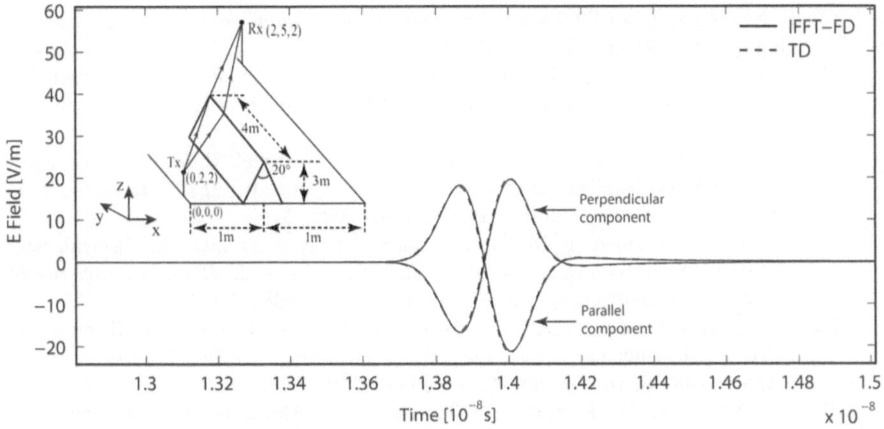

FIGURE 16.12 Different field components of corner diffraction by 3-D wedge scenario.

polarizations respectively [33]. The TD term $c_n^0(t;\phi+\phi')$ that corresponds to $C_n^0(\gamma)$ (expressed in (16.29)) can be written as [36]

$$c_n^0(t;\phi+\phi') = -\frac{Ln}{\pi\sqrt{2}\sqrt{\gamma_n^0 c}\,\sin\beta_o}\sin(2a_n^0)tan^{-1}\left(\sqrt{\frac{t}{\gamma_n^0}}\right) \times \left| F\left[\frac{2Ln^2sin^2a_n^0}{2\pi L_c a\left(\pi+\beta_{oc}-\beta_c\right)}\right]\right| \qquad (16.36)$$

where $\gamma_n^0 = \frac{2Ln^2sin^2a_n^0}{c}$ with $a_0 = \frac{[\pi-(\phi+\phi')]}{2n}$ and $a_n = \frac{[\pi+(\phi+\phi')]}{2n}$.

Now considering the above FD and TD formulations for corner diffraction, Figure 16.12 shows the corner diffracted field for UWB corner diffraction by a dielectric wedge made up of dry concrete. The TD results are observed as in good agreement with the IFFT-FD counterparts. Further as has been discussed earlier also, the shape distortion is observed due to the nature of the UWB channel that depends on the frequency.

Confirming the TD accuracy, it is summarized here that the TD approach can prove significant to study the UWB diffraction (either edge diffraction or corner diffraction) by lossy obstacles in realistic scenarios.

REFERENCES

[1]. FCC first report and order: In the matter of revision of part 15 of the comparison's rules regarding ultra-wideband transmission system, FCC 02-48, April 2002.

[2]. F. Nekoogar, *Ultra-Wideband Communications: Fundamentals and Applications*. Upper Saddle River, NJ, USA: Prentice Hall, 2005.
[3]. V. Yajnanarayana and P. Handel, "Joint estimation of TOA and PPM symbols using sub-Nyquist sampled IR-UWB signal," *IEEE Communications Letters*, vol. 21, no. 4, pp. 949–952, 2017.
[4]. K. Yu, K. Wen, Y. Li, S. Zhang, and K. Zhang, "A novel NLOS mitigation algorithm for UWB localization in harsh indoor environments," *IEEE Transactions on Vehicular Technology*, vol. 68, no. 1, pp. 686–699. 2019.
[5]. S. Sangodoyin and A. F. Molisch, "Impact of body mass index on ultrawideband MIMO BAN channels–measurements and statistical model," *IEEE Transactions on Wireless Communications*, vol. 17, no. 9, pp. 6067–6081, 2018.
[6]. S. Sangodoyin and A. F. Molisch, "A measurement-based model of BMI impact on UWB multi-antenna PAN and B2B channels," *IEEE Transactions on Communications*, vol. 66, no. 12, pp. 6494–6510, 2018.
[7]. P. Leelatien, K. Ito, K. Saito, M. Sharma, and A. Alomainy, "Channel characteristics and wireless telemetry performance of transplanted organ monitoring system using ultrawideband communication," *IEEE Journal of Electromagnetics, RF, and Microwaves in Medicine and Biology*, vol. 2, no. 2, pp. 94–101, 2018.
[8]. A. Sharma, I. J. G. Zuazola, R. Martnez, A. Perallos, and J. C. Batchelor, "Channel-based antenna synthesis for improved in-vehicle UWB MB-OFDM communications," *IET Microwaves, Antennas & Propagation*, vol. 13, no. 9, pp. 1358–1367, 2019.
[9]. C. Briso, C. Calvo, and Y. Xu, "UWB propagation measurements and modeling in large indoor environments," *IEEE Access*, vol. 7, pp. 41913–41920, 2019.
[10]. R. C. Qiu, C. Zhou, and Q. Liu, "Physics-based pulse distortion for ultra-wideband signals," *IEEE Transactions on Vehicular Technology*, vol. 54, no. 5, pp. 1546–1555, 2005.
[11]. B. Hu and N. C. Beaulieu, "Pulse shapes for ultrawideband communication systems," *IEEE Transactions on Wireless Communications*, vol. 4, no. 4, pp. 1789–1797, 2005.
[12]. M. Ghavami, L. B. Michael, and R. Kohno, *Ultra Wideband Signals and Systems in Communication Engineering*. Chichester, U.K.: Wiley, 2004.
[13]. A. F. Molisch, "Ultrawideband propagation channels-theory, measurement, modeling," *IEEE Transactions on Vehicular Technology*, vol. 54, no. 5, pp. 1528–1545, 2005.
[14]. APT report on ultra wideband (UWB), No. APT/AWF/REP-01, Edition: Sep. 2012. https://www.apt.int/sites/default/files/APT-AWF-REP-01Report_on_UWB.pdf
[15]. C. C. Chong, F. Watanabe, and H. Inamura, "Potential of UWB technology for the next generation wireless communications," in *2006 IEEE Ninth International Symposium on Spread Spectrum Techniques and Applications*, Manaus-Amazon, Brazil, pp. 422–429, 2006.
[16]. "Digital Land Mobile Radio Communications," *Commission of the European Communities*, COST 207, 1984–1988.
[17]. D. Tholl, M. Fattouche, R. J. C. Bultitude, P. Melancon, and H. Zaghloul, "A comparison of two radio propagation channel impulse response determination techniques," *IEEE Transactions on Antennas and Propagation*, vol. 41, no. 4, pp. 515–517, 1993.
[18]. J. D. Taylor, *An Introduction to Ultra-Wideband Radar Technology*. Boca Raton, FL: CRC, 1995.
[19]. T. S. Rappaport, *Wireless Communications: Principles and Practice. Piscataway*, NJ, USA: IEEE Press, 1996.
[20]. J. Jemai, P. C. F. Eggers, G. F. Pedersen, and T. Ku¨rner, "Calibration of a UWB sub-band channel model using simulated annealing," *IEEE Transactions on Antennas and Propagation*, vol. 57, no. 10, pp. 3439–3443, 2009.

[21]. R. C. Qiu, "A Generalized time domain multipath channel and its application in ultra-wideband (UWB) wireless optimal receiver design—part II: physics-based system analysis," *IEEE Transactions on Wireless Communications*, vol. 3, no. 6, pp. 2312–2324, 2004.
[22]. A. F. Molisch, "Status of models for UWB propagation channel," *IEEE 802.15.4a Channel Model (Final Report)*, 2004.
[23]. R. C. Qiu, "A study of the ultra-wideband wireless propagation channel and optimum UWB receiver design," *IEEE Journal on Selected Areas in Communications*, vol. 20, no. 9, pp. 1628–1637, 2002.
[24]. R. C. Qiu, "A generalized time domain multipath channel and its application in ultra-wideband (UWB) wireless optimal receiver—part III: system performance analysis," *IEEE Transactions on Wireless Communications*, vol. 5, no. 10, pp. 2685–2695, 2006.
[25]. I. A. Hemadeh, K. Satyanarayana, M. E.-Hajjar, and L. Hanzo, "Millimeter-wave communications: physical channel models, design considerations, antenna constructions, and link-budget," *IEEE Communications Surveys & Tutorials*, vol. 20, no. 2, pp. 870–913, 2018.
[26]. M. E. Rasekh, A. A. Shishegar, and F. Farzaneh, "A study of the effect of diffraction and rough surface scatter modeling on ray tracing results in an urban environment at 60 GHz," *in Proceedings of theIEEE First Conference on Millimeter-Wave and Terahertz Technologies (MMWaTT)*, Tehran, Iran, 2009, pp. 27–31.
[27]. M. Jacob, S. Priebe, R. Dickhoff, T. K.-Ostmann, T. Schrader, and T. Kurner, "Diffraction in mm and Sub-mm wave indoor propagation channels," *IEEE Transactions on Microwave Theory and Techniques*, vol. 60, no. 3, pp. 833–844, 2012.
[28]. R. G. Kouyoumjian and P. H. Pathak, "A uniform geometrical theory of diffraction for an edge in a perfectly conducting surface," *Proceedings of the IEEE*, vol. 62, no. 11, pp. 1448–1461, 1974.
[29]. A. Karousos and C. Tzaras, "Multiple time-domain diffraction for UWB signals," *IEEE Transactions on Antennas and Propagation*, vol. 56, pp. 5, 1420–1427, 2008.
[30]. P. J. Joseph, A. D. Tyson, and W. D. Burnside, "An absorber tip diffraction coefficient," *IEEE Transactions on Electromagnetic Compatibility*, vol. 36, no. 4, pp. 372–379, 1974.
[31]. B. Bansal, S. Soni, V. K. Mishra, A. Gupta, and A. Agrawal, "A novel heuristic time-domain diffraction model for UWB diffraction by lossy wedges and buildings," *Physical Communication*, vol. 34, pp. 80–89, 2019.
[32]. R. J. Luebbers, "A heuristic UTD slope diffraction coefficient for rough lossy wedges," *IEEE Transactions on Antennas and Propagation*, vol. 37, no. 2, pp. 206–211, 1989.
[33]. P. B. Barnes and F. M. Tesche, "On the direct calculation of a transient plane wave reflected from a finitely conducting half space," *IEEE Transactions on Electromagnetic Compatibility*, vol. 33, no. 2, pp. 90–96, 1991.
[34]. P. Liu, J. Wang, and Y. Long, "Time-domain double diffraction for UWB signals," in *PIERS Proceedings*, Beijing China, pp. 848–852, 2009.
[35]. W. D. Burnside and K. W. Burgener, "High frequency scattering by a thin lossless dielectric slab," *IEEE Transactions on Antennas and Propagation*, vol. AP-31, no. 1, pp. 104–110, 1983.
[36]. B. Bansal and S. Soni, "A new time-domain corner diffraction coefficient for metallic and dielectric objects for UWB signals," *Microwave and Optical Technology Letters*, vol. 57, no. 7, pp. 1760–1765, 2015.

17 Adopting Artificial Neural Network Modelling Technique to Analyze and Design Microstrip Patch Antenna for C-Band Applications

Shuchismita Pani, Parnika Saxena, and Yogesh
Department of Electronics & Communication Engineering, Amity University, Noida, India

17.1 INTRODUCTION

Due to the fact that the size of microstrip patch antennas is very small, the parameters for the design of these antennas need to be measured with high accuracy that provides high gain, fewer losses, etc. The dielectric constant and the dimensions of the antenna are related to the frequency at which the antenna radiates. The microstrip patch antenna (MPA) having features like good accuracy, low production, and small size have caused it gained a lot of attention in this area. They are utilized in various applications which include satellite applications, high-performance aircraft, and where the bandwidth and size is the constraint, etc. Research on optimization of antenna parameters using ANN and analyzing their performances has been carried out [1–3]. The commonly used microstrip antennas are circular, triangular, rectangular, and elliptical [4–6]. Microstrip antennas are widely used for C-Band applications. The C-Band lies in the range of 4–8 GHz. Gigahertz frequency finds applications in radio astronomy, W-LAN, most modern radars, amateur radio, satellite radio, etc. The SHF (super high frequency) lies in the range of 3–30 GHz frequency. It is also known as the centimeter wave. Further SHF consists of various bands out of which the C-band lies in the range of 4–8 GHz. C-Band mainly finds its use in satellite communication, radar systems, etc. A lot of work has previously been done for C-Band applications. Wi-Fi antenna design and modeling

DOI: 10.1201/9781003187325-17

using artificial neural networks have been proposed [7]. In [8] an I-slot-shaped antenna design, the substrate used is glass. The neural network training algorithms are used in the simulation of results for training the samples to minimize the error and to obtain results for different inputs. The artificial neural network has the ability to respond accurately to the inputs that are in the interval which defines the network [9–11]. For complex classification tasks, ANN forms arbitrary nonlinear design boundaries. The weights are optimized by the NN between its neurons by using a training process. After the training, the network gives the results for different inputs. A non-invasive blood glucose measurement and their sensitivity optimization based on ANN is proposed by designing a novel antenna presented in [12]. ANN is widely being used in various applications. Due to growing applications in the context of electromagnetics ANN is now being used in antenna designing and the synthesis or optimization of radiation patterns etc. Previously a lot of work has been done on the antenna and neural networks. Neural networks have been used to determine the bandwidth of antenna [13]. In [14] neural network technique is used to design an elliptical patch antenna using an adaptive-network that is based on a fuzzy inference system (ANFIS). In [15] an ANN is proposed for modeling of U-slot antennas, for increasing the accuracy of modeling, the use of ANN is suggested. ANN possesses various features that make them a suitable choice. Earlier ANN modeling techniques have been used for multi-antenna systems to analyze and design. ANN models are fast and give response much earlier and faster [16]. In [17], for millimeter-wave slot Patch antennas, ANN has been used. For the ANN Model, there are input and output parameters. For analyzing and designing a microstrip patch antenna with ANN model, the parameters taken into consideration are the operating frequency, dielectric constant, and the size of the antenna. Several shapes of patches have been used to determine the relationship between the parameters [18].

The proposed work presents an efficient modeling technique (i.e., artificial neural network [ANN] modeling) to analyze the microstrip patch antenna for C-Band applications. C-Band lies in the range of 4–8 GHz. The artificial neural network is used for analysis. The software used for the analysis is MATLAB. Once the analysis has been done and results have been obtained, an antenna is designed from the analysis results. The antenna is designed using the dimension obtained from the network's best validation result. The designed antenna is simulated on HFSS.

17.2 METHODS

17.2.1 Microstrip Antenna

Microstrip antennas as the name suggests are an antenna whose fabrication is done by adopting microstrip methods or techniques on a printed circuit board. They are also known as the "Printed Antennas". These antennas have become mostly used because of

- Small size
- Simple in design

- Cost of production is low
- Available in a variety of shapes

Patch antennas are the most widely or commonly used type of microstrip antennas. In general, this kind of antenna is narrowband wide-beam antennas. In simple words, on one side of the microstrip patch antennas (MPA), there is a radiating patch, and on the other side, there is a ground plane. The patch is made on the substrate. The results obtained in the simulation also depend on the choice of the substrate. If the dielectric constant of the substrate is high and there is a bandwidth issue, then a lower value substrate is chosen. Thus, the choice of the substrate is done accordingly. Due to the various advantages that these antennas have, they are used in a lot of applications.

17.2.2 Artificial Neural Network

The functionality of artificial neural network (ANN) is similar to that of the structure of the human brain. In short, the neural network is a framework that contains several nodes called 'neurons', interconnected to each other. These neurons are arranged in the form of layers. The function of the layers is the processing of information by using the dynamic state responses to the external inputs. Thus, the algorithm in use is very helpful in recognizing patterns that are too complex to be manually be extracted and difficult for the machine to be taught to identify the same.

Neural networks have gained huge attention since the last few years. They are of great use since they provide alternative approaches for solving and understanding of the various problems. Rather than adopting conventional methods of solving a problem, neural networks take a different approach to such a problem. Conventional here refers to the fact that to a particular problem there is a particular solution, i.e., there are a set of instructions that the computer follows in order to solve a particular problem. If the answer to a particular problem is known, only then is a computer able to solve the given problem. Neural networks are here beneficial since for any kind of problem they have a solution, they just have to be trained with particular problem and their solution.

17.2.3 ANN Characteristics

The method through which artificial neural networks (ANN) process the data is similar to how the human brain processes the data. ANN structure is based on biological neurons. Just like the brain is able to make the decisions on its own, ANN is capable of making decisions. Its structure consists of various processing units which communicate through each other. In the proposed method the artificial neural network consists of three elements in the model:

- Synapses: It is represented by weights
- Function: Responsible for the combination of weighted input signals

- An activation function

where

$$u_k = \sum_{j=1}^{p} w_{kj} x_j \tag{17.1}$$

$$y_k = \varnothing(u_k - \theta_k) \tag{17.2}$$

The denotation of the variables is mentioned in Table 17.1.

The flow chart of the proposed method is shown in Figure 17.1 that includes the input data, creation of the neural network object, configuration of the inputs and outputs, initialization of the weight and bias, network training, validation, and output prediction of the length and width of the patch antenna. In [19], a method proposed to extract the local features used for model with less processing time. The model is optimized by changing the weight when applied to the NN [20].

17.2.4 ANN Learning Methods

The learning methods of ANN are:

- Supervised learning method: In supervised learning, the designed network is trained using the database. This database gives the standard output. The network weights are adjusted according to the error generated between the output data and the target data.
- Unsupervised learning method: In this type of learning, there is no database that is provided in order to train the network or there is no source that provides the desired values.
- Reinforced: In this learning, the network is not fed the information about the expected output, the network is used only to represent the correctness of the output.

TABLE 17.1
Denotation of the Variables

Variable	Meaning
$x_1, x_2, \ldots\ldots\ldots x_p$	Input signal
$w_{k1}, w_{k2}, \ldots\ldots\ldots\ldots w_{kp}$	Synaptic weight
u_k	Linear accumulator
θ_k	Threshold function
$\varnothing$	Sigmoid function
y_k	Output signal

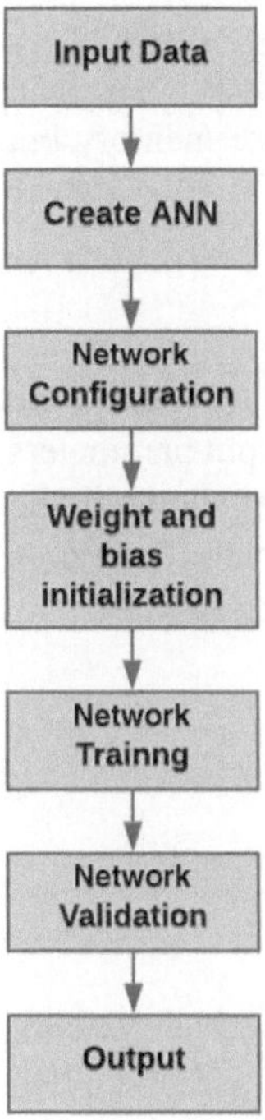

FIGURE 17.1 The flow of the proposed method.

17.2.5 Backpropagation Learning Algorithm

This training algorithm for errors being propagated back is based on the concept of supervised learning. If a dataset is composed then,

The input vector, $x = [x_1, x_2, \ldots x_I]$

Answer vector, $d = [d_1, d_2, \ldots d_k]$

The response of the network, $y = [y_1, y_2, \ldots y_k]$

The more the value of vector y is near to vector d, better is the result, i.e., lesser are the errors. The generated error at the output of the neuron k is given as

$$e_k = d_k(n) - y_k(n) \tag{17.3}$$

The main focus of using this algorithm is to determine the smallest error which is done by using a search on the error surface provided by the gradient descent. There are various types of backpropagation learning algorithms.

17.2.6 The Levenberg-Marquardt (LM)

The functionality of the Levenberg Marquardt algorithm lies between gradient descent method and the GNA (Gauss-Newton algorithm). In this algorithm, if the present solution is distant to the minimum, i.e., local minimum and there is a large value of H, LMA acts the steepest descent method; it is slow but achieves convergence. Gauss-Newton method is applied when the solution is near the local minimum, and there is fast convergence. Levenberg-Marquardt backpropagation algorithm is used. This algorithm has a certain feature which makes it an ideal choice. Some of the key features of this algorithm are: requires more memory but less time, among the most effective NN and more precise.

17.3 DATABASE GENERATION

To create the database for the training, the input parameters are the frequency, the permittivity of substrate and dielectric or substrate height, and the output parameters are the dimensions of antenna, i.e., length and width of the antenna as shown in the ANN model in Figure 17.2. The data set is composed of 194 samples. The range taken into consideration for creating the database is:

Frequency: 4–8 GHz
Dielectric Constant values: 2.2–4.4
Dielectric Height: 0.1 mm–2.5 mm

17.4 ANN MODEL AND RESULTS

For the proposed work, the ANN model is designed on MATLAB. The network architecture consists of input, output, and hidden layers. The number of hidden nodes is 5. MATLAB consists of various toolboxes for the creation of the ANN model. Various tools like nntool, nftool, nctool, etc. are available which is be used to design the same. Out of the available toolboxes, nntool is used for the proposed work, and nntool allows the created database to be imported and be used as well as allows the data and the neural networks to be exported. In the designed model Levenberg Marquardt algorithm is implemented.

17.4.1 Network Analysis Results

Regression Plot: The use of regression is done when the database is divided into 3 sections, i.e., training data, validation data, and testing data. It is used for recognizing/determining the relationship between the variables. When R approaches 1, the fitting of the model to the data is better and better and is the designed or the proposed model. The regression plots obtained for the training, validation, and

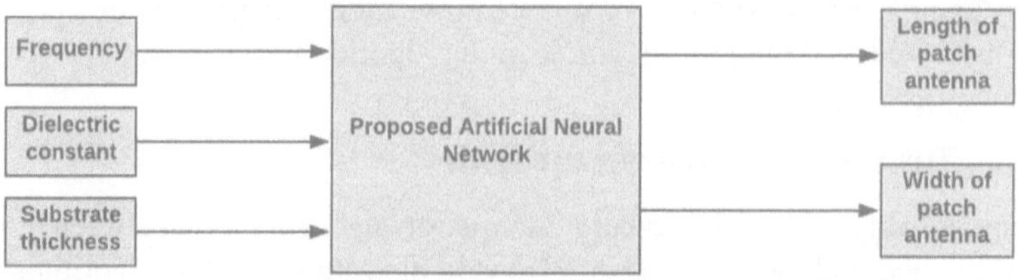

FIGURE 17.2 Input and output parameters.

testing having the value of R as 0.98546, 0.9966, and 0.99427 respectively. The value of R for all is 0.98845 which is near the value 1.

Gradient Plot: The gradient plot shows the performance of the training state. The performance is obtained with regard to the gradient, number of validation checks, and Mu. The number of validation checks, Mu, and Gradient values obtained at an epoch 103 are 6, 0.8133, and 0.001 respectively.

Performance Plot: A plot of the number of epochs versus the mean squared error (MSE) is used to analyze the system performance. The plots are monitored to observe the trend using the epoch versus mean square error (MSE) from the plot, and as the number of epochs increases, the value of MSE decreases. The slope of the three is almost the same, having a small difference between them. The best validation performance is 0.25181 at epoch 97.

17.4.2 Networks Best Results

The designed network predicts the length, width of a patch for the input parameters given. For an input frequency of 7.55 GHz, the dielectric constant of 4.4 and substrate thickness of 2.1 mm, the network predicts the length and the width. The value of length, width of the microstrip patch antenna predicted is 8.528768 mm and 12.10997 mm respectively. The error in the prediction of the parameters is –0.00049 and –0.01896 for length and width respectively.

17.5 ANTENNA DESIGN AND RESULTS

From the dimensions obtained for the patch we gather the network's best performance. The substrate material used is FR4_epoxy which has the value of dielectric permittivity as 4.4. FR4 possesses attractive features like low cost and has excellent mechanical properties. It is widely used, as a choice of the substrate in antenna design and various other applications. The value of 'h', i.e., the height of the substrate is 2.1 mm. The designed antenna operates at a frequency of 7.55 GHz.

17.5.1 Gain

For the designed antenna a gain of 6.2991 dB is obtained. The gain of the proposed antenna with the optimized solution predicted by ANN shows the estimation of intensity in the direction shown in Figure 17.3.

17.5.2 Return Loss

The return loss is also known as the reflection coefficient and is the amount of power being lost due to the reflected signal. The proportion of radio waves arriving at the antenna input that are rejected as a ratio against those that are accepted. For the designed antenna the S_{11} parameter is obtained as shown in Figure 17.4. The value of S_{11} or the return loss obtained is –11.2822 dB at an operating frequency of 7.55 GHz.

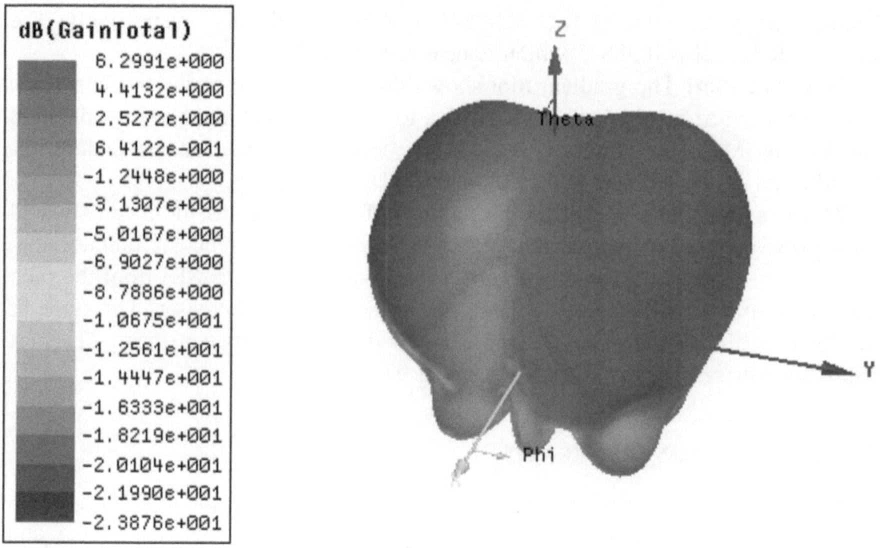

FIGURE 17.3 Three dimension polar plot.

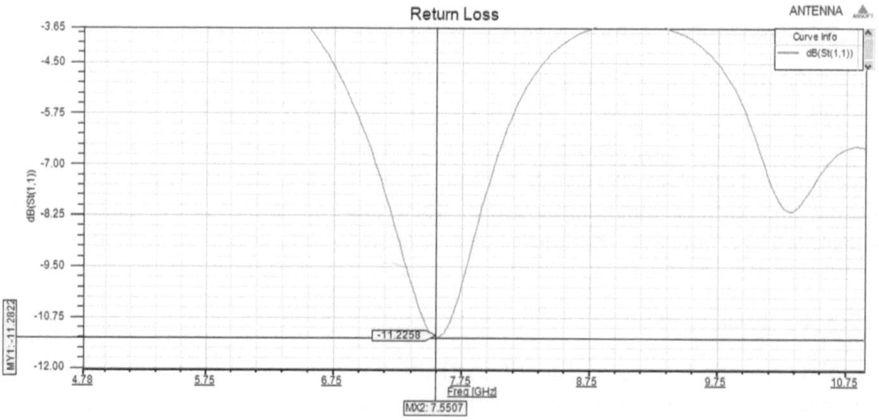

FIGURE 17.4 Return loss plot.

17.5.3 Radiation Pattern

Radiation patterns are used to depict how the radiated energy is distributed in space. Different kinds of radiation patterns are obtained for different antennas. Figure 17.5 shows the radiation pattern obtained for the designed antenna for an operational frequency of 7.55 GHz.

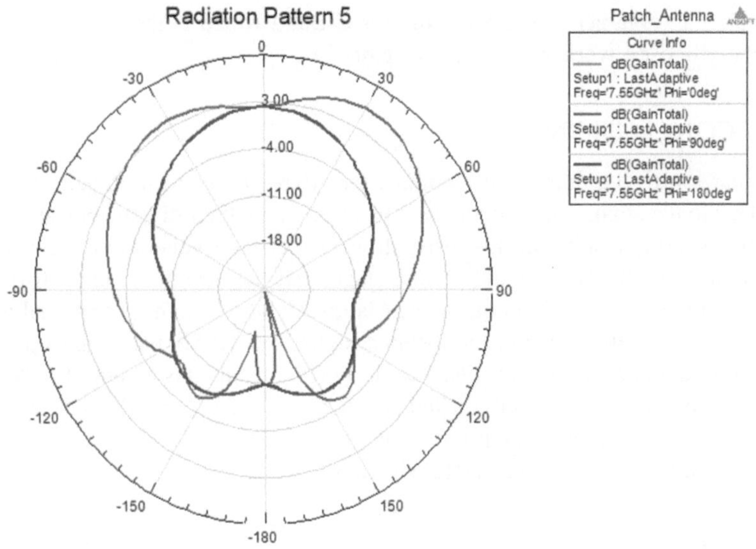

FIGURE 17.5 Radiation pattern.

17.5.4 Current Distribution Plot

Figure 17.6 predicts the surface current distribution, i.e., Jsurf Plot of the designed antenna. The maximum value of the current intensity is obtained at the feed points and along the horizontal direction.

The plots obtained from the designed microstrip patch antenna are good. It shows a gain of 6.2991 dB. The gain obtained is greater than the one obtained in [8]. In [8], and I- slot-shaped antenna was designed operating for C-Band and X-Band.

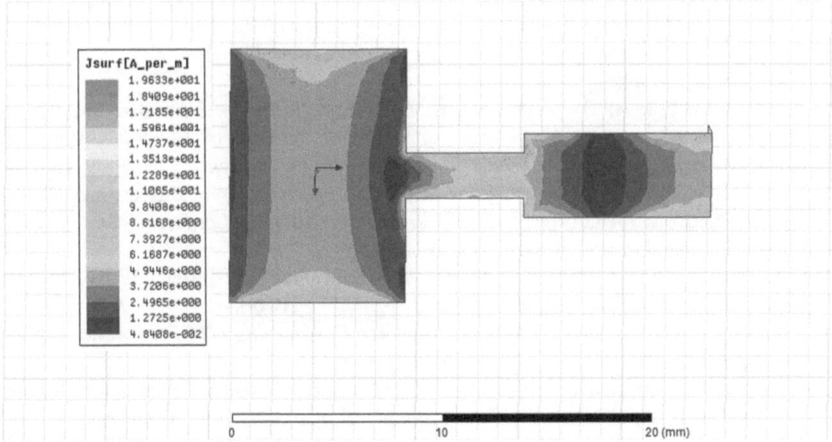

FIGURE 17.6 Surface current distribution of designed antenna.

The peak gain obtained was 2.9483 dB for C-Band which is less than 6.2991 dB. Thus the proposed work gives a higher gain.

17.6 CONCLUSIONS

In comparison to the numerical methods, the ANN proves to be much more efficient. Due to the small size of microstrip patch antenna, their dimensions are correct to obtain accurate results. In this research work, database generation is considered as the first stage, analysis as the second stage, and the design of microstrip patch antenna as the last stage. The Levenberg-Marquardt algorithm has been used for the analysis. Further, the antenna has been designed on HFSS using the optimized dimensions and shows good results. The designed antenna has a gain of 6.2991 dB which is a good result, and analysis and design of MPA for C-Band applications has been successfully completed. Optimization of microstrip antennas parameters using artificial neural networks gives more accuracy.

REFERENCES

1. M. Chetioui, A. Boudkhil, N. Benabdallah and N. Benahmed, "Design and Optimization of SIW Patch Antenna for Ku Band Applications Using ANN Algorithms," *2018 4th International Conference on Optimization and Applications (ICOA)*, Mohammedia, 2018, pp. 1–4, doi: 10.1109/ICOA.2018.8370530.
2. A. I. Hammoodi, F. Al-Azzo, M. Milanova and H. Khaleel, "Bayesian Regularization Based ANN for the Design of Flexible Antenna for UWB Wireless Applications," *2018 IEEE Conference on Multimedia Information Processing and Retrieval (MIPR)*, Miami, FL, 2018, pp. 174–177, doi: 10.1109/MIPR.2018.00039.
3. M. Sarevska, Z. Stankovic, N. Doncov, I. Milovanovic and B. Milovanovic, "Design of Well-Matched UHF Planar Bowtie Dipole Antenna using Neural Model," *2019 14th International Conference on Advanced Technologies, Systems and Services in Telecommunications (TELSIKS)*, Nis, Serbia, 2019, pp. 331–334, doi: 10.1109/TELSIKS46999.2019.9002319.
4. Nurhan Turker, Filiz Gunes, Tulay Yildirim, "Artificial Neural Design of Microstrip Antennas," *The Turkish Journal of Electrical Engineering & Computer Sciences*, 2006, 14(3), pp. 445–453.
5. Abhilasa Mishra, A. B. Nandgaonkar, V. D. Bhagile, S. C. Mehrotra "Design of Square and Rectangular Microstrip Antenna with the use of FFBP algorithm of Artificial Neural Network," *Applied Electromagnetics Conference*, Dec 2009, pp. 1–4
6. B. K. Singh, "Design of Rectangular Microstrip Patch Antenna Based on Artificial Neural Network Algorithm," *2015 2nd International Conference on Signal Processing and Integrated Networks*, 2015, pp. 6–9
7. P. k. Abbassi, N. M. Badra, A. M. M. A. Allam and A. El-Rafei, "WiFi Antenna Design and Modeling using Artificial Neural Networks," *2019 International Conference on Innovative Trends in Computer Engineering (ITCE)*, Aswan, Egypt, 2019, pp. 270–274, doi: 10.1109/ITCE.2019.8646616.
8. P. Verma, K. Sharma and A. Sharma, "I-Slot Loaded Microstrip Antenna for C-Band and X-Band Applications," *International Conference for Convergence for Technology-2014*, Pune, 2014, pp. 1–4, doi: 10.1109/I2CT.2014.7092276.

9. K. Güney, S. Sagiroglu and M. Erler, "Generalized Neural Method to Determine Resonant Frequencies of Various Microstrip Antennas," *International Journal of RF and Microwave Computer-Aided Engineering*, 2001, 12(1), pp. 131–139.
10. S. Haykin, *Redes Neurais: Princípios e prática.* Porto Alegre: Bookman, 2nd Edition, 2001.
11. D. Karaboga, K. Guney, S. Sagiroglu, M.Erler, "Neural Computation of Resonant Frequency of Electrically Thin and Thick Rectangular Microstrip Antennas," *IEEE Proceedings on Microwave Antennas and Propagation*, 1999, 126(2), pp. 155–159.
12. S. Raj, P. Tripathi, N. Kishore, S. S. Tripathi and V. S. Tripathi, "A Novel Antenna Design for Non-Invasive Blood Glucose Measurement and its Sensitivity Optimization using ANN," *2020 International Conference on Electrical and Electronics Engineering (ICE3)*, Gorakhpur, India, 2020, pp. 355–358, doi: 10.1109/ICE348803.2020.9122876.
13. Thakare, V. V., Singhal, P., & Das, K., "Calculation of Microstrip Antenna Bandwidth Using Artificial Neural Network," *2008 IEEE International RF and Microwave Conference*, 2008
14. A. Gehani, J. Ghadiya and D. Pujara, "Design of a Circularly Polarized Elliptical Patch Antenna Using Artificial Neural Networks and Adaptive Neuro-Fuzzy Inference System," *2013 Proceedings of the International Symposium on Antennas & Propagation*, Nanjing, 2013, pp. 388–390.
15. L. Xiao, W. Shao, T. Liang and B. Wang, "Artificial Neural Network With Data Mining Techniques for Antenna Design," *2017 IEEE International Symposium on Antennas and Propagation & USNC/URSI National Radio Science Meeting*, San Diego, CA, 2017, pp. 159–160.
16. R. Addaci and R. Staraj, "Artificial Neural Networks Modeling Technique for Fast Analysis and Design of Multi-Antenna Systems," *2013 7th European Conference on Antennas and Propagation (EuCAP), Gothenburg*, 2013, pp. 2034–2037.
17. Aliakbari, Hanieh, Abdipour, Abdolali, Costanzo, Alessandra, Masotti, Diego, Mirzavand, Rashid, Mousavi, Pedram, "ANN-Based Design of a Versatile Millimetre-Wave Slotted Patch Multi-Antenna Configuration for 5G Scenarios," *IET Microwaves, Antennas & Propagation*, 2017, 11(9), pp. 1288–1295
18. F. N. dos Santos, S. S. Nascimento, V. F. Rodriguez-Esquerre and F. G. S. Filho, "Analysis and Design of Microstrip Antennas by Artificial Neural Networks," *2011 SBMO/IEEE MTT-S International Microwave and Optoelectronics Conference (IMOC 2011)*, Natal, 2011, pp. 23–226.
19. Yogesh and A. K. Dubey, "Fruit Defect Detection Based on Speeded Up Robust Feature Technique," *2016 5th International Conference on Reliability, Infocom Technologies and Optimization (Trends and Future Directions) (ICRITO)*, Noida, 2016, pp. 590–594, doi: 10.1109/ICRITO.2016.7785023.
20. T. Makkar, Y. Kumar, A. K. Dubey, A. Rocha and A. Goyal, "Analogizing Time Complexity of KNN and CNN in Recognizing Handwritten Digits," *2017 Fourth International Conference on Image Information Processing (ICIIP)*, Shimla, 2017, pp. 1–6, doi: 10.1109/ICIIP.2017.8313707.

18 Microstrip Antenna in IoT
From Basic to Applications

Arun Kumar Singh[1] *and Vikas Pandey*[2]
[1]Saudi Electronic University, Saudi Arabia-KSA
[2]Babu Banarasi Das University, Lucknow, India

18.1 INTRODUCTION

For short range communication general use of buzzword is IoT. Most of the IoT applications need narrow bandwidth. Usually bandwidth range from 10 MHz to 100 MHz. Internet of Things (IoT) is technology where devices get a particular identity to be able to connect and communicate with one another through internet networks without human-to-human or human-to-machine interactions. A collection of IoT devices connected by a network is called IoT infrastructure. It is grouped into four layers, i.e., sensors and actuators, internet gateway and data acquisition system, edge handler, and data center. Sensors and actuators are used to collect data from the environment or physically observed objects. Units of sensors and actuator are what we called nodes. Each node can communicate with each other using specific protocols to produce this useful set of data. Analog data from sensors and actuators are converted into a digital form by data acquisition devices, which is then forwarded by the internet gateway to Edge handler layer. Edge handler function is to prevent data from the edge to consume data center bandwidth. It also can process raw data into data that is ready to be processed. The last layer is Data Centre and Cloud, and at this layer, data is processed and analyzed in-depth for later use by its users. Shortrange, low power, omnidirectional, and seamless connectivity for short time duration are some of the general specifications of Internet of Things. As there is exponential growth in demand for Internet of Things, the spectrum regulatory authorities world-wide have been releasing new frequency bands at various spectrum range from sub GHz to several GHz. The antenna design for IOT applications should fit in the range of spectrum released for the purpose. This paper makes an attempt to describe an antenna for IOT needs in the allocated frequency bands 830 to 840 MHz, 850 to 890 MHz, and 1190 to 1200 MHz range.

In the 21st century graphene has been termed as one of the significant material in wireless terahertz communications due to amazing electrical and optical properties. According to Edholm's law, wireless data rates tend to be doubled in eighteen months and frequently approaching to the wired communication system. Higher

DOI: 10.1201/9781003187325-18

channel capacity 100 Gbps or more required for the next generation communication system demands the exploitation of higher frequency band for data transfer. To meet high speed data transmission, one of the possible solutions is terahertz (THz) frequency band which alleviates the spectrum scarcity of current wireless system. THz frequency band lies between 0.1 THz and 10 THz. Although the frequency band below and above these bands have extensively investigated but THz band is one of the least explored frequency bands for future wireless communication. High data rates, less attenuation compared to optical signal, and higher bandwidth may be achieved by THz frequency band. This frequency band has different applications like medical imaging, defence and security based technology, ultra-first spectroscopy of materials and Internet of Thing (IoT) devices may also be benefitted from THz spectrum. In addition, high performance antenna is required to support THz band wireless communication [1].

Microstrip patch antenna introduced in the 1950s may be one possible solution to support THz communication due to low cost, low profile, planner configuration, easy to fabricate and feeding, superior probability, and easy to integrate with antenna elements like Monolithic Microwave Integrated Circuits (MMICs). Microstrip patch antenna has variety of applications, i.e., mobile communication, personal wireless communications, radar, Radio Frequency Identification (RFID), Surveillance systems, aerospace telecommunications, weapons and missile, Global Positioning System (GPS), and many others. Graphene is an allotrope of carbon packed into a two dimensional (2D) honeycomb and hexagonal lattice structure are widely used in Nano-photonics, Nano-electronics and THz wireless communication. It has better electromagnetic conductivity, chemical, mechanical, and optical properties. At room temperature, graphene carrier mobility is very high and may vary from 8000–200000 cm^2 V^{-1} s^{-1}. The carrier interband transition and carrier intraband transition are the two parts of graphene conductivity for smaller size. The total conductivity of graphene is the summation of intraband transition and interband transition which can be calculated from the Kubo formula, where e is the electron charge, T is temperature, ω is radian frequency, $\hbar$ is the reduced Planck's constant. μ_c is the chemical potential, Γ is a phenomenological scattering rate that is assumed to be independent of energy ε.

Summation of intraband and interband transition: calculated from the Kubo formula

Total Transition

$$\sigma_{\text{Total}} = \sigma_{\text{Inter}} + \sigma_{\text{Intra}}$$

The contribution of intraband

$$\sigma_{intra}(\omega) = \frac{2e^{2k_BT_i}}{\pi\hbar^2(\omega + i\tau^{-1})} In\left[2\cos h\left(\frac{\mu_c}{2k_BT}\right)\right]$$

The contribution by interband

$$\sigma_{intra}(\omega) = \frac{e^2}{4\hbar}[H\left(\frac{\omega}{2}\right) + i\frac{4\omega}{\pi}\int_0^\infty \frac{H(\varepsilon) - H\left(\frac{\omega}{2}\right)}{\omega^2 - 4\varepsilon^2}\partial\varepsilon$$

$$H(\varepsilon) = \frac{\sin h\left(\frac{\hbar\varepsilon}{k_BT}\right)}{\cos h\left(\frac{\mu_c}{k_BT}\right) + \cos h\left(\frac{\hbar\varepsilon}{k_BT}\right)}$$

Different researchers have been studying an antenna that works in terahertz frequency for wireless communication for various application. Intrapezoidal microstrip patch antenna for THz wireless application has been presented with

Photonic Band Gap (PBG) based substrate. They have shown that PBG substrate improves the performance of the antenna, like return loss, gain, and bandwidth. Graphene based patch antenna on polyimide substrate in the frequency range 0.725–0.775 THz has been analyzed and investigated. In hexagonal slotted antenna with microstrip feedline has been described in terms of VSWR, input impedance, realized gain and radiation properties for 0 THz–12 THz frequency using polyimide substrate. Graphene based dipole antenna with tunable resonant frequency has been presented on. An elliptical microstrip patch antenna with polyimide substrate for THz wireless application reported on. Their proposed antenna result has been described in terms of directivity, peak realized gain, radiation efficiency, and VSWR. Rectangular microstrip patch antenna for 1 THz resonant frequency has been proposed in for wireless communication [2–5].

In contrast to the current paradigm on the Internet, which bases on human-to-human relations, Gutiérrez mention IoT to have a paradigm as the future internet, where every physical or virtual object that can be identified with unique identifiers will be considered to be interconnected. So, keeping this in mind, although IoT uses distributed networks in nature, IoT has driven combinations with other technologies, such as short-range communication, real-time localization, embedded sensors, and ad-hoc networks, as a way to turn everyday things into smart things. Combining IoT with an ad-hoc network provides benefits because of the ad hoc properties as self-organized networks, they are built spontaneously by several connected devices [6,7]. on a router or base station, so they are suitable for implementation where the deployment of new fixed infrastructure is not feasible.

18.2 WHAT IS AN IOT SOLUTION AND PLATFORM

An IoT solution is more than an embedded system that has a connectivity feature. An IoT solution is a set of devices and sensors that are connected to a cloud platform through a gateway. The cloud platform provides the infrastructure that is necessary to manage, store, secure, and analyze a large amount of data to extract valuable information and insights from it. IoT is the result of the convergence of several technologies, including wireless communications, micro-electrical systems, and the internet. A "thing" is any object with embedded electronics that can transfer data over a network without any human interaction. Some examples are wearable devices, environmental sensors, machinery in factories, devices in homes and buildings, or components in a vehicle. IoT can make life easier for all. For example, smart homes and connected devices in your home can make your life easier. You can turn off lights, lock doors, turn on or off heat or air-conditioners, and do other actions through voice commands or mobile apps. The truly transformative use of IoT is to combine structured and unstructured data with cognitive analytics. IoT in the era of AI technologies makes it possible for you to make sense of the vast amounts of IoT data to understand your data better and deeper. Infusing intelligence into systems and processes can help you increase efficiency, improve customer satisfaction, uncover new business opportunities, and mitigate risks and threats proactively [8–10].

An IoT platform provides a set of ready-to-use features that greatly speed up the development of applications for connected devices and provide scalability and cross-device compatibility. An IoT platform is an integrated service that provides the capabilities that you need to bring physical objects online. IoT platforms originated as IoT middleware, which functions as a mediator between the hardware and application layers. The primary tasks of this middleware include data collection from devices over different protocols and network topologies, remote device configuration and control, device management, and over-the-air firmware updates [11].

18.3 DIFFERENT ANTENNA NEEDS FOR LOW BANDWIDTH APPLICATIONS

IoT is disparate tremendously high throughput wireless networks such as 5G, high-efficiency wireless (HEW also known as IEEE 802.11ax), and WiGig that enhancement speeds on the direction of 10 Gbps through exploitation vast continuous spectrum space in the millimeter-wave (mmWave) bands or through the use of high order advanced modulation schemes such as carrier accumulation, 64-QAM OFDM and multi-user multiple-input and multiple-output (MU-MIMO). Networks such as these will often use highly classy antenna assemblies like Active Electronically Scanning Antennas (AESAs) such as phased array antennas or switched beam arrays in Massive MIMO and microcell installations [12]. IoT networks and devices in its place influence the licensed and unlicensed sub-6 GHz bands using minute amounts of bandwidth (<5 MHz) whereas LTE-A uses up to 20 MHz with carrier aggregation and beyond that with mmWave spectrum utilization for 5G. Instead of leveraging beamforming algorithms with AESAs, IoT networks use star, mesh, or point-to-point topologies to smartly guide uplinks and downlinks between gateways and end-devices. Oftentimes, this means the job of establishing a link is accomplished through the use of relatively simple omni-directional antenna structures such as chip, PCB, whip, rubber duck, patch, and wire antennas. Many IoT-based development kits and radio modules such as Qualcomm's Internet-of-Everything (IoE) development platform or the Arduino GSM come with GPS, Bluetooth, and WiFi antennas [13].

18.4 ANTENNA DESIGN AND CONFIGURATION

The geometry of the proposed graphene based Microstrip patch antenna with PBG substrate at 1.10 THz resonant frequency has been shown in Figure 18.1. The antenna has been simulated using Silicon substrate with dielectric constant/relative permittivity 11.9 having thickness of 30 μm with the antenna dimension of 100 × 100 μm^2. Photonic crystal has many applications in microwave circuit, optical communication, antenna, and so forth. The Photonic Band Gap (PBG) structure is created on the substrate by drilling periodic circular cylinder. Due to non-transmission of PBG structure, it reduces the substrate absorption compared to the conventional patch antenna. Physical mechanism of photonic band gap suppresses the surface waves propagating along the surface of the substrate and reflects most of electromagnetic wave energy radiating to the substrate significantly. The dimension of the proposed antenna substrate is 100 μm × 100 μm × 30 μm.

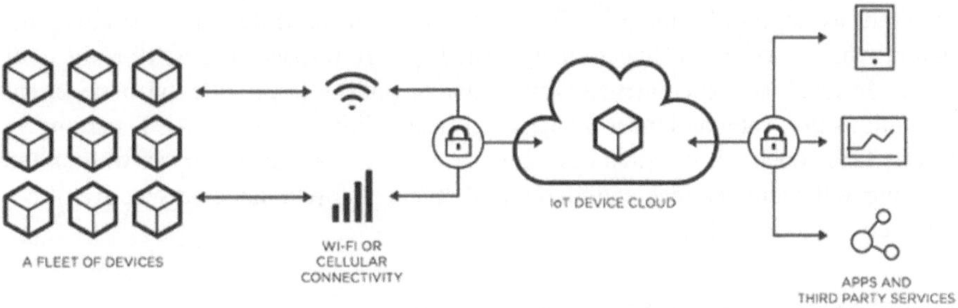

FIGURE 18.1 IoT Platform.

The radiating patch of 50 μm × 50 μm is etched in the top surface of Silicon substrate which is excited by microstrip feed line. The width of the microstrip feedline is 8 μm and has been used to feed the power to the proposed antenna having input impedance 50 Ω. All the antenna dimension for the proposed antenna at 1.10 THz center frequency have been tabulated on Table 18.1[14,15].

18.5 ANTENNAS PARAMETERS

There are usually two essential parameters for antennas: gain and directivity. Directivity expresses the attentiveness of a beam radiation in a specific direction. Omni-directional antennas are, therefore, somewhat evenly concentrated in all three dimensions, while a directional antenna displays narrower radiation patterns [16,17]. This is frequently proficient by combining multiple radiating elements. Gain—a description most often seen on datasheets—is a measurement of the total power radiated from an antenna. IoT networks that exploit the unlicensed ISM bands must be within the Equivalent (or Effective) Isotropic Radiated Power (EIRP)

TABLE 18.1
Geometric Parameters of the Graphene-Based Antenna

Parameter	Dimensions (μm)
Length of the patch (Lp)	50
Width of the patch (Wp)	50
Length of the substrate (Ls)	100
Width of the substrate (Ws)	100
Length of the ground (Lg)	100
Width of the ground (Wg)	100
Height of the substrate (Hs)	30
Microstrip feed width (Wf)	8

required, as definite by the FCC in order to be a licensed digital transmitter, mitigating the risk of any nosiness on congested spectrum space. The EIRP is the total power an ideal isotropic antenna would have to put out to provide the same signal strength as the Antenna Under Test (AUT) in the direction of the AUT's strongest beam. This constraint considers the gain, transmitter output power, as well as the loss due to the antenna feed and therefore delivers a more holistic perspective of the transceiver module [18].

18.6 ANTENNAS USED IN IOT DEVICES

The characteristic antenna leveraged in IoT end-devices are wire, whip, rubber duck, paddle, chip, and PCB. IoT devices with embedded antennas like chip and PCB have the benefit of fitting into small spaces, shrinking a sensor node's dimensions. Composed of conductive traces, PCB antennas often exhibit higher gains than their chip-based counterparts. There are various antenna topologies for PCB antennas including inverted-F, L, and folded monopole. The ground plane is of particular importance in the generation of PCB antennas as a smaller ground plane can constrain design significantly with a much narrower functional bandwidth, lessened antenna radiation efficiency, and modified radiation pattern. As with any antenna, the radiative element's volume is directly proportional to its gain and (oftentimes) embedded PCB antennas take up lots of space on the board. Nonetheless, sensor nodes implementing a PCB antenna have the characteristic of maintaining a relatively flat shape that could allow for them to be more readily encased and mounted in any environment [19].

There are variants with dual/multi-band antennas and flexible antenna structures for applications that require the antenna be routed along the body of the device (e.g., UAV). The frequency of the given IoT application must fall into the antenna's bandwidth. Table 18.2 depicts some common IoT applications and their corresponding wireless networking technologies, along with the frequency bands in which they function. While utmost of these applications purpose in the unlicensed ISM bands, Medical Body Area Networks (MBAN), and Wireless Avionics Intra-Communications both have dedicated spectra [20].

It is not a secret that the Internet of Things (IoT) is geared to boom in the coming decades. With the number of short-range IoT gadgets projected to surpass mobile devices, IoT will fill industrial, commercial, and consumer applications. The landscape of IoT end devices, gateways, modems, and base stations is fraught with many standards organizations with respective protocols attempting to alleviate the limitations that come with custom proprietary hardware and a lack of vendor neutrality. These consortiums also attempt to optimize the hardware and software specifications for a particular application, so that (while there is no one-size-fits-all IoT platform) there is likely a standard physical (PHY) and medium access control (MAC) layer that would best fit an application of choice. For instance, the Wireless HART standard was developed by the HART communication foundation with low latency, high reliability, decent battery life (3–5 years), and medium throughput (~150 Mbps) transmissions for the Industrial IoT (IIoT). The up-and-coming Low Power Wide Area Network (LPWAN) architectures allow for ultra-high link

TABLE 18.2
IoT Applications and Technologies

Applications	IoT Technologies	Frequency
General-Smart Office, Smart Apartment, Smart Home, etc.	Zigbee	915 MHz, 2.4 GHz
	GPS	1575.42 MHz, 1227.6 MHz, 1176.45 MHz
	Z-Wave	2.4 GHz
	Bluetooth	2.4 GHz
	Wi-Fi	2.4 GHz, 3.6 GHz, 5 GHz, 5.9 GHz
Low Power WAN-Smart Farming, Smart City, etc.	Sigfox	868 MHz, 902 MHz
	LoRa	433 MHz for ASIA, 868 MHz for Europe, and 915 MHz for the USA
Medical Use	IEEE802.15.6-MBAN	2360 MHz–2400 MHz
	IEEE802.15.6-WBAN	2.4 GHz, 400 MHz, 800 MHz, 900 MHz
Avionics Use	WAIC	4200 MHz–4400 MHz
Industrial IoT	Wireless HART, ISA100.11a	2.4 GHz

distances (<1 km), extremely high battery lifetimes (~10 years), and low throughput on the order of bit per second (bps) transmissions for agricultural, industrial, medical, and smart city applications [19,21,22].

Along with this diverse ecosystem of IoT alliances come a myriad of end-devices utilizing all sorts of sensors/actuators, MEMS devices, batteries/cells, energy harvest techniques, radio modules, and finally, antennas. In any wireless application the choice and design of the antenna varies depending on the space available, transmission strength, and frequency range. Still, there are significant differences between the various wireless network topologies out there. In advanced wireless communication, Terahertz (THz) band spectrum plays an important role in ultra-wideband and high speed secured data transmission. A high performance based antenna has been designed for THz application using CST microwave studio simulation tool. The performance of proposed antenna is discussed in details with respect to some important parameter. The result of this designed antenna has been described in return loss (s11), antenna gain, directivity, VSWR, E-plane, H-Plane, and radiation efficiency. The return loss (s11) of the proposed graphene based microstrip patch antenna with PBG structure [23–26].

18.6 CONCLUSION

Internet of Things is described as an infinitude of embedded devices and small sensors, integrated in a wide network with a permanent access to the user. One of the major application of IoT is in the Smart home/office concept, allowing more convenience, efficiency (with at additional aspects), and security. Nowadays, with numbers of devices, it is mandatory for these devices to be small, low-power, and at the same time have more capabilities and efficiencies. IoT is disparate tremendously high throughput wireless networks such as 5G, high-efficiency wireless (HEW also known as IEEE 802.11ax), and WiGig that enhancement speeds on the direction of 10 Gbps through exploitation vast continuous spectrum space in the millimeter-wave (mmWave) bands or through the use of high order advanced modulation schemes such as carrier accumulation, 64-QAM OFDM, and multi-user multiple-input and multiple-output (MU-MIMO). This antenna has reduced dimensions, ideal to be integrated in most of IoT sensors. For future wireless communication, low cost, low profile, minimal weight, and high performance antenna is required to support high speed data transmission. This paper presents a graphene based wideband microstrip patch antenna at 1.10 THz resonate frequency. Graphene has been used in the proposed antenna for higher electrical conductivity, mobility, and saturation severity in THz band regime. Microstrip patch antenna has a variety of applications, i.e., mobile communication, personal wireless communications, radar, Radio Frequency Identification (RFID), surveillance systems, aerospace tele-communications, weapons and missile, Global Positioning System (GPS), and many others. Graphene is an allotrope of carbon packed into a two dimensional (2D) honeycomb and hexagonal lattice structure are widely used in Nano-photonics, Nano-electronics, and THz wireless communication. It has better electromagnetic conductivity, chemical, mechanical, and optical properties.

REFERENCES

1. T. Varum, M. Duarte, J. N. Matos and P. Pinho (2018). Microstrip Antenna for IoT/ WLAN Applications in Smart Homes at 17 GHz. In *12th European Conference on Antennas and Propagation (EuCAP 2018)*, London, pp. 1–4, doi: 10.1049/cp.2018.04 75.
2. Dhasarathan, V., Bilakhiya, N., Parmar, J., Ladumor, M., & Patel, S. K. (2020). Numerical Investigation of Graphene-Based Metamaterial Microstrip Radiating Structure. Materials Research Express, 7(1), 016203.
3. Hanson, G. W. (2008). Dyadic Green's Functions for an Anisotropic, Non-Local Model of Biased Graphene. IEEE Transactions on Antennas and Propagation, 56(3), 747–757.
4. Singh, A., & Singh, S. (2015). A Trapezoidal Microstrip Patch Antenna on Photonic Crystal Substrate for High Speed Thz Applications. Photonics and Nanostructures-Fundamentals and Applications, 14, 52–62.
5. Anand, S., Kumar, D. S., Wu, R. J., & Chavali, M. (2014). Graphene Nanoribbon Based Terahertz Antenna on Polyimide Substrate. Optik, 125(19), 5546–5549.
6. Reina, D. G., Toral, S. L., Barrero, F., Bessis, N., Asimakopoulou, E. (2013). The Role of Ad Hoc Networks in the Internet of Things: A Case Scenario for Smart Environments. In: Bessis N, Xhafa F, Varvarigou D, Hill R, Li M, editors. Internet of Things and Inter-cooperative Computational Technologies for Collective Intelligence. Berlin, Heidelberg: Springer, [cited 2019 Aug 17]. p. 89–113. (Studies in Computational Intelligence).

7. Lu Tan, Neng Wang (2010). Future internet: The Internet of Things. In: *2010 3rd International Conference on Advanced Computer Theory and Engineering (ICACTE)*. 2010. p. V5–376.
8. . IoT Architectures, Models, and Platforms for Smart City Applications, Copyright: © 2020, DOI: 10.4018/978-1-7998-1253-1.ch009
9. https://hackernoon.com/how-to-choose-the-right-iot-platform-the-ultimate-checklist-47b5575d4e20
10. https://www.5gtechnologyworld.com/specifying-antennas-for-various-iot-applications/
11. Khan, M. A. K., Shaem, T. A., Alim, M. A. (2020). Graphene Patch Antennas with Different Substrate Shapes and Materials. Optik, 202, 163700.
12. Cherry, S. (2004). Edholm's Law of Bandwidth. IEEE Spectrum, 41(7), 58–60.
13. Koenig, S., Lopez-Diaz, D., Antes, J., Boes, F., Henneberger, R., Leuther, A.,... & Zwick, T. (2013). Wireless Sub-Thz Communication System with High Data Rate. Nature Photonics, 7(12), 977.
14. Bansal, G., Marwaha, A., & Singh, A. (2020). A Graphene-Based Multiband Antipodal Vivaldi Nanoantenna for UWB Applications. Journal of Computational Electronics, 19, 709–718. https://doi.org/10.1007/s10825-020-01460-2
15. Akyildiz, I. F., Jornet, J. M., & Han, C. (2014). Terahertz Band: Next Frontier for Wireless Communications. Physical Communication, 12, 16–32.
16. Ji, Y. B., Oh, S. J., Kang, S. G., Heo, J., Kim, S. H., Choi, Y.,... &Haam, S. J. (2016). Terahertz Reflectometry Imaging For Low and High Grade Gliomas. Scientific Reports, 6(1), 1–9.
17. Corsi, C., & Sizov, F. (Eds.). (2014). *Thz and Security Applications: Detectors, Sources and Associated Electronics for Thz Applications*. New York: Springer. McIntosh, A. I., Yang, B., Goldup, S. M., Watkinson, M., &Donnan, R. S. (2012). Terahertz Spectroscopy: A Powerful New Tool for The Chemical Sciences?. Chemical Society Reviews, 41(6), 2072-2082.
18. Nagatsuma, T., Ducournau, G., & Renaud, C. C. (2016). Advances in Terahertz CommunicationsAccelerated by Photonics. Nature Photonics, 10(6), 371.
19. Prabu, R. T., Benisha, M., Bai, V. T., & Ranjeetha, R. (2019). Design of 5G mm-Wave Antenna Using Line Feed and Corporate Feed Techniques. In *Smart Intelligent Computing and Applications* (pp. 367–380). Singapore: Springer.
20. Kumar, P., Thakur, N., &Sanghi, A. (2013). Micro Strip Patch Antenna for 2.4 GHZ WirelessApplications. International Journal of Engineering Trends and Technology (IJETT), 4(8).
21. Khraisat, Y. S. (2012). Design of 4 Elements Rectangular Microstrip Patch Antenna with High Gain for2.4 Ghz Applications. Modern Applied Science, 6(1), 68.
22. Wang, C., Yao, Y., Yu, J., & Chen, X. (2019). 3d Beam Reconfigurable THz Antenna with GrapheneBased High-Impedance Surface. Electronics, 8(11), 1291.
23. Singhal, S., &Budania, J. (2019). Hexagonal Fractal Antenna for Super Wideband Terahertz Applications. Optik, 206, 163615.
24. Tripathi, S. K., Kumar, M., & Kumar, A. (2019). Graphene Based Tunable and Wideband TerahertzAntenna for Wireless Network Communication. Wireless Networks, 25(7), 4371–4381.
25. Singhal, S. (2019). Ultrawideband Elliptical Microstrip Antenna for Terahertz Applications. Microwaveand Optical Technology Letters, 61(10), 2366–2373.
26. Nickpay, M. R., Danaie, M., & Shahzadi, A. (2019). Wideband Rectangular Double-Ring NanoribbonGraphene-Based Antenna for Terahertz Communications. IETE Journal of Research, 1–10.

19 UWB-MIMO Antenna with Band-Notched Characteristic

Chandrasekhar Rao Jetti[1] *and*
Venkateswara Rao Nandanavanam[2]
[1]Associate Professor, Department of ECE, Bapatla Engineering College, Bapatla, Andhra Pradesh, India
[2]Professor, Department of ECE, Bapatla Engineering College, Bapatla, Andhra Pradesh, India

19.1 INTRODUCTION

Higher data rates and improved quality of service are the primary concerns of future wireless communication systems like 4G and 5G. Since the Federal Communications Commission (FCC) allocated the unlicensed frequency spectrum from 3.1–10.6 GHz for commercial applications in 2002 [1], ultra-wideband (UWB) technology has attained considerable attention because of its inherent features like high data rate communications, extremely less power consumption, and low cost. However, multipath fading and frequency interference with other communication systems are the important problems that should be well solved for UWB systems. In an indoor communication application, like other wireless communication systems, the UWB system performance is also restricted by multipath fading due to rich scattering environments which cause inter-symbol interference. In present times, digital communication using multiple input multiple output (MIMO) technology has emerged as a breakthrough for a wireless system. The MIMO system employs multiple antennas at the transmitter and receiver. It makes use of the rich multipath environment to mitigate the multipath fading effect. And it improves the range of communication and system capacity (data rate) without the need for additional bandwidth or transmitted signal power [2,3]. Hence, the UWB system with MIMO technology is a viable solution to reduce the multipath fading effect and to improve the quality of service, the range of communication, and system capacity [4].

The electromagnetic interaction between the radiating (antenna) elements in multiple antenna or MIMO system is known as mutual coupling. The closely spaced antennas, especially in portable devices, inevitably cause strong mutual coupling between antennas. The mutual coupling is undesirable which causes fluctuations in the input impedance of individual antenna element, i.e., impedance mismatch which

DOI: 10.1201/9781003187325-19

degrades the radiation efficiency, deviations in antenna radiation pattern due to the high correlation between antenna signals, and decreases the channel capacity of MIMO systems. Since the mutual coupling has a considerable impact on the MIMO system performance, the reduction of mutual coupling between antennas and enhancement of isolation between ports is imperative. However, placing multiple antennas in a space-limited portable wireless device is a big challenge for antenna designers [5]. Hence, designing compact UWB-MIMO antenna exhibiting band-notch function and less mutual coupling is very much needed. Various designs were proposed in the recent years to suppress the effects of mutual coupling in UWB MIMO antennas [6–15]. Methods include placing radiating elements perpendicular to each other and adding two long protruding stubs to ground [6], use of tree-like structure on the ground plane [7], etching a T-shaped slot and a line slot on the ground [8], adding a Y-shaped slot on the T-shaped protruded ground plane [9], placing two shorts at 45 degrees between the microstrip lines and in the opposite direction [10], protruding ground structure [11], T-shaped metallic stub [12,13], adopting wideband neutralization line [14], and using modified ground structure along with T-shaped slot on the ground [15].

Ultra-wideband is an emerging technology for short distance low power communications. It makes use of short duration pulses which have very low power spectral density for transmission of data. Since the UWB system is operating from 3.1 to 10.6 GHz, it could easily interfere with existing narrowband communication systems such as Wireless Local Area Network (WLAN-5.15–5.825 GHz). So, UWB antenna with integrated frequency notching function at the interfering frequency band is a feasible solution to mitigate the frequency interference [16].

The frequency interference produced by a UWB transmitter to a narrowband system is very negligible because of the transmitted signal emission power (power spectral density) is very less compared with narrowband systems. But, when a UWB receiver is located near to the narrowband interferer, the interference caused is high. So, a notch at the interfering frequency is needed to suppress its effect. The traditional RF filter circuits using lumped elements can be used to implement this frequency notching feature but it increases the system complexity, cost, and occupies more space when integrated with other microwave circuits in the portable device. Another viable solution is to design a UWB antenna with an integrated band-notched feature to mitigate the frequency interference which decreases the complexity and cost of the UWB system. The idea of designing the UWB antenna with band-notched characteristic is given in [17,18].

The band notched UWB antenna does not interfere with existing communication systems which are operating at notch frequency. Hence, the design of UWB antenna with band-notched function is needed. The band-notch characteristics for UWB systems can be obtained by etching slots or split ring resonators (SRR) on the radiating element or ground plane or feed line. This can also be done by adding strips or stubs adjacent to the radiator or ground or feed line. Each slot or strip can function as a resonator and the slot or strip length controls the notch center frequency. However, the desired notch band is obtained by proper tuning of length and width of the slot or strip. The width of the slot or strip has a negligible effect on the notch-band position, but it has a significant impact on the notch bandwidth. The total length of etched slots or added strips should be λ/2 or λ/4 corresponding to the notched-band center frequency as given in equations (19.1) and (19.2) [19], where λ is the guided wavelength.

$$L_{Notch} = \frac{c}{2f_{Notch}\sqrt{\varepsilon_{eff}}}, \quad (19.1)$$

or

$$L_{Notch} = \frac{c}{4f_{Notch}\sqrt{\varepsilon_{eff}}}, \quad (19.2)$$

$$\lambda = \frac{c}{f_{Notch}\sqrt{\varepsilon_{eff}}}, \quad (19.3)$$

$$\varepsilon_{eff} = \frac{(1 + \varepsilon_r)}{2}, \quad (19.4)$$

where c denotes light speed, f_{Notch} represents notch center frequency, L_{Notch} is the total length of slot or strip, ε_{eff} indicates effective dielectric constant, and ε_r denotes dielectric constant.

Several investigations were reported earlier to create band notch function at WLAN band for UWB systems in [19–30]. Methods include inserting λ/4 and λ/2 slot resonators on the ground plane [19], using a pair of ground stubs locating along the edge of the ground plane [20], inserting open stub in the printed folded monopole [21], etching folded U-shaped slots in the feed line of the antenna [22,23], incorporating SRR slots on radiating element [24], quarter-wave stub connected to the ground [25], adding protruding two rectangular stubs on the ground plane [26], with a slot of length 1.0 λ in the radiator [27], open-ended quarter-wavelength L-shaped slots were etched on the rectangular radiating patches [28], using C-shaped and Z-shaped slot resonators on the ground [29], and employing elliptical SRR on the radiating element [30].

The antenna designs presented in the above literature exhibit acceptable isolation and notching characteristics, but some designs were not compact enough and few are a bit complex. So, the design of simple and compact band notched UWB MIMO antenna with low mutual coupling is needed. In this chapter, we have presented a compact isolation-enhanced planar UWB-MIMO antenna with single band-notched characteristics[31]. Ansoft HFSS v.13 is used to carry out the proposed antenna design, optimization, and simulations. For validating the simulation results, all the proposed antenna has been fabricated and tested using the Agilent N5224A PNA, Anritsu MS2037C vector network analyzer, and an anechoic chamber. In Section 19.2, UWB-MIMO antenna with single band-notched characteristic is discussed. Finally, conclusions of the work are given in Section 19.3.

19.2 Single Band-Notched UWB-MIMO Antenna

Compact ultra-wideband MIMO antenna displaying band-notch features at WLAN band (5 to 5.9 GHz) for portable wireless applications is discussed in this section [31]. The subsequent sections discuss the complete description of the proposed antenna.

19.2.1 ANTENNA DESIGN

The proposed UWB-MIMO antenna geometry and photo of the fabricated antenna are displayed in Figure 19.1(a) and 19.1(b), respectively. The antenna is produced on an FR4-epoxy substrate material having relative permittivity (ε_r) of 4.4, a height of 0.8 mm, and a loss tangent of 0.02. The total dimension of the proposed design is $L \times W \times h$ mm^3. The proposed design contains two similar rectangular monopole antenna elements, designated as PM1 and PM2 with dimensions $L_R \times W_R$. Both PM1 and PM2 elements are excited with 50-Ω coplanar waveguide with sizes $F_{L1} \times W_F$. And the shared ground is designed by assembling $L_G \times W_G$ and $L_G \times L$ defected ground parts. The elements PM1 and PM2 are placed perpendicularly to each other to decrease the mutual coupling among the elements and to enhance the isolation between the ports. A long rectangular strip with dimensions $S_L \times S_W$ is protruded on the ground plane between the elements to further improve the isolation and to expand the bandwidth of the antenna. To generate a band-notch characteristic at 5–5.9 GHz, an inverted U-slot resonator is located on the feed line.

The optimized antenna sizes are specified as follows: (unit: mm): $D_1 = 5.1$, $D_2 = 6.1$, $D_3 = 11.2$, $F_{L1} = 9.5$, $F_{L2} = 1.5$, $F_{L3} = 0.4$, $L = 26$, $L_G = 8$, $L_R = 10$, $S_L = 18$, $S_W = 1$, $W = 40$, $W_F = 1.8$, $W_G = 3.2$, $W_R = 11$, $U_1 = 7.8$, $U_2 = 0.4$, and $U_W = 0.3$.

And Figure 19.2(a) and 19.2(b) shows the simulated S-parameters such as S_{11} and S_{21} of the Antenna 1 (UWB-MIMO antenna without ground strip), Antenna 2 (UWB-MIMO with a ground strip), and the proposed antenna. It can be observed that the proposed UWB MIMO antenna is operating from 2.2 to 11.4 GHz with

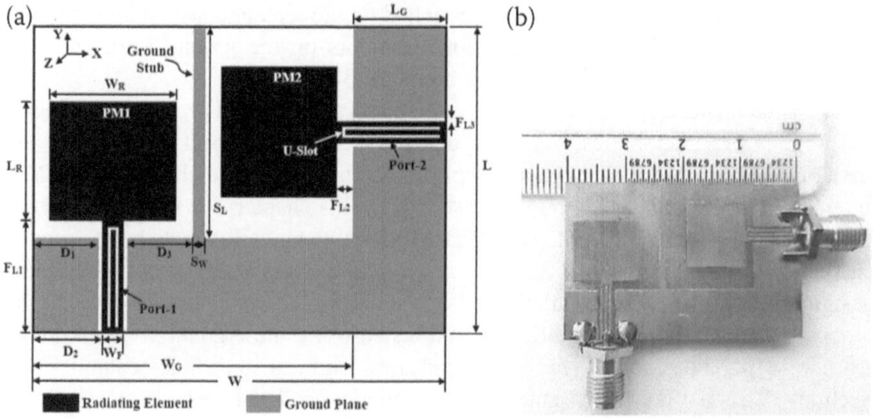

FIGURE 19.1 (a) Geometry of the proposed antenna and (b) fabricated antenna [31].

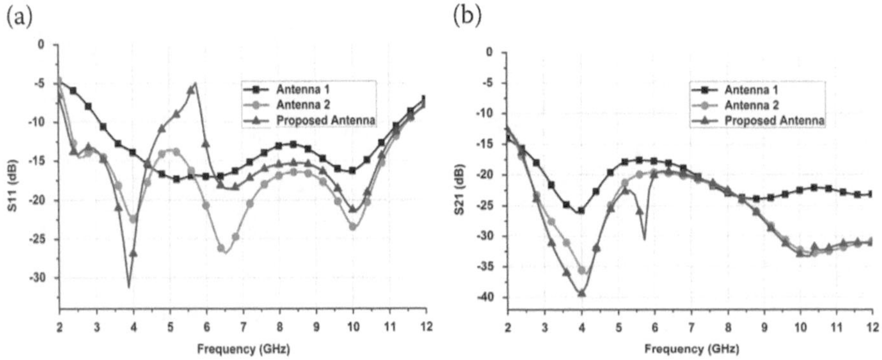

FIGURE 19.2 (a) Simulated S_{11} parameter and (b) simulated S_{21} parameter [31].

good impedance bandwidth except at notch band from 5 to 5.9 GHz. Also, the mutual coupling of less than −20 dB is obtained over the entire UWB band.

19.2.2 STUDY OF MIMO ANTENNA

Since the ground and antenna elements are smaller in sizes, the movement of surface currents and near-field radiation causes insufficient impedance matching and low isolation, which limits the MIMO antenna performance. Figure 19.3(a) and 19.3(b) presents the UWB MIMO antenna without and with ground strip. And the ground strip effects on the bandwidth and isolation are shown in Figure 19.4(a), 19.4(b), and 19.4(c), respectively.

With ground strip among the PM1 and PM2 (Antenna 2), the first resonance is produced at 2.5 GHz having a lower cutoff frequency of 2.3 GHz and offers bandwidth from 2.3 to 11.4 GHz. And the isolation of more than 20 dB is found in the whole UWB band. The ground strip acts as a reflector so that the path of the surface currents is changed. Thus, the distance among the ports is enlarged.

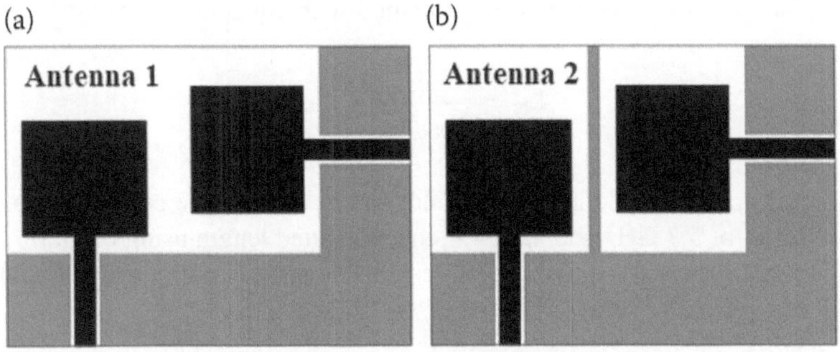

FIGURE 19.3 (a) UWB-MIMO antenna without a ground strip (Antenna 1) and (b) UWB-MIMO antenna with a ground strip (Antenna 2) [31].

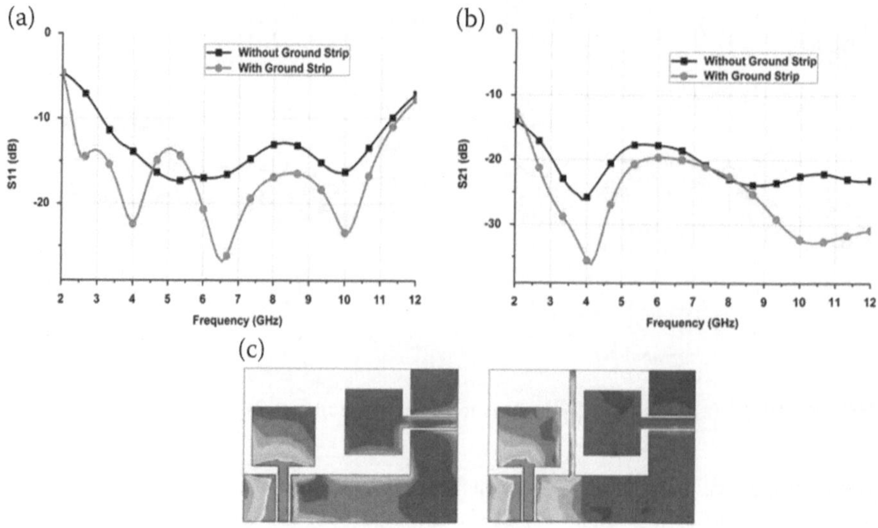

FIGURE 19.4 (a) S_{11} without and with a ground strip, (b) S_{21} without and with a ground strip, and (c) surface current distribution at 3.8 GHz when port 1 excited without and with a ground strip (Dark blue indicates no radiation and dark red indicates more radiation from antenna) [31].

Therefore, the mutual coupling between the antenna elements is drastically reduced. Also, the ground strip will increase impedance matching properties of the antenna.

To produce the band-notch characteristics for ultra-wideband systems, slots of several forms or split-ring resonators (SRRs) or strips can be placed on or beside the feed line or the antenna element or the ground plane as described previously. The slot or SRR or strip can work as a band-notch resonator. The center frequency of the notch band is altered by the resonator length and bandwidth is regulated by the resonator width. In this proposed antenna, an inverted U-shaped slot is adopted as a band-notch resonator and is imprinted on the feed line of Antenna 2 which makes the proposed band-notch antenna as displayed in Figure 19.5(a) to 19.5(d).

The U-shaped resonator length is computed with equation (19.5) [19]:

$$L_N = \frac{c}{2f_N\sqrt{\varepsilon_{eff}}} \approx \frac{\lambda}{2}, \tag{19.5}$$

where L_N means the whole length of U-slot and f_N denotes the center frequency of notch. With f_N of 5.7 GHz and ε_r of 4.4, the computed length using equation (19.5) is eqaul to 16.01 mm. The simulated or designed length is 16 mm and is found with the help of equation (19.6).

$$L_{U\text{-}Slot} = 2\,U_1 + U_2 \approx \frac{\lambda}{2}. \tag{19.6}$$

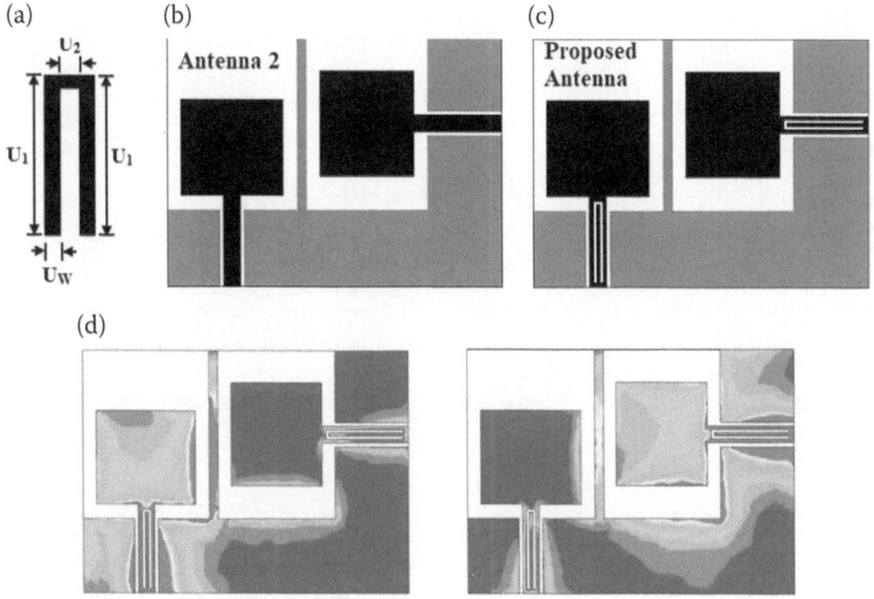

FIGURE 19.5 (a) an inverted U-slot resonator, (b) Antenna 2, (c) proposed antenna, and (d) current distribution at 5.7 GHz when port 1 and port 2 excited (dark blue indicates no radiation and dark red indicates more radiation from antenna) [31].

It is identified the computed (theoretical) length is almost equal to simulated (practical) length. Figure 19.5(d) displays the surface currents without and with inverted U-slot resonator. It is observed that substantial current is focused around the slot at 5.7 GHz. Therefore the flow of current on the antenna is jammed and thus the antenna does not work at the designed notch band.

19.3 RESULTS AND DISCUSSION

The proposed design attains impedance bandwidth ($|S_{11}|<-10$ dB) from 2.2 to 11.4 GHz having notch band at 5–5.9 GHz as confirmed in Figure 19.6(a). Therefore, the proposed structure is able to suppress the frequency interference from WLAN band. In addition, the simulation and measurement isolation (S_{21}) is more than 20 dB in the working band demonstrating good isolation between the ports as observed from Figure 19.6(b).

The simulation and measurement 2-D E-plane and H-plane patterns of the proposed structure at 3.8, 6.5, and 10 GHz when port 1 is excited and port 2 is loaded with 50-ohmare depicted in Figure 19.7. It is found that there is good promise between the simulation and measurement radiation patterns. The elements PM1 and PM2 have relatively omnidirectional patterns in H-planes, i.e., the XZ plane and the YZ plane, respectively, at 3.8 and 6.5 GHz frequencies. But, at 10 GHz, the H-plane patterns are less omnidirectional due to higher-order resonances. Also, at 6.5 and 10 GHz, PM1 and PM2 providing bidirectional E-plane patterns, i.e., the YZ plane

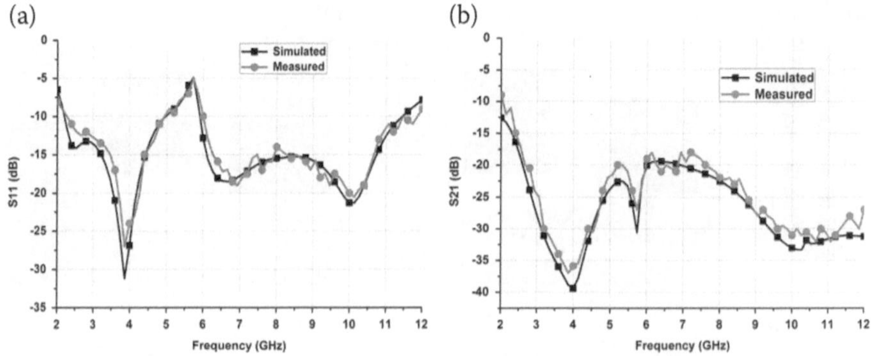

FIGURE 19.6 The simulation and measurement: (a) S_{11}-parameter and (b) S_{21}-parameter [31].

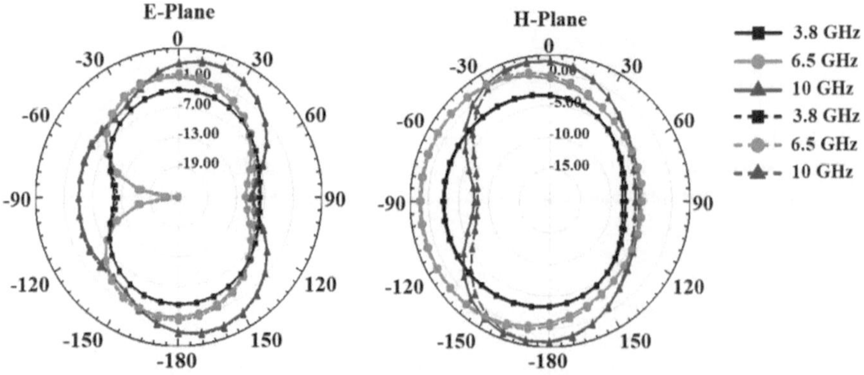

FIGURE 19.7 Simulation (solid line) and measurement (dashed line) 2-D radiation patterns[31].

and the XZ plane, respectively. Nevertheless, PM1 and PM2 do not offer the dumbbell shaped E-plane patterns at 3.8 GHz since the ground plane strip deviates from the current distributions. Moreover, the H-plane patterns of port 1 and port 2 are approximately mirror images proving the good pattern diversity.

Figure 19.8(a) shows the simulation and measurement peak gain of the proposed structure. It is found that the peak gain of 2.4 to 7.5 dBi in the entire band apart from the 5–5.9 GHz notch band. The measured peak gain drops to –2.2 dBi at the notch. The simulation and measurement radiation efficiency is plotted in Figure 19.8(b). The radiation efficiency is more than 90% excepting at notch band. the efficiency falls to 12% at 5–5.9 GHz. It is manifested that the proposed structure is able to suppress the interference from 5–5.9 GHzWLAN band more effectively.

The envelope correlation coefficient (ECC) is a vital parameter to analyse the antenna diversity. The ECC is computed through S-parameters by using equation (19.7) of a two-port MIMO antenna system stated in [32]. To ensure the diversity

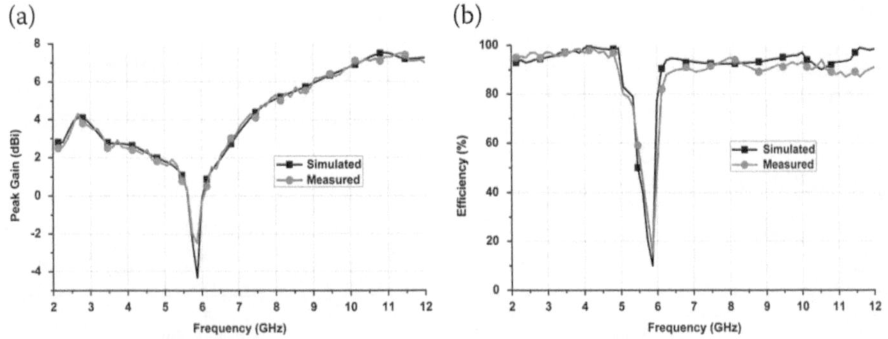

FIGURE 19.8 Simulation and measurement (a) peak gain and (b) radiation efficiency[31].

for MIMO antenna system, the ECC should be less than 0.5. The simulation and measurement ECC is shown in Figure 19.9(a). The simulation ECC is around 0.005 and measurement ECC is less than 0.008 from 2.2 to 11.4 GHz.

$$\mathrm{ECC} = \frac{|S_{11}^{*}S_{12} + S_{21}^{*}S_{22}|^{2}}{(1 - (|S_{11}|^{2} + |S_{21}|^{2}))(1 - (|S_{22}|^{2} + |S_{12}|^{2}))}, \tag{19.7}$$

In addition to ECC, the diversity gain (DG) and total active reflection coefficient (TARC) are also essential parameters to study the diversity behaviour of MIMO antenna. The diversity gain and total active reflection coefficient of the proposed antennas can be estimated by using the equations (19.8) and (19.9) [33] as follows:

$$\mathrm{DG} = 10\sqrt{1 - \mathrm{ECC}^{2}} \tag{19.8}$$

$$\mathrm{TARC} = \sqrt{\frac{(S_{11} + S_{12})^{2} + (S_{21} + S_{22})^{2}}{2}} \tag{19.9}$$

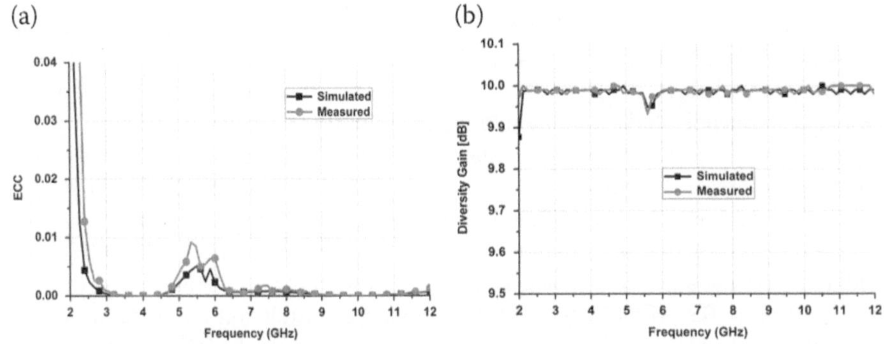

FIGURE 19.9 (a) Simulated and measured ECC [31] and (b) simulated and measured diversity gain.

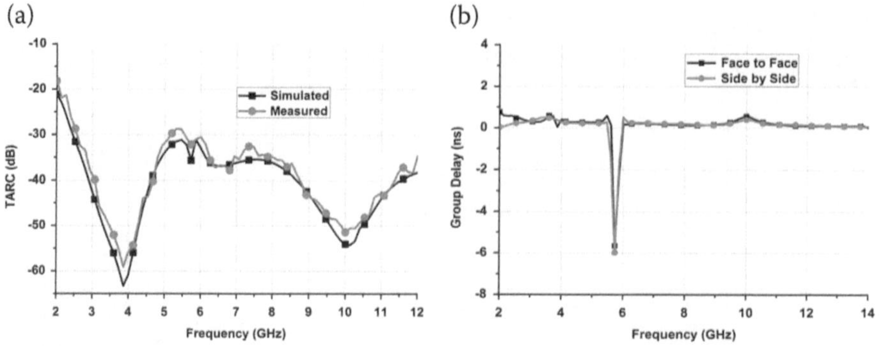

FIGURE 19.10 (a) Simulated and measured TARC and (b) Group delay of the proposed antenna.

The simulated and measured diversity gain plots are given in Figure 19.9(b). The diversity gain of >9.95 dB is found in the UWB band.

And Figure 19.10(a) shows the simulated and measured TARC. It is observed that the TARC of less than −28 dB is obtained in the whole UWB band. The group delay of the proposed antenna is measured in face to face and side by side situations with the space of 30 cm shown in Figure 19.10(b). The group delay is almost uniform and is below 1 ns in the complete working band except at stopband. At the notch band, i.e., at 5.7 GHz, the group delay of 5.8 ns in the face to face orientation and 6 ns in the side by side orientation ensures that the proposed antenna can transmit the UWB signal with minimum distortion. It is observed from the above results that there is good agreement between the simulated and measured S-parameters except for some deviations due to fabrication and soldering imperfections, losses in dielectrics and conductors, effects of SMA connector, and measurement tolerances.

19.4 CONCLUSION

To mitigate the frequency interference from narrowband systems like WLAN, a compact planar UWB antenna design with singleband-notched characteristics for portable wireless devices applications is discussed in this chapter. The monopoles are placed perpendicularly to mitigate mutual coupling. A long strip is added to the ground to advance the impedance matching properties and to the improve the isolation. To create the notch function at 5–5.9 GHz, an inverted U-shaped slot is etched on feed line. The simulation and measurement results demonstrate that the proposed design is providing bandwidth of $S_{11} \leq -10$ dB from 3.1–10.6 GHz excluding at the intended notch band and offering low mutual coupling (S_{21}) of less than −20 dB in the whole band. The less envelope correlation coefficient, nearly constant gain, stable radiation patterns, more directive gain, TARC and less group delay validate that the proposed MIMO antenna is a suitable choice for UWB systems.

REFERENCES

1. Docket, E. T. (2002). *Docket 98-153, FCC, Revision of Part 15 of the Commissions Rules Regarding Ultra-Wideband Transmission Systems*. Technical Report.
2. Kaiser, T., Zheng, F., & Dimitrov, E. (2009). An overview of ultra-wide-band systems with MIMO. *Proceedings of the IEEE, 97*(2), 285–312.
3. Mabrouk, I. B., Talbi, L., Nedil, M., & Hettak, K. (2012). MIMO-UWB channel characterization within an underground mine gallery. *IEEE Transactions on Antennas and Propagation, 60*(10), 4866–4874.
4. Tran, V. P., & Sibille, A. (2006). Spatial multiplexing in UWB MIMO communications. *Electronics Letters*, *42*(16), 1.
5. Zheng, L., & Tse, D. N. C. (2003). Diversity and multiplexing: A fundamental tradeoff in multiple-antenna channels. *IEEE Transactions on Information Theory*, *49*(5), 1073–1096.
6. Liu, L., Cheung, S. W., & Yuk, T. I. (2013). Compact MIMO antenna for portable devices in UWB applications. *IEEE Transactions on Antennas and Propagation*, *61*(8), 4257–4264.
7. Zhang, S., Ying, Z., Xiong, J., & He, S. (2009). Ultrawideband MIMO/diversity antennas with a tree-like structure to enhance wideband isolation. *IEEE Antennas and Wireless Propagation Letters*, *8*, 1279–1282.
8. Luo, C. M., Hong, J. S., & Zhong, L. L. (2015). Isolation enhancement of a very compact UWB-MIMO slot antenna with two defected ground structures. *IEEE Antennas and Wireless Propagation Letters*, *14*, 1766–1769.
9. Tao, J., & Feng, Q. (2016). Compact ultrawideband MIMO antenna with half-slot structure. *IEEE Antennas and Wireless Propagation Letters*, *16*, 792–795.
10. Gallo, M., Antonino-Daviu, E., Ferrando-Bataller, M., Bozzetti, M., Molina-Garcia-Pardo, J. M., & Juan-Llacer, L. (2011). A broadband pattern diversity annular slot antenna. *IEEE Transactions on Antennas and Propagation*, *60*(3), 1596–1600.
11. Rao, J. C., & Rao, N. V. (2016). CPW-fed compact ultra wideband MIMO antenna for portable devices. *Indian Journal of Science and Technology*, *9*(17), 1–9.
12. Bassi, M., Caruso, M., Khan, M. S., Bevilacqua, A., Capobianco, A. D., & Neviani, A. (2013). An integrated microwave imaging radar with planar antennas for breast cancer detection. *IEEE Transactions on Microwave Theory and Techniques*, *61*(5), 2108–2118.
13. Capobianco, A. D., Khan, M. S., Caruso, M., & Bevilacqua, A. (2014). 3–18 GHz compact planar antenna for short-range radar imaging. *Electronics Letters*, *50*(14), 1016–1018.
14. Zhang, S., & Pedersen, G. F. (2015). Mutual coupling reduction for UWB MIMO antennas with a wideband neutralization line. *IEEE Antennas and Wireless Propagation Letters*, *15*, 166–169.
15. Li, H., Liu, J., Wang, Z., & Yin, Y. Z. (2017). Compact 1× 2 and 2× 2 MIMO antennas with enhanced isolation for ultrawideband application. *Progress in Electromagnetics Research*, *71*, 41–49.
16. Kerkhoff, A., & Ling, H. (2003, June). Design of a planar monopole antenna for use with ultra-wideband (UWB) having a band-notched characteristic. In *IEEE Antennas and Propagation Society International Symposium. Digest. Held in conjunction with: USNC/CNC/URSI North American Radio Sci. Meeting (Cat. No. 03CH37450)* (Vol. 1, pp. 830–833). IEEE.
17. Weng, Y. F., Cheung, S. W., Yuk, T. I., & Liu, L. (2012). Creating band-notched characteristics for compact UWB monopole antennas. In *Ultra Wideband-Current Status and Future Trends*. IntechOpen.

18. Liu, L. (2014). Compact planar UWB antennas for wireless device applications. *HKU Theses Online (HKUTO)*.
19. Zheng, Z. A., Chu, Q. X., & Tu, Z. H. (2010). Compact band-rejected ultrawideband slot antennas inserting with $\lambda/2$ and $\lambda/4$ resonators. *IEEE Transactions on Antennas and Propagation*, *59*(2), 390–397.
20. Weng, Y. F., Cheung, S. W., & Yuk, T. I. (2011). Compact ultra-wideband antennas with single band-notched characteristic using simple ground stubs. *Microwave and Optical Technology Letters*, *53*(3), 523–529.
21. Lee, J. M., Kim, K. B., Ryu, H. K., & Woo, J. M. (2012). A compact ultrawideband MIMO antenna with WLAN band-rejected operation for mobile devices. *IEEE Antennas and wireless propagation letters*, *11*, 990–993.
22. Sayidmarie, K. H., & Najm, T. A. (2013). Performance evaluation of band notch techniques for printed dual band monopole antennas. *International Journal of Electromagnetics and Applications*, *3*(4), 70–80.
23. Majeed, A. H., Abdullah, A. S., Sayidmarie, K. H., Abd-Alhameed, R. A., Elmegri, F., & Noras, J. M. (2015). Compact dielectric resonator antenna with band-notched characteristics for ultra-wideband applications. *Progress in Electromagnetics Research*, *57*, 137–148.
24. Gao, P., He, S., Wei, X., Xu, Z., Wang, N., & Zheng, Y. (2014). Compact printed UWB diversity slot antenna with 5.5-GHz band-notched characteristics. *IEEE Antennas and Wireless Propagation Letters*, *13*, 376–379.
25. Khan, M. S., Capobianco, A. D., Naqvi, A., Shafique, M. F., Ijaz, B., & Braaten, B. D. (2015). Compact planar UWB MIMO antenna with on-demand WLAN rejection. *Electronics Letters*, *51*(13), 963–964.
26. Liu, L., Cheung, S. W., & Yuk, T. I. (2015). Compact MIMO antenna for portable UWB applications with band-notched characteristic. *IEEE Transactions on Antennas and Propagation*, *63*(5), 1917–1924.
27. Sipal, D., Abegaonkar, M. P., & Koul, S. K. (2016). Compact band-notched UWB antenna for MIMO applications in portable wireless devices. *Microwave and Optical Technology Letters*, *58*(6), 1390–1394.
28. Tao, J., & Feng, Q. Y. (2016). Compact isolation-enhanced UWB MIMO antenna with band-notch character. *Journal of Electromagnetic Waves and Applications*, *30*(16), 2206–2214.
29. Liu, Z., Wu, X., Zhang, Y., Ye, P., Ding, Z., & Hu, C. (2017). Very compact 5.5 GHz band-notched UWB-MIMO antennas with high isolation. *Progress in Electromagnetics Research*, *76*, 109–118.
30. Tripathi, S., Mohan, A., & Yadav, S. K. (2017). A compact MIMO/diversity antenna with WLAN band-notch characteristics for portable UWB applications. *Progress in Electromagnetics Research*, *77*, 29–38.
31. Jetti, C. R., & Nandanavanam, V. R. (2018). Compact MIMO antenna with WLAN band-notch characteristics for portable UWB systems. *Progress in Electromagnetics Research*, *88*, 1–12.
32. Blanch, S., Romeu, J., & Corbella, I. (2003). Exact representation of antenna system diversity performance from input parameter description. *Electronics Letters*, *39*(9), 705–707.
33. Najam, A. I., Duroc, Y., & Tedjini, S. (2012). Multiple-input multiple-output antennas for ultra wideband communications. *IntechOpen*, *10*, 209–236.

Index